Springer Undergraduate Mathematics Series

Advisory Board

Other books in this series

Marek Capiński and Ekkehard Kopp

Measure, Integral and Probability

Second Edition

With 23 Figures

 Springer

Marek Capiński, PhD
Institute of Mathematics, Jagiellonian University, Reymonta 4, 30-059 Kraków, Poland

Peter Ekkehard Kopp, DPhil
Department of Mathematics, University of Hull, Cottingham Road, Hull, HU6 7RX, UK

Cover illustration elements reproduced by kind permission of:
Aptech Systems, Inc., Publishers of the GAUSS Mathematical and Statistical System, 23804 S.E. Kent-Kangley Road, Maple Valley, WA 98038, USA. Tel: (206) 432 - 7855 Fax: (206) 432 - 7832 email: info@aptech.com URL: www.aptech.com.
American Statistical Association: Chance Vol 8 No 1 1995 article by KS and KW Heiner "Tree Rings of the Northern Shawangunks" page 32 fig. 2.
Springer-Verlag: Mathematica in Education and Research Vol 4 Issue 3 1995 article by Roman E. Maeder, Beatrice Amrhein and Oliver Gloor "Illustrated Mathematics: Visualization of Mathematical Objects" page 9 fig 11, originally published as a CD ROM "Illustrated Mathematics" by TELOS: ISBN 0-387-14222-3, German edition by Birkhauser: ISBN-13:978-1-85233-781-0
Mathematica in Education and Research Vol 4 Issue 3 1995 article by Richard J Gaylord and Kazume Nishidate "Traffic Engineering with Cellular Automata" page 35 fig. 2. Mathematica in Education and Research Vol 5 Issue 2 1996 article by Michael Trott "The Implicitization of a Trefoil Knot" page 14.
Mathematica in Education and Research Vol 5 Issue 2 1996 article by Lee de Cola "Coins, Trees, Bars and Bells: Simulation of the Binomial Process" page 19 fig. 3. Mathematica in Education and Research Vol 5 Issue 2 1996 article by Richard Gaylord and Kazume Nishidate "Contagious Spreading" page 33 fig 1. Mathematica in Education and Research Vol 5 Issue 2 1996 article by Joe Buhler and Stan Wagon "Secrets of the Madelung Constant" page 50 fig 1.

British Library Cataloguing in Publication Data
Capiński, Marek, 1951–
    Measure, integral and probability. - 2nd ed. - (Springer
    undergraduate mathematics series)
    1. Lebesgue integral   2. Measure theory   3. Probabilities
    I. Title   II. Kopp, P. E., 1944–
    515.4'2

Library of Congress Cataloging-in-Publication Data
Capiński, Marek, 1951–
    Measure, integral and probability / Marek Capiński and Ekkehard Kopp.—2nd ed.
        p. cm. — (Springer undergraduate mathematics series, ISSN 1615-2085)
    Includes bibliographical references and index.
    ISBN-13:978-1-85233-781-0
        1. Measure theory.   2. Integrals, Generalized.   3. Probabilities.   I. Kopp, P. E., 1944–   II.
    Title.   III. Series.
    QA312.C36 2004
    515'.42—dc22   2004040204

Springer Undergraduate Mathematics Series ISSN 1615-2085

ISBN-13:978-1-85233-781-0   e-ISBN-13:978-1-4471-0645-6

DOI: 10.1007/978-1-4471-0645-6

Springer Science+Business Media
springeronline.com

© Springer-Verlag London Limited 2004
2nd printing 2005

Reprint of the original edition 2005

Typesetting: Electronic files provided by the authors

12/3830-54321   Printed on acid-free paper   SPIN 11494904

To our children; grandchildren:
Piotr, Maciej, Jan, Anna; Łukasz,
Anna, Emily

Preface to the second edition

After five years and six printings it seems only fair to our readers that we should respond to their comments and also correct errors and imperfections to which we have been alerted in addition to those we have discovered ourselves in reviewing the text. This second edition also introduces additional material which earlier constraints of time and space had precluded, and which has, in our view, become more essential as the make-up of our potential readership has become clearer. We hope that we manage to do this in a spirit which preserves the essential features of the text, namely providing the material rigorously and in a form suitable for directed self-study. Thus the focus remains on accessibility, explicitness and emphasis on concrete examples, in a style that seeks to encourage readers to become directly involved with the material and challenges them to prove many of the results themselves (knowing that solutions are also given in the text!).

Apart from further examples and exercises, the new material presented here is of two contrasting kinds. The new Chapter 7 adds a discussion of the comparison of general measures, with the Radon–Nikodym theorem as its focus. The proof given here, while not new, is in our view more constructive and elementary than the usual ones, and we utilise the result consistently to examine the structure of Lebesgue–Stieltjes measures on the line and to deduce the Hahn–Jordan decomposition of signed measures. The common origin of the concepts of variation and absolute continuity of functions and measures is clarified. The main probabilistic application is to conditional expectations, for which an alternative construction via orthogonal projections is also provided in Chapter 5. This is applied in turn in Chapter 7 to derive elementary properties of martingales in discrete time.

The other addition occurs at the end of each chapter (with the exception of Chapters 1 and 5). Since it is clear that a significant proportion of our current readership is among students of the burgeoning field of mathematical finance, each relevant chapter ends with a brief discussion of ideas from that subject. In these sections we depart from our aim of keeping the book self-contained, since we can hardly develop this whole discipline afresh. Thus we neither define nor explain the origin of the finance concepts we address, but instead seek to locate them mathematically within the conceptual framework of measure and probability. This leads to conclusions with a mathematical precision that sometimes eludes authors writing from a finance perspective.

To avoid misunderstanding we repeat that the purpose of this book remains the development of the ideas of measure and integral, especially with a view to their applications in probability and (briefly) in finance. This is therefore neither a textbook in probability theory nor one in mathematical finance. Both of these disciplines have a large specialist literature of their own, and our comments on these areas of application are intended to assist the student in understanding the mathematical framework which underpins them.

We are grateful to those of our readers and to colleagues who have pointed out many of the errors, both typographical and conceptual, of the first edition. The errors that inevitably remain are our sole responsibility. To facilitate their speedy correction a webpage has been created for the notification of errors, inaccuracies and queries, at http://www.springeronline.com/1-85233-781-8 and we encourage our readers to use it mercilessly. Our thanks also go to Stephanie Harding and Karen Borthwick at Springer-Verlag, London, for their continuing care and helpfulness in producing this edition.

Kraków, Poland   Marek Capiński
Hull, UK   Ekkehard Kopp
October 2003

Preface to the first edition

The central concepts in this book are Lebesgue measure and the Lebesgue integral. Their role as standard fare in UK undergraduate mathematics courses is not wholly secure; yet they provide the principal model for the development of the abstract measure spaces which underpin modern probability theory, while the Lebesgue function spaces remain the main source of examples on which to test the methods of functional analysis and its many applications, such as Fourier analysis and the theory of partial differential equations.

It follows that not only budding analysts have need of a clear understanding of the construction and properties of measures and integrals, but also that those who wish to contribute seriously to the applications of analytical methods in a wide variety of areas of mathematics, physics, electronics, engineering and, most recently, finance, need to study the underlying theory with some care.

We have found remarkably few texts in the current literature which aim explicitly to provide for these needs, at a level accessible to current undergraduates. There are many good books on modern probability theory, and increasingly they recognise the need for a strong grounding in the tools we develop in this book, but all too often the treatment is either too advanced for an undergraduate audience or else somewhat perfunctory. We hope therefore that the current text will not be regarded as one which fills a much-needed gap in the literature!

One fundamental decision in developing a treatment of integration is whether to begin with measures or integrals, i.e. whether to start with sets or with functions. Functional analysts have tended to favour the latter approach, while the former is clearly necessary for the development of probability. We have decided to side with the probabilists in this argument, and to use the (reasonably) systematic development of basic concepts and results in probability theory as the principal field of application – the order of topics and the terminology we use reflect this choice, and each chapter concludes with further development of the relevant probabilistic concepts. At times this approach may seem less 'efficient' than the alternative, but we have opted for direct proofs and explicit constructions, sometimes at the cost of elegance. We hope that it will increase understanding.

The treatment of measure and integration is as self-contained as we could make it within the space and time constraints: some sections may seem too pedestrian for final-year undergraduates, but experience in testing much of the material over a number of years at Hull University teaches us that familiarity and confidence with basic concepts in analysis can frequently seem somewhat shaky among these audiences. Hence the preliminaries include a review of Riemann integration, as well as a reminder of some fundamental concepts of elementary real analysis.

While probability theory is chosen here as the principal area of application of measure and integral, this is not a text on elementary probability, of which many can be found in the literature.

Though this is not an advanced text, it is intended to be studied (not skimmed lightly) and it has been designed to be useful for directed self-study as well as for a lecture course. Thus a significant proportion of results, labelled 'Proposition', are not proved immediately, but left for the reader to attempt before proceeding further (often with a hint on how to begin), and there is a generous helping of exercises. To aid self-study, proofs of the propositions are given at the end of each chapter, and outline solutions of the exercises are given at the end of the book. Thus few mysteries should remain for the diligent.

After an introductory chapter, motivating and preparing for the principal definitions of measure and integral, Chapter 2 provides a detailed construction of Lebesgue measure and its properties, and proceeds to abstract the axioms appropriate for probability spaces. This sets a pattern for the remaining chapters, where the concept of independence is pursued in ever more general contexts, as a distinguishing feature of probability theory.

Chapter 3 develops the integral for non-negative measurable functions, and introduces random variables and their induced probability distributions, while Chapter 4 develops the main limit theorems for the Lebesgue integral and compares this with Riemann integration. The applications in probability lead to a discussion of expectations, with a focus on densities and the role of characteristic functions.

In Chapter 5 the motivation is more functional-analytic: the focus is on the Lebesgue function spaces, including a discussion of the special role of the space L^2 of square-integrable functions. Chapter 6 sees a return to measure theory, with the detailed development of product measure and Fubini's theorem, now leading to the role of joint distributions and conditioning in probability. Finally, following a discussion of the principal modes of convergence for sequences of integrable functions, Chapter 7 adopts an unashamedly probabilistic bias, with a treatment of the principal limit theorems, culminating in the Lindeberg–Feller version of the central limit theorem.

The treatment is by no means exhaustive, as this is a textbook, not a treatise. Nonetheless the range of topics is probably slightly too extensive for a one-semester course at third-year level: the first five chapters might provide a useful course for such students, with the last two left for self-study or as part of a reading course for students wishing to continue in probability theory. Alternatively, students with a stronger preparation in analysis might use the first two chapters as background material and complete the remainder of the book in a one-semester course.

Kraków, Poland   Marek Capiński
Hull, UK   Ekkehard Kopp
May 1998

Contents

1
Motivation and preliminaries

Life is an uncertain business. We can seldom be sure that our plans will work out as we intend, and are thus conditioned from an early age to think in terms of the *likelihood* that certain events will occur, and which are 'more likely' than others. Turning this vague description into a *probability model* amounts to the construction of a rational framework for thinking about uncertainty. The framework ought to be a general one, which enables us to handle equally situations where we have to sift a great deal of prior information, and those where we have little to go on. Some degree of judgement is needed in all cases; but we seek an orderly theoretical framework and methodology which enables us to formulate general laws in quantitative terms.

This leads us to mathematical models for probability, that is to say, idealised abstractions of empirical practice, which nonetheless have to satisfy the criteria of wide applicability, accuracy and simplicity. In this book our concern will be with the construction and use of generally applicable probability models in which we can also consider infinite sample spaces and infinite sequences of trials: that such are needed is easily seen when one tries to make sense of apparently simple concepts such as 'drawing a number at random from the interval $[0, 1]$' and in trying to understand the limit behaviour of a sequence of identical trials. Just as elementary probabilities are computed by finding the comparative sizes of sets of outcomes, we will find that the fundamental problem to be solved is that of measuring the 'size' of a set with infinitely many elements. At least for sets on the real line, the ideas of basic real analysis provide us with a convincing answer, and this contains all the ideas needed for the abstract axiomatic framework on which to base the theory of probability. For

1

this reason the development of the concept of *measure*, and *Lebesgue measure* on $\mathbb{R}$ in particular, has pride of place in this book.

1.1 Notation and basic set theory

In measure theory we deal typically with families of subsets of some arbitrary given set and consider functions which assign real numbers to sets belonging to these families. Thus we need to review some basic set notation and operations on sets, as well as discussing the distinction between countably and uncountably infinite sets, with particular reference to subsets of the real line $\mathbb{R}$. We shall also need notions from analysis such as limits of sequences, series, and open sets. Readers are assumed to be largely familiar with this material and may thus skip lightly over this section, which is included to introduce notation and make the text reasonably self-contained and hence useful for self-study. The discussion remains quite informal, without reference to foundational issues, and the reader is referred to basic texts on analysis for most of the proofs. Here we mention just two introductory textbooks: [8] and [11].

1.1.1 Sets and functions

In our operations with sets we shall always deal with collections of subsets of some universal set Ω; the nature of this set will be clear from the context – frequently Ω will be the set $\mathbb{R}$ of real numbers or a subset of it. We leave the concept of 'set' as undefined and given, and concern ourselves only with set membership and operations. The empty set is denoted by $\emptyset$; it has no members. Sets are generally denoted by capital letters.

Set membership is denoted by $\in$, so $x \in A$ means that the element x is a member of the set A. Set inclusion, $A \subset B$, means that every member of A is a member of B. This includes the case when A and B are equal; if the inclusion is strict, i.e. $A \subset B$ and B contains elements which are not in A (written $x \notin A$), this will be stated separately. The notation $\{x \in A : P(x)\}$ is used to denote the set of elements of A with property P. The set of all subsets of A (its *power set*) is denoted by $\mathcal{P}(A)$.

We define the *intersection* $A \cap B = \{x : x \in A \text{ and } x \in B\}$ and *union* $A \cup B = \{x : x \in A \text{ or } x \in B\}$. The *complement* A^c of A consists of the elements of Ω which are not members of A; we also write $A^c = \Omega \setminus A$, and, more generally, we have the *difference* $B \setminus A = \{x \in B : x \notin A\} = B \cap A^c$ and the *symmetric difference* $A \triangle B = (A \setminus B) \cup (B \setminus A)$. Note that $A \triangle B = \emptyset$ if and only if $A = B$.

Intersection (resp. union) gives expression to the logical connective 'and' (resp. 'or') and, via the logical symbols $\exists$ (there exists) and $\forall$ (for all), they have extensions to arbitrary collections. Indexed by some set Λ these are given by

$$\bigcap_{\alpha \in \Lambda} A_\alpha = \{x : x \in A_\alpha \text{ for all } \alpha \in \Lambda\} = \{x : \forall \alpha \in \Lambda, \ x \in A_\alpha\};$$

$$\bigcup_{\alpha \in \Lambda} A_\alpha = \{x : x \in A_\alpha \text{ for some } \alpha \in \Lambda\} = \{x : \exists \alpha \in \Lambda, \ x \in A_\alpha\}.$$

These are linked by *de Morgan's laws*:

$$(\bigcup_\alpha A_\alpha)^c = \bigcap_\alpha A_\alpha^c; \qquad (\bigcap_\alpha A_\alpha)^c = \bigcup_\alpha A_\alpha^c.$$

If $A \cap B = \varnothing$ then A and B are *disjoint*. A family of sets $(A_\alpha)_{\alpha \in \Lambda}$ is *pairwise disjoint* if $A_\alpha \cap A_\beta = \varnothing$ whenever $\alpha \neq \beta$ $(\alpha, \beta \in \Lambda)$.

The *Cartesian product* $A \times B$ of sets A and B is the set of ordered pairs $A \times B = \{(a, b) : a \in A, b \in B\}$. As already indicated, we use $\mathbb{N}, \mathbb{Z}, \mathbb{Q}, \mathbb{R}$ for the basic number systems of natural numbers, integers, rationals and reals respectively. Intervals in $\mathbb{R}$ are denoted via each endpoint, with a square bracket indicating its inclusion, an open bracket exclusion, e.g. $[a, b) = \{x \in \mathbb{R} : a \leq x < b\}$. We use ∞ and $-\infty$ to describe unbounded intervals, e.g. $(-\infty, b) = \{x \in \mathbb{R} : x < b\}$, $[0, \infty) = \{x \in \mathbb{R} : x \geq 0\} = \mathbb{R}^+$. The set $\mathbb{R}^2 = \mathbb{R} \times \mathbb{R}$ denotes the plane, more generally, $\mathbb{R}^n$ is the n-fold Cartesian product of $\mathbb{R}$ with itself, i.e. the set of all n-tuples $(x_1, \ldots, x_n)$ composed of real numbers. Products of intervals, called *rectangles*, are denoted similarly.

Formally, a *function* $f : A \to B$ is a subset of $A \times B$ in which each first coordinate determines the second: if $(a, b), (a, c) \in f$ then $b = c$. Its *domain* $\mathcal{D}_f = \{a \in A : \exists b \in B, (a, b) \in f\}$ and *range* $\mathcal{R}_f = \{b \in B : \exists a \in A, (a, b) \in f\}$ describe its scope. Informally, f associates elements of B with those of A, such that each $a \in A$ has at most one *image* $b \in B$. We write this as $b = f(a)$. The set $X \subset A$ has *image* $f(X) = \{b \in B : b = f(a) \text{ for some } a \in X\}$ and the *inverse image* of a set $Y \subset B$ is $f^{-1}(Y) = \{a \in A : f(a) \in Y\}$. The *composition* $f_2 \circ f_1$ of $f_1 : A \to B$ and $f_2 : B \to C$ is the function $h : A \to C$ defined by $h(a) = f_2(f_1(a))$. When $A = B = C$, $x \mapsto (f_1 \circ f_2)(x) = f_1(f_2(x))$ and $x \mapsto (f_2 \circ f_1)(x) = f_2(f_1(x))$ both define functions from A to A. In general, these will not be the same: for example, let $f_1(x) = \sin x$, $f_2(x) = x^2$, then $x \mapsto \sin(x^2)$ and $x \mapsto (\sin x)^2$ are not equal.

The function g *extends* f if $\mathcal{D}_f \subset \mathcal{D}_g$ and $g = f$ on $\mathcal{D}_f$; alternatively we say that f *restricts* g to $\mathcal{D}_f$. These concepts will be used frequently for real-valued set functions, where the domains are collections of sets and the range is a subset of $\mathbb{R}$.

The algebra of real functions is defined pointwise, i.e. the *sum* $f + g$ and *product* fg are given by $(f + g)(x) = f(x) + g(x)$, $(fg)(x) = f(x)g(x)$.

The *indicator* function $\mathbf{1}_A$ of the set A is the function

$$\mathbf{1}_A(x) = \left\{ \begin{array}{ll} 1 & \text{for } x \in A \\ 0 & \text{for } x \notin A. \end{array} \right.$$

Note that $\mathbf{1}_{A \cap B} = \mathbf{1}_A \cdot \mathbf{1}_B$, $\mathbf{1}_{A \cup B} = \mathbf{1}_A + \mathbf{1}_B - \mathbf{1}_A \mathbf{1}_B$, and $\mathbf{1}_{A^c} = 1 - \mathbf{1}_A$.

We need one more concept from basic set theory, which should be familiar: for any set E, an *equivalence relation* on E is a relation (i.e. a subset R of $E \times E$, where we write $x \sim y$ to indicate that $(x, y) \in R$) with the following properties:

1. *reflexive*: for all $x \in E$, $x \sim x$,

2. *symmetric*: $x \sim y$ implies $y \sim x$,

3. *transitive*: $x \sim y$ and $y \sim z$ implies $x \sim z$.

An equivalence relation $\sim$ on E partitions E into disjoint *equivalence classes*: given $x \in E$, write $[x] = \{z : z \sim x\}$ for the equivalence class of x, i.e. the set of all elements of E that are equivalent to x. Thus $x \in [x]$, hence $E = \bigcup_{x \in E} [x]$. This is a disjoint union: if $[x] \cap [y] \neq \varnothing$, then there is $z \in E$ with $x \sim z$ and $z \sim y$, hence $x \sim y$, so that $[x] = [y]$. We shall denote the set of all equivalence classes so obtained by $E/\sim$.

1.1.2 Countable and uncountable sets in $\mathbb{R}$

We say that a set A is *countable* if there is a one–one correspondence between A and a subset of $\mathbb{N}$, i.e. a function $f : A \to \mathbb{N}$ that takes distinct points to distinct points. Informally, A is *finite* if this correspondence can be set up using only an initial segment $\{1, 2, ..., N\}$ of $\mathbb{N}$ (for some $N \in \mathbb{N}$), while we call A *countably infinite* or *denumerable* if all of $\mathbb{N}$ is used. It is not difficult to see that countable unions of countable sets are countable; in particular, the set $\mathbb{Q}$ of rationals is countable.

Cantor showed that the set $\mathbb{R}$ *cannot* be placed in one–one correspondence with (a subset of) $\mathbb{N}$; thus it is an example of an *uncountable* set. Cantor's proof assumes that we can write each real number uniquely as a decimal (always choosing the non-terminating version). We can also restrict ourselves (why?) to showing that the interval $[0,1]$ is uncountable.

If this set were countable, then we could write its elements as a sequence $(x_n)_{n \geq 1}$, and since each x_n has a unique decimal expansion of the form

$$x_n = 0.a_{n1}a_{n2}a_{n3}....a_{nn}...$$

for digits a_{ij} chosen from the set $\{0, 1, 2..., 9\}$, we could therefore write down the array

$$x_1 = 0.a_{11}a_{12}a_{13} \ldots$$

$$x_2 = 0.a_{21}a_{22}a_{23} \ldots$$

$$x_3 = 0.a_{31}a_{32}a_{33} \ldots$$

$$\ldots$$

Now let $y = 0.b_1 b_2 b_3 \ldots$, where the digits b_n are chosen to differ from a_{nn}. Such a decimal expansion defines a number $y \in [0, 1]$ that differs from each of the x_n (since its expansion differs from that of x_n in the nth place). Hence our sequence does not exhaust [0,1], and the contradiction shows that [0,1] cannot be countable.

Since the union of two countable sets must be countable, and since $\mathbb{Q}$ is countable, it follows that $\mathbb{R}\backslash\mathbb{Q}$ is uncountable, i.e. there are far 'more' irrationals than rationals! One way of making this seem more digestible is to consider the problem of choosing numbers at random from an interval in $\mathbb{R}$.

Recall that rational numbers are precisely those real numbers whose decimal expansion recurs (we include 'terminates' under 'recurs'). Now imagine choosing a real number from [0,1] at random: think of the set $\mathbb{R}$ as a pond containing all real numbers, and imagine you are 'fishing' in this pond, pulling out one number at a time.

How likely is it that the first number will be rational, i.e. how likely are we to find a number whose expansion recurs? It would be like rolling a ten-sided die infinitely many times and expecting, after a finite number of throws, to say with certainty that *all* subsequent throws will give the same digit. This does not seem at all likely, and we should therefore not be too surprised to find that countable sets (including $\mathbb{Q}$) will be among those we can 'neglect' when measuring sets on the real line in the 'unbiased' or uniform way in which we have used the term 'random' so far. Possibly more surprising, however, will be the discovery that even some uncountable sets can be 'negligible' from the point of view adopted here.

1.1.3 Topological properties of sets in $\mathbb{R}$

Recall the definition of an *open set* $O \subset \mathbb{R}$:

Definition 1.1

A subset O of the real line $\mathbb{R}$ is *open* if it is a union of open intervals, i.e. for intervals $(I_\alpha)_{\alpha \in \Lambda}$, where Λ is some index set (countable or not)

$$O = \bigcup_{\alpha \in \Lambda} I_\alpha.$$

A set is *closed* if its complement is open. Open sets in $\mathbb{R}^n$ ($n > 1$) can be defined as unions of n-fold products of open intervals.

This definition seems more general than it actually is, since, on $\mathbb{R}$, countable unions will always suffice – though the freedom to work with general unions will be convenient later on. If Λ is an index set and I_α is an open interval for each $\alpha \in \Lambda$, then there exists a countable collection $(I_{\alpha_k})_{k \geq 1}$ of these intervals whose union equals $\cup_{\alpha \in \Lambda} I_\alpha$. What is more, the sequence of intervals can be chosen to be pairwise disjoint.

It is easy to see that a finite intersection of open sets is open; however, a countable intersection of open sets need not be open: let $O_n = (-\frac{1}{n}, 1)$ for $n \geq 1$, then $E = \bigcap_{n=1}^{\infty} O_n = [0, 1)$ is not open.

Note that $\mathbb{R}$, unlike $\mathbb{R}^n$ or more general spaces, has a *linear order*, i.e. given $x, y \in \mathbb{R}$ we can decide whether $x \leq y$ or $y \leq x$. Thus u is an *upper bound* for a set $A \subset \mathbb{R}$ if $a \leq u$ for all $a \in A$, and a *lower bound* is defined similarly. The *supremum* (or least upper bound) is then the minimum of all upper bounds and written $\sup A$. The *infimum* (or greatest lower bound) $\inf A$ is defined as the maximum of all lower bounds. The *completeness property* of $\mathbb{R}$ can be expressed by the statement that every set which is bounded above has a supremum.

A real function f is said to be *continuous* if $f^{-1}(O)$ is open for each open set O. Every continuous real function defined on a closed bounded set *attains its bounds* on such a set, i.e. has a minimum and maximum value there. For example, if $f : [a, b] \to \mathbb{R}$ is continuous, $M = \sup\{f(x) : x \in [a, b]\} = f(x_{\max})$, $m = \inf\{f(x) : x \in [a, b]\} = f(x_{\min})$ for some points $x_{\max}, x_{\min} \in [a, b]$. The Intermediate Value Theorem says that a continuous function takes all intermediate values between the extreme ones, i.e. for each $y \in [m, M]$ there is a $\theta \in [a, b]$ such that $y = f(\theta)$.

Specialising to real sequences (x_n), we can further define the *upper limit* $\limsup_n x_n$ as

$$\inf\{\sup_{m \geq n} x_m : n \in \mathbb{N}\}$$

and the *lower limit* $\liminf_n x_n$ as

$$\sup\{\inf_{m \geq n} x_m : n \in \mathbb{N}\}.$$

The sequence x_n converges if and only if these quantities coincide and their common value is then its limit. Recall that a sequence (x_n) *converges* and the real number x is its *limit*, written $x = \lim_{x \to \infty} x_n$, if for every $\varepsilon > 0$ there is an $N \in \mathbb{N}$ such that $|x_n - x| < \varepsilon$ whenever $n \geq N$. A series $\sum_{n \geq 1} a_n$ converges if the sequence $x_m = \sum_{n=1}^{m} a_n$ of its partial sums converges, and its limit is then the *sum* $\sum_{n=1}^{\infty} a_n$ of the series.

1.2 The Riemann integral: scope and limitations

In this section we give a brief review of the Riemann integral, which forms part of the staple diet in introductory analysis courses, and consider some of the reasons why it does not suffice for more advanced applications.

Let $f : [a, b] \to \mathbb{R}$ be a bounded real function, where a, b, with $a < b$, are real numbers. A *partition* of $[a, b]$ is a finite set $P = \{a_0, a_1, a_2, \ldots, a_n\}$ with

$$a = a_0 < a_1 < a_2 < \cdots < a_n = b.$$

The partition P gives rise to the *upper* and *lower Riemann sums*

$$U(P, f) = \sum_{i=1}^{n} M_i \Delta a_i, \qquad L(P, f) = \sum_{i=1}^{n} m_i \Delta a_i$$

where $\Delta a_i = a_i - a_{i-1}$,

$$M_i = \sup_{a_{i-1} \leq x \leq a_i} f(x)$$

and

$$m_i = \inf_{a_{i-1} \leq x \leq a_i} f(x)$$

for each $i \leq n$. (Note that M_i and m_i are well-defined real numbers since f is bounded on each interval $[a_{i-1}, a_i]$.)

In order to define the Riemann integral of f, one first shows that for any given partition P, $L(P, f) \leq U(P, f)$, and next that for any *refinement*, i.e. a partition $P' \supset P$, we must have $L(P, f) \leq L(P', f)$ and $U(P', f) \leq U(P, f)$. Finally, since for any two partitions P_1 and P_2, their union $P_1 \cup P_2$ is a refinement of both, we see that $L(P, f) \leq U(Q, f)$ for *any* partitions P, Q. The set $\{L(P, f) : P \text{ is a partition of } [a, b]\}$ is thus bounded above in $\mathbb{R}$, and we call its supremum the *lower integral* $\underline{\int_a^b} f$ of f on $[a, b]$. Similarly, the infimum of the set of upper sums is the *upper integral* $\overline{\int_a^b} f$. The function f is now said to be

Riemann-integrable on $[a, b]$ if these two numbers coincide, and their common value is the *Riemann integral* of f, denoted by $\int_a^b f$ or, more commonly,

$$\int_a^b f(x)\, dx.$$

This definition does not provide a convenient criterion for checking the integrability of particular functions; however, the following formulation provides a useful criterion for integrability – see [8] for a proof.

Theorem 1.2 (Riemann's criterion)

$f : [a, b] \to \mathbb{R}$ is Riemann-integrable if and only if for every $\varepsilon > 0$ there exists a partition P_ε such that $U(P_\varepsilon, f) - L(P_\varepsilon, f) < \varepsilon$.

Example 1.3

We calculate $\int_0^1 f(x)\, dx$ when $f(x) = \sqrt{x}$: our immediate problem is that square roots are hard to find except for perfect squares. Therefore we take partition points which are perfect squares, even though this means that the interval lengths of the different intervals do not stay the same (there is nothing to say that they should do, even if it often simplifies the calculations). In fact, take the sequence of partitions

$$P_n = \{0, (\frac{1}{n})^2, (\frac{2}{n})^2, \ldots, (\frac{i}{n})^2, \ldots, 1\}$$

and consider the upper and lower sums, using the fact that f is increasing:

$$U(P_n, f) = \sum_{i=1}^n (\frac{i}{n})\{(\frac{i}{n})^2 - (\frac{i-1}{n})^2\} = \frac{1}{n^3} \sum_{i=1}^n (2i^2 - i)$$

$$L(P_n, f) = \sum_{i=1}^n (\frac{i-1}{n})\{(\frac{i}{n})^2 - (\frac{i-1}{n})^2\} = \frac{1}{n^3} \sum_{i=1}^n (2i^2 - 3i + 1).$$

Hence

$$U(P_n, f) - L(P_n, f) = \frac{1}{n^3} \sum_{i=1}^n (2i - 1) = \frac{1}{n^3}\{n(n+1) - n\} = \frac{1}{n}.$$

By choosing n large enough, we can make this difference less than any given $\varepsilon > 0$, hence f is integrable. The integral must be $\frac{2}{3}$, since both $U(P_n, f)$ and $L(P_n, f)$ converge to this value, as is easily seen.

Riemann's criterion still does not give us a precise picture of the *class* of Riemann-integrable functions. However, it is easy to show (see [8]) that any bounded *monotone* function belongs to this class, and only a little more difficult to see that any *continuous* function $f : [a, b] \to \mathbb{R}$ (which is of course automatically bounded) will be Riemann-integrable.

This provides quite sufficient information for many practical purposes, and the tedium of calculations such as that given above can be avoided by proving the following theorem:

Theorem 1.4 (fundamental theorem of calculus)

If $f : [a, b] \to \mathbb{R}$ is continuous and the function $F : [a, b] \to \mathbb{R}$ has derivative f (i.e. $F' = f$ on (a, b)) then

$$F(b) - F(a) = \int_a^b f(x)\, dx.$$

This result therefore links the Riemann integral with differentiation, and displays F as a *primitive* (also called 'anti-derivative') of f:

$$F(x) = \int_{-a}^x f(x)\, dx$$

up to a constant, thus justifying the elementary techniques of integration that form part of any calculus course.

We can relax the continuity requirement. A trivial step is to assume f bounded and continuous on $[a, b]$ except at finitely many points. Then f is Riemann-integrable. To see this, split the interval into pieces on which f is continuous. Then f is integrable on each and hence one can derive integrability of f on the whole interval. As an example consider a function f equal to zero for all $x \in [0, 1]$ except $a_1, \ldots, a_n$ where it equals 1. It is integrable with integral over $[0, 1]$ equal to 0.

Taking this further, however, will require the power of the Lebesgue theory: in Theorem 4.33 we show that f is Riemann-integrable if and only if it is continuous at 'almost all' points of $[a, b]$. This result is by no means trivial, as you will discover if you try to prove directly that the following function f, due to *Dirichlet*, is Riemann-integrable over $[0, 1]$:

$$f(x) = \begin{cases} \frac{1}{n} & \text{if } x = \frac{m}{n} \in \mathbb{Q} \\ 0 & \text{if } x \notin \mathbb{Q}. \end{cases}$$

In fact, it is not difficult (see [8]) to show that f is continuous at each irrational and discontinuous at every rational point, hence (as we will see) is continuous at 'almost all' points of $[0, 1]$.

Since the purpose of this book is to present *Lebesgue's* theory of integration, we should discuss *why* we need a new theory of integration at all: what, if anything, is wrong with the simple Riemann integral described above?

First, scope: it doesn't deal with all the kinds of functions that we hope to handle.

The results that are most easily proved rely on continuous functions on bounded intervals. In order to handle integrals over unbounded intervals, e.g.

$$\int_{-\infty}^{\infty} e^{-x^2}\, dx$$

or the integral of an unbounded function

$$\int_0^1 \frac{1}{\sqrt{x}}\, dx,$$

we have to resort to 'improper' Riemann integrals, defined by a limit process. For example consider the integrals

$$\int_{-n}^{n} e^{-x^2}\, dx, \qquad \int_{\varepsilon}^{1} \frac{1}{\sqrt{x}}\, dx,$$

and let $n \to \infty$ or $\varepsilon \to 0$ respectively. This isn't all that serious a flaw.

Second, dependence on intervals: we have no easy way of integrating over more general sets, or of integrating functions whose values are distributed 'awkwardly' over sets that differ greatly from intervals. For example, consider the upper and lower sums for the indicator function $\mathbf{1}_{\mathbb{Q}}$ of $\mathbb{Q}$ over $[0,1]$; however we partition $[0,1]$, each subinterval must contain both rational and irrational points; thus each upper sum is 1 and each lower sum 0. Hence we cannot calculate the Riemann integral of f over the interval $[0,1]$; it is simply 'too discontinuous'. (You may easily convince yourself that f is discontinuous at all points of $[0,1]$.)

Third, lack of completeness: rather more importantly from the point of view of applications, the Riemann integral doesn't interact well with taking the limit of a sequence of functions. One may expect results of the following form: if a sequence (f_n) of Riemann-integrable functions converges (in some appropriate sense) to f, then $\int_a^b f_n\, dx \to \int_a^b f\, dx$.

We give two counterexamples showing what difficulties can arise if the functions (f_n) converge to f pointwise, i.e. $f_n(x) \to f(x)$ for all x.

1. The limit need not be Riemann-integrable, and so the convergence question does not even make sense. Here we may take $f = \mathbf{1}_{\mathbb{Q}}$, $f_n = \mathbf{1}_{A_n}$ where

$A_n = \{q_1, \ldots, q_n\}$, and the sequence (q_n), $n \geq 1$ is an enumeration of the rationals, so that (f_n) is even monotone increasing.

2. The limit is Riemann-integrable, but the convergence of Riemann integrals does not hold. Let $f = 0$, consider $[a, b] = [0, 1]$, and put

$$f_n(x) = \begin{cases} 4n^2 x & \text{if } 0 \leq x < \frac{1}{2n} \\ 4n - 4n^2 x & \text{if } \frac{1}{2n} \leq x < \frac{1}{n} \\ 0 & \text{if } \frac{1}{n} \leq x \leq 1. \end{cases}$$

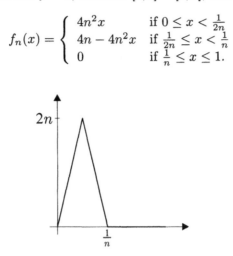

Figure 1.1 Graph of f_n

This is a continuous function with integral 1. On the other hand, the sequence $f_n(x)$ converges to $f = 0$ since for all x, $f_n(x) = 0$ for n sufficiently large (such that $\frac{1}{n} < x$). See Figure 1.1.

To avoid problems of this kind, we can introduce the idea of *uniform* convergence: a sequence (f_n) in $C[0, 1]$ converges uniformly to f if the sequence $a_n = \sup\{|f_n(x) - f(x)| : 0 \leq x \leq 1\}$ converges to 0. In this case one can easily prove the convergence of the Riemann integrals:

$$\int_0^1 f_n(x)\, dx \to \int_0^1 f(x)\, dx.$$

However, the 'distance' $\sup\{|f(x) - g(x)| : 0 \leq x \leq 1\}$ has nothing to do with integration as such and the uniform convergence is too restrictive for many applications. A more natural concept of 'distance', given by $\int_0^1 |f(x) - g(x)|\, dx$, leads to another problem. By defining

$$g_n(x) = \begin{cases} 0 & \text{if } 0 \leq x \leq \frac{1}{2} \\ n(x - \frac{1}{2}) & \text{if } \frac{1}{2} < x < \frac{1}{2} + \frac{1}{n} \\ 1 & \text{otherwise,} \end{cases}$$

it can be shown that $\int_0^1 |g_n(x) - g_m(x)|\, dx \to 0$ as $m, n \to \infty$; in Figure 1.2 the shaded area vanishes. (We say that (f_n) is a *Cauchy sequence* in this distance.)

Yet there is no continuous function f to which this sequence converges since the pointwise limit is $f(x) = 1$ for $x > \frac{1}{2}$ and 0 otherwise, so that $f = \mathbf{1}_{(\frac{1}{2}, 1]}$. So the space $C([0, 1])$ of all continuous functions $f \colon [0, 1] \to \mathbb{R}$ is too small from this point of view.

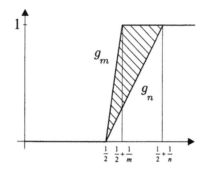

Figure 1.2 Graphs of g_n, g_m

This is rather similar to the situation which leads one to work with $\mathbb{R}$ rather than just with the set of rationals $\mathbb{Q}$ (there are Cauchy sequences without limits in $\mathbb{Q}$, for example a sequence of rational approximations of $\sqrt{2}$). Recalling the crucial importance of completeness in the case of $\mathbb{R}$, we naturally look for a theory of integration which does not have this shortcoming. In the process we shall find that our new theory, which will include the Riemann integral as a special case, also solves the other problems listed.

1.3 Choosing numbers at random

Before we start to develop the theory of *Lebesgue measure* to make sense of the 'length' of a general subset of $\mathbb{R}$, let us pause to consider some practical motivation. The simplicity of elementary probability with finite sample spaces vanishes rapidly when we have an *infinite* number of outcomes, such as when we 'pick a number between 0 and 1 at random'. We face making sense of the 'probability' that a given $x \in [0, 1]$ is chosen. A similar, slightly more general question, is the following: what is the probability that the number we pick is rational?

First a prior question: what do we mean by saying that we pick the number x at random? 'Random' plausibly means that in each such trial, each real number is 'equally likely' to be picked, so that we impose the uniform probability distribution on $[0, 1]$. But the 'number' of possible choices is infinite. Hence the event A_x that a fixed x is chosen ought to have zero probability. On the other hand, since *some* number between 0 and 1 *is* chosen, and it is not impossible that it could be our x. Thus a set $A_x \neq \emptyset$ can have $P(A_x) = 0$. Our way of 'measuring' probabilities need not, therefore, be able to distinguish completely between sets – we cannot really expect this in general if we want to handle infinite sets.

We can go slightly further: the probability that any one of a *finite* set of reals $A = \{x_1, x_2, \ldots, x_n\}$ is selected should also be 0, since it seems natural that this probability $P(A)$ should equal $\sum_{i=1}^{n} P(\{x_i\})$. We can extend this to claim the finite additivity property of the probability function $A \mapsto P(A)$, i.e. that if $A_1, A_2, \ldots, A_n$ are *disjoint* sets, then $P(\bigcup_{i=1}^{n} A_i) = \sum_{i=1}^{n} P(A_i)$. This claim looks very plausible, and we shall see that it becomes an essential feature of any sensible basis for a calculus of probabilities.

Less obvious is the claim that, under the uniform distribution, any *countably infinite* set, such as $\mathbb{Q}$, must also carry probability 0 – yet that is exactly what an analysis of the 'area under the graph' of the function $\mathbf{1}_{\mathbb{Q}}$ suggests. We can reinterpret this as a result of a 'continuity property' of the mapping $A \mapsto P(A)$ when we let $n \to \infty$ in the above: if the sequence (A_i) of subsets of $\mathbb{R}$ is disjoint then we would like to have

$$P(\bigcup_{i=1}^{\infty} A_i) = \lim_{n \to \infty} \sum_{i=1}^{n} P(A_i) = \sum_{i=1}^{\infty} P(A_i).$$

We shall see in Chapter 2 that this condition is indeed satisfied by Lebesgue measure on the real line $\mathbb{R}$, and it will be used as the defining property of abstract measures on arbitrary sets.

There is much more to probability than is developed in this book: for example, we do not discuss finite sample spaces and the elegant combinatorial ideas that characterise good introductions to probability, such as [6] and [10]. Our focus throughout remains on the essential role played by Lebesgue measure in the description of probabilistic phenomena based on infinite sample spaces. This leads us to leave to one side many of the interesting examples and applications which can be found in these texts, and provide, instead, a consistent development of the theoretical underpinnings of random variables with densities.

2

Measure

2.1 Null sets

The idea of a 'negligible' set relates to one of the limitations of the Riemann integral, as we saw in the previous chapter. Since the function $f = 1_{\mathbb{Q}}$ takes a non-zero value only on $\mathbb{Q}$, and equals 1 there, the 'area under its graph' (if such makes sense) must be very closely linked to the 'length' of the set $\mathbb{Q}$. This is why it turns out that we cannot integrate f in the Riemann sense: the sets $\mathbb{Q}$ and $\mathbb{R} \setminus \mathbb{Q}$ are so different from intervals that it is not clear how we should measure their 'lengths' and it is clear that the 'integral' of f over $[0,1]$ should equal the 'length' of the set of rationals in $[0,1]$. So how *should* we define this concept for more general sets?

The obvious way of defining the 'length' of a set is to start with intervals nonetheless. Suppose that I is a bounded interval of any kind, i.e. $I = [a,b]$, $I = [a,b)$, $I = (a,b]$ or $I = (a,b)$. We simply define the length of I as $l(I) = b - a$ in each case.

As a particular case we have $l(\{a\}) = l([a,a]) = 0$. It is then natural to say that a one-element set is 'null'. Before we extend this idea to more general sets, first consider the length of a finite set. A finite set is not an interval but since a single point has length 0, adding finitely many such lengths together should still give 0. The underlying concept here is that if we decompose a set into a finite number of disjoint intervals, we compute the length of this set by adding the lengths of the pieces.

As we have seen, in general it may not be always possible to decompose a set into non-trivial intervals. Therefore, we consider systems of intervals that cover a given set. We shall generalise the above idea by allowing a countable number of covering intervals. Thus we arrive at the following more general definition of sets of 'zero length':

Definition 2.1

A *null set* $A \subseteq \mathbb{R}$ is a set that may be covered by a sequence of intervals of arbitrarily small total length, i.e. given any $\varepsilon > 0$ we can find a sequence $\{I_n : n \geq 1\}$ of intervals such that

$$A \subseteq \bigcup_{n=1}^{\infty} I_n$$

and

$$\sum_{n=1}^{\infty} l(I_n) < \varepsilon.$$

(We also say simply that '*A is null*'.)

Exercise 2.1

Show that we get an equivalent notion if in the above definition we replace the word 'intervals' by any of these: 'open intervals', 'closed intervals', 'the intervals of the form $(a, b]$', 'the intervals of the form $[a, b)$'.

Note that the intervals do not need to be disjoint. It follows at once from the definition that the empty set is null.

Next, any one-element set $\{x\}$ is a null set. For, let $\varepsilon > 0$ and take $I_1 = (x - \frac{\varepsilon}{4}, x + \frac{\varepsilon}{4})$, $I_n = [0, 0]$ for $n \geq 2$. (Why take $I_n = [0, 0]$ for $n \geq 2$? Well, why not! We could equally have taken $I_n = (0, 0) = \emptyset$, of course!) Now

$$\sum_{n=1}^{\infty} l(I_n) = l(I_1) = \frac{\varepsilon}{2} < \varepsilon.$$

More generally, any countable set $A = \{x_1, x_2, ...\}$ is null. The simplest way to show this is to take $I_n = [x_n, x_n]$, for all n. However, as a gentle introduction to the next theorem we will cover A by open intervals. This way it is more fun.

For, let $\varepsilon > 0$ and cover A with the following sequence of intervals:

$$I_1 = (x_1 - \tfrac{\varepsilon}{8}, x_1 + \tfrac{\varepsilon}{8}) \quad l(I_1) = \frac{1}{2}\varepsilon \cdot \frac{1}{2^1}$$

$$I_2 = (x_2 - \tfrac{\varepsilon}{16}, x_2 + \tfrac{\varepsilon}{16}) \quad l(I_2) = \frac{1}{2}\varepsilon \cdot \frac{1}{2^2}$$

$$I_3 = (x_3 - \tfrac{\varepsilon}{32}, x_3 + \tfrac{\varepsilon}{32}) \quad l(I_3) = \frac{1}{2}\varepsilon \cdot \frac{1}{2^3}$$

$$\cdots \qquad\qquad \cdots$$

$$I_n = (x_n - \tfrac{\varepsilon}{2\cdot 2^n}, x_n + \tfrac{\varepsilon}{2\cdot 2^n}) \; l(I_n) = \frac{1}{2}\varepsilon \cdot \frac{1}{2^n}$$

Since $\sum_{n=1}^{\infty} \frac{1}{2^n} = 1$,

$$\sum_{n=1}^{\infty} l(I_n) = \frac{\varepsilon}{2} < \varepsilon,$$

as needed.

Here we have the following situation: A is the union of countably many one-element sets. Each of them is null and A turns out to be null as well.

We can generalise this simple observation:

Theorem 2.2

If $(N_n)_{n \geq 1}$ is a sequence of null sets, then their union

$$N = \bigcup_{n=1}^{\infty} N_n$$

is also null.

Proof

We assume that all N_n, $n \geq 1$, are null and to show that the same is true for N we take any $\varepsilon > 0$. Our goal is to cover the set N by countably many intervals with total length less than ε.

The proof goes in three steps, each being a little bit tricky.

Step 1. We carefully cover each N_n by intervals.

'Carefully' means that the lengths have to be small. 'Small' means that we are going to add them up later to end up with a small number (and 'small' here means less than ε).

Since N_1 is null, there exist intervals I_k^1, $k \geq 1$, such that

$$\sum_{k=1}^{\infty} l(I_k^1) < \frac{\varepsilon}{2}, \quad N_1 \subseteq \bigcup_{k=1}^{\infty} I_k^1.$$

For N_2 we find a system of intervals I_k^2, $k \geq 1$, with

$$\sum_{k=1}^{\infty} l(I_k^2) < \frac{\varepsilon}{4}, \quad N_2 \subseteq \bigcup_{k=1}^{\infty} I_k^2.$$

You can see a cunning plan of making the total lengths smaller at each step at a geometric rate. In general, we cover N_n with intervals I_k^n, $k \geq 1$, whose total length is less than $\frac{\varepsilon}{2^n}$:

$$\sum_{k=1}^{\infty} l(I_k^n) < \frac{\varepsilon}{2^n}, \quad N_n \subseteq \bigcup_{k=1}^{\infty} I_k^n.$$

Step 2. The intervals I_k^n are formed into a sequence.

We arrange the countable family of intervals $\{I_k^n\}_{k \geq 1, n \geq 1}$ into a sequence J_j, $j \geq 1$. For instance we put $J_1 = I_1^1$, $J_2 = I_2^1$, $J_3 = I_1^2$, $J_4 = I_3^1$, etc. so that none of the I_k^n are skipped. The union of the new system of intervals is the same as the union of the old one and so

$$N = \bigcup_{n=1}^{\infty} N_n \subseteq \bigcup_{n=1}^{\infty} \bigcup_{k=1}^{\infty} I_k^n = \bigcup_{j=1}^{\infty} J_j.$$

Step 3. Compute the total length of J_j.

This is tricky because we have a series of numbers with two indices:

$$\sum_{j=1}^{\infty} l(J_j) = \sum_{n=1, k=1}^{\infty} l(I_k^n).$$

Now we wish to write this as a series of numbers, each being the sum of a series. We can rearrange the double sum because the components are non-negative (a fact from elementary calculus)

$$\sum_{n=1, k=1}^{\infty} l(I_k^n) = \sum_{n=1}^{\infty} \left(\sum_{k=1}^{\infty} l(I_k^n) \right) < \sum_{n=1}^{\infty} \frac{\varepsilon}{2^n} = \varepsilon,$$

which completes the proof. $\qquad\qquad\qquad\qquad\qquad\qquad\qquad\qquad\qquad\qquad\square$

Thus any countable set is null, and null sets appear to be closely related to countable sets – this is no surprise as any proper interval is uncountable, so any countable subset is quite 'sparse' when compared with an interval, hence makes no real contribution to its 'length'. (You may also have noticed the similarity between Step 2 in the above proof and the 'diagonal argument' which is commonly used to show that $\mathbb{Q}$ is a countable set.)

However, uncountable sets *can* be null, provided their points are sufficiently 'sparsely distributed', as the following famous example, due to Cantor, shows:

1. Start with the interval $[0, 1]$, remove the 'middle third', i.e. the interval $(\frac{1}{3}, \frac{2}{3})$, obtaining the set C_1, which consists of the two intervals $[0, \frac{1}{3}]$ and $[\frac{2}{3}, 1]$.

2. Next remove the middle third of each of these two intervals, leaving C_2, consisting of four intervals, each of length $\frac{1}{9}$, etc. (See Figure 2.1.)

3. At the nth stage we have a set C_n, consisting of 2^n disjoint closed intervals, each of length $\frac{1}{3^n}$. Thus the total length of C_n is $\left(\frac{2}{3}\right)^n$.

Figure 2.1 Cantor set construction (C_3)

We call

$$C = \bigcap_{n=1}^{\infty} C_n$$

the *Cantor set*.

Now we show that C is null as promised.

Given any $\varepsilon > 0$, choose n so large that $\left(\frac{2}{3}\right)^n < \varepsilon$. Since $C \subseteq C_n$, and C_n consists of a (finite) sequence of intervals of total length less than ε, we see that C is a null set.

All that remains is to check that C is an uncountable set. This is left for you as

Exercise 2.2

Prove that C is uncountable.

Hint Adapt the proof of the uncountability of $\mathbb{R}$: begin by expressing each x in $[0, 1]$ in ternary form:

$$x = \sum_{k=1}^{\infty} \frac{a_k}{3^k} = 0.a_1 a_2 \ldots$$

with $a_k = 0$, 1 or 2. Note that $x \in C$ iff all its a_k equal 0 or 2.

Why is the Cantor set null, even though it is uncountable? Clearly it is the distribution of its points, the fact that it is 'spread out' all over [0,1], which causes the trouble. This makes it the source of many examples which show that intuitively 'obvious' things are not always true! For example, we can use the Cantor set to define a function, due to Lebesgue, with very odd properties:

If $x \in [0, 1]$ has ternary expansion (a_n), i.e. $x = 0.a_1 a_2 \ldots$ with $a_n = 0, 1$ or 2, define N as the first index n for which $a_n = 1$, and set $N = \infty$ if none of the a_n are 1 (i.e. when $x \in C$). Now set $b_n = \frac{a_n}{2}$ for $n < N$ and $b_N = 1$, and let $F(x) = \sum_{n=1}^{N} \frac{b_n}{2^n}$ for each $x \in [0, 1]$. Clearly, this function is monotone increasing and has $F(0) = 0$, $F(1) = 1$. Yet it is constant on the middle thirds (i.e. the complement of C), so all its increase occurs on the Cantor set. Since we have shown that C is a null set, F 'grows' from 0 to 1 entirely on a 'negligible' set. The following exercise shows that it has no jumps!

Exercise 2.3

Prove that the Lebesgue function F is continuous and sketch its partial graph.

2.2 Outer measure

The simple concept of null sets provides the key to our idea of length, since it tells us what we can 'ignore'. A quite general notion of 'length' is now provided by:

Definition 2.3

The (Lebesgue) *outer measure* of any set $A \subseteq \mathbb{R}$ is given by

$$m^*(A) = \inf Z_A$$

where

$$Z_A = \{\sum_{n=1}^{\infty} l(I_n) : I_n \text{ are intervals, } A \subseteq \bigcup_{n=1}^{\infty} I_n\}.$$

We say the $(I_n)_{n \geq 1}$ *cover* the set A. So the outer measure is the infimum of lengths of all possible covers of A. (Note again that some of the I_n may be empty; this avoids having to worry whether the sequence (I_n) has finitely or infinitely many different members.)

Clearly $m^*(A) \geq 0$ for any $A \subseteq \mathbb{R}$. For some sets A, the series $\sum_{n=1}^{\infty} l(I_n)$ may diverge for any covering of A, so $m^*(A)$ may by equal to ∞. Since we wish to be able to add the outer measures of various sets we have to adopt a convention to deal with infinity. An obvious choice is $a + \infty = \infty$, $\infty + \infty = \infty$ and a less obvious but quite practical assumption is $0 \times \infty = 0$, as we have already seen.

The set Z_A is bounded from below by 0, so the infimum always exists. If $r \in Z_A$, then $[r, +\infty] \subseteq Z_A$ (clearly, we may expand the first interval of any cover to increase the total length by any number). This shows that Z_A is either $\{+\infty\}$ or the interval $(x, +\infty]$ or $[x, +\infty]$ for some real number x. So the infimum of Z_A is just x.

First we show that the concept of null set is consistent with that of outer measure:

Theorem 2.4

$A \subseteq \mathbb{R}$ is a null set if and only if $m^*(A) = 0$.

Proof

Suppose that A is a null set. We wish to show that $\inf Z_A = 0$. To this end we show that for any $\varepsilon > 0$ we can find an element $z \in Z_A$ such that $z < \varepsilon$.

By the definition of null set we can find a sequence (I_n) of intervals covering A with $\sum_{n=1}^{\infty} l(I_n) < \varepsilon$ and so $\sum_{n=1}^{\infty} l(I_n)$ is the required element z of Z_A.

Conversely, if $A \subseteq \mathbb{R}$ has $m^*(A) = 0$, then by the definition of inf, given any $\varepsilon > 0$, there is $z \in Z_A$, $z < \varepsilon$. But a member of Z_A is the total length of some covering of A. That is, there is a covering (I_n) of A with total length less than ε, so A is null.   □

This combines our general outer measure with the special case of 'zero measure'. Note that $m^*(\varnothing) = 0$, $m^*(\{x\}) = 0$ for any $x \in \mathbb{R}$, and $m^*(\mathbb{Q}) = 0$ (and in fact, for any countable X, $m^*(X) = 0$).

Next we observe that m^* is monotone: the bigger the set, the greater its outer measure.

Proposition 2.5

If $A \subset B$ then $m^*(A) \leq m^*(B)$.

Hint Show that $Z_B \subset Z_A$ and use the definition of inf.

The second step is to relate outer measure to the length of an interval. This innocent result contains the crux of the theory, since it shows that the formal definition of m^*, which is applicable to *all* subsets of $\mathbb{R}$, coincides with the intuitive idea for intervals, where our thought processes began. We must therefore expect the proof to contain some hidden depths, and we have to tackle these in stages: the hard work lies in showing that the length of the interval cannot be greater than its outer measure: for this we need to appeal to the famous Heine–Borel theorem, which states that every closed, bounded subset B of $\mathbb{R}$ is *compact*: given any collection of open sets O_α covering B (i.e. $B \subset \bigcup_\alpha O_\alpha$), there is a finite subcollection $(O_{\alpha_i})_{i \leq n}$ which still covers B, i.e. $B \subset \bigcup_{i=1}^n O_{\alpha_i}$ (for a proof see [1]).

Theorem 2.6

The outer measure of an interval equals its length.

Proof

If I is unbounded, then it is clear that it cannot be covered by a system of intervals with finite total length. This shows that $m^*(I) = \infty$ and so $m^*(I) = l(I) = \infty$.

So we restrict ourselves to bounded intervals.

Step 1. $m^*(I) \leq l(I)$.

We claim that $l(I) \in Z_I$. Take the following sequence of intervals: $I_1 = I$, $I_n = [0, 0]$ for $n \geq 2$. This sequence covers the set I, and the total length is equal to the length of I, hence $l(I) \in Z_I$. This is sufficient since the infimum of Z_I cannot exceed any of its elements.

Step 2. $l(I) \leq m^*(I)$.

(i) $I = [a, b]$. We shall show that for any $\varepsilon > 0$,

$$l([a, b]) \leq m^*([a, b]) + \varepsilon. \tag{2.1}$$

This is sufficient since we may obtain the required inequality passing to the limit, $\varepsilon \to 0$. (Note that if $x, y \in \mathbb{R}$ and $y > x$ then there is an $\varepsilon > 0$ with $y > x + \varepsilon$, e.g. $\varepsilon = \frac{1}{2}(y - x)$.)

So we take an arbitrary $\varepsilon > 0$. By the definition of outer measure we can find a sequence of intervals I_n covering $[a, b]$ such that

$$\sum_{n=1}^\infty l(I_n) \leq m^*([a, b]) + \frac{\varepsilon}{2}. \tag{2.2}$$

We shall slightly increase each of the intervals to an open one. Let the endpoints of I_n be a_n, b_n, and we take

$$J_n = \left(a_n - \frac{\varepsilon}{2^{n+2}}, b_n + \frac{\varepsilon}{2^{n+2}}\right).$$

It is clear that

$$l(I_n) = l(J_n) - \frac{\varepsilon}{2^{n+1}},$$

so that

$$\sum_{n=1}^{\infty} l(I_n) = \sum_{n=1}^{\infty} l(J_n) - \frac{\varepsilon}{2}.$$

We insert this in (2.2) and we have

$$\sum_{n=1}^{\infty} l(J_n) \le m^*([a,b]) + \varepsilon. \tag{2.3}$$

The new sequence of intervals of course covers $[a,b]$, so by the Heine–Borel theorem we can choose a finite number of J_n to cover $[a,b]$ (the set $[a,b]$ is compact in $\mathbb{R}$). We can add some intervals to this finite family to form an initial segment of the sequence (J_n) – just for simplicity of notation. So for some finite index m we have

$$[a,b] \subseteq \bigcup_{n=1}^{m} J_n. \tag{2.4}$$

Let $J_n = (c_n, d_n)$. Put $c = \min\{c_1, \ldots, c_m\}$, $d = \max\{d_1, \ldots, d_m\}$. The covering (2.4) means that $c < a$ and $b < d$, hence $l([a,b]) < d - c$.

Next, the number $d - c$ is certainly smaller than the total length of J_n, $n = 1, \ldots, m$ (some overlapping takes place) and

$$l([a,b]) < d - c < \sum_{n=1}^{m} l(J_n). \tag{2.5}$$

Now it is sufficient to put (2.3) and (2.5) together in order to deduce (2.1) (the finite sum is less than or equal to the sum of the series since all terms are non-negative).

(ii) $I = (a,b)$. As before, it is sufficient to show (2.1). Let us fix any $\varepsilon > 0$.

$$l((a,b)) = l([a + \frac{\varepsilon}{2}, b - \frac{\varepsilon}{2}]) + \varepsilon$$
$$\le m^*([a + \frac{\varepsilon}{2}, b - \frac{\varepsilon}{2}]) + \varepsilon \quad \text{(by (1))}$$
$$\le m^*((a,b)) + \varepsilon \quad \text{(by Proposition 2.5)}.$$

(iii) $I = [a, b)$ or $I = (a, b]$.

$$l(I) = l((a,b)) \leq m^*((a,b)) \quad \text{(by (2))}$$
$$\leq m^*(I) \quad \text{(by Proposition 2.5)},$$

which completes the proof. □

Having shown that outer measure coincides with the natural concept of length for intervals, we now need to investigate its properties. The next theorem gives us an important technical tool which will be used in many proofs.

Theorem 2.7

Outer measure is countably subadditive, i.e. for any sequence of sets $\{E_n\}$,

$$m^*\left(\bigcup_{n=1}^{\infty} E_n\right) \leq \sum_{n=1}^{\infty} m^*(E_n).$$

(Note that both sides may be infinite here.)

Proof (a warm-up)

Let us prove first a simpler statement:

$$m^*(E_1 \cup E_2) \leq m^*(E_1) + m^*(E_2).$$

Take an $\varepsilon > 0$ and we show an even easier inequality

$$m^*(E_1 \cup E_2) \leq m^*(E_1) + m^*(E_2) + \varepsilon.$$

This is however sufficient because taking $\varepsilon = \frac{1}{n}$ and letting $n \to \infty$ we get what we need.

So for any $\varepsilon > 0$ we find covering sequences $(I_k^1)_{k\geq 1}$ of E_1 and $(I_k^2)_{k\geq 1}$ of E_2 such that

$$\sum_{k=1}^{\infty} l(I_k^1) \leq m^*(E_1) + \frac{\varepsilon}{2},$$

$$\sum_{k=1}^{\infty} l(I_k^2) \leq m^*(E_2) + \frac{\varepsilon}{2};$$

hence, adding up,

$$\sum_{k=1}^{\infty} l(I_k^1) + \sum_{k=1}^{\infty} l(I_k^2) \leq m^*(E_1) + m^*(E_2) + \varepsilon.$$

The sequence of intervals $(I_1^1, I_1^2, I_2^1, I_2^2, I_3^1, I_3^2, \ldots)$ covers $E_1 \cup E_2$, hence

$$m^*(E_1 \cup E_2) \le \sum_{k=1}^{\infty} l(I_k^1) + \sum_{k=1}^{\infty} l(I_k^2),$$

which combined with the previous inequality gives the result.   □

Proof of the theorem

If the right-hand side is infinite, then the inequality is of course true. So, suppose that $\sum_{n=1}^{\infty} m^*(E_n) < \infty$. For each given $\varepsilon > 0$ and $n \ge 1$ find a covering sequence $(I_k^n)_{k \ge 1}$ of E_n with

$$\sum_{k=1}^{\infty} l(I_k^n) \le m^*(E_n) + \frac{\varepsilon}{2^n}.$$

The iterated series converges:

$$\sum_{n=1}^{\infty} \left(\sum_{k=1}^{\infty} l(I_k^n) \right) \le \sum_{n=1}^{\infty} m^*(E_n) + \varepsilon < \infty$$

and since all its terms are non-negative,

$$\sum_{n=1}^{\infty} \left(\sum_{k=1}^{\infty} l(I_k^n) \right) = \sum_{n,k=1}^{\infty} l(I_k^n).$$

The system of intervals $(I_k^n)_{k,n \ge 1}$ covers $\bigcup_{n=1}^{\infty} E_n$, hence

$$m^*\left(\bigcup_{n=1}^{\infty} E_n \right) \le \sum_{n,k=1}^{\infty} l(I_k^n) \le \sum_{n=1}^{\infty} m^*(E_n) + \varepsilon.$$

To complete the proof we let $\varepsilon \to 0$.   □

A similar result is of course true for a finite family $(E_n)_{n=1}^{m}$:

$$m^*\left(\bigcup_{n=1}^{m} E_n \right) \le \sum_{n=1}^{m} m^*(E_n).$$

It is a corollary to Theorem 2.7 with $E_k = \emptyset$ for $k > m$.

Exercise 2.4

Prove that if $m^*(A) = 0$ then for each B, $m^*(A \cup B) = m^*(B)$.

Hint Employ both monotonicity and subadditivity of outer measure.

Exercise 2.5

Prove that if $m^*(A \Delta B) = 0$, then $m^*(A) = m^*(B)$.

Hint Note that $A \subseteq B \cup (A \Delta B)$.

We conclude this section with a simple and intuitive property of outer measure. Note that the length of an interval does not change if we shift it along the real line: $l([a, b]) = l([a+t, b+t]) = b - a$ for example. Since the outer measure is defined in terms of the lengths of intervals, it is natural to expect it to share this property. For $A \subset \mathbb{R}$ and $t \in \mathbb{R}$ we put $A + t = \{a + t : a \in A\}$.

Proposition 2.8

Outer measure is translation-invariant, i.e.

$$m^*(A) = m^*(A + t)$$

for each A and t.

Hint Combine two facts: the length of interval does not change when the interval is shifted and outer measure is determined by the length of the coverings.

2.3 Lebesgue-measurable sets and Lebesgue measure

With outer measure, subadditivity (as in Theorem 2.7) is as far as we can get. We wish, however, to ensure that if sets (E_n) are pairwise disjoint (i.e. $E_i \cap E_j = \emptyset$ if $i \neq j$), then the inequality in Theorem 2.7 becomes an equality. It turns out that this will not in general be true for outer measure, although examples where it fails are quite difficult to construct (we give such examples in the Appendix). But our wish is an entirely reasonable one: any 'length function' should at least be finitely additive, since decomposing a set into finitely many disjoint pieces should not alter its length. Moreover, since we constructed our length function via approximation of complicated sets by 'simpler' sets (i.e. intervals) it seems fair to demand a *continuity property*: if pairwise disjoint (E_n) have union E, then the lengths of the sets $B_n = E \setminus \bigcup_{k=1}^{n} E_k$ may be expected to decrease to 0 as $n \to \infty$. Combining this with finite additivity leads quite naturally to the demand that 'length' should be *countably additive*, i.e. that

$$m^* \left(\bigcup_{n=1}^{\infty} E_n \right) = \sum_{n=1}^{\infty} m^*(E_n) \quad \text{when } E_i \cap E_j = \emptyset \text{ for } i \neq j.$$

We therefore turn to the task of finding the class of sets in $\mathbb{R}$ which have this property. It turns out that it is also the key property of the abstract concept of measure, and we will use it to provide mathematical foundations for probability.

In order to define the 'good' sets which have this property, it also seems plausible that such a set should apportion the outer measure of *every* set in $\mathbb{R}$ properly, as we state in Definition 2.9 below. Remarkably, this simple demand will suffice to guarantee that our 'good' sets have all the properties we demand of them!

Definition 2.9

A set $E \subseteq \mathbb{R}$ is (Lebesgue-) *measurable* if for every set $A \subseteq \mathbb{R}$ we have

$$m^*(A) = m^*(A \cap E) + m^*(A \cap E^c) \tag{2.6}$$

where $E^c = \mathbb{R} \setminus E$, and we write $E \in \mathcal{M}$.

We obviously have $A = (A \cap E) \cup (A \cap E^c)$, hence by Theorem 2.7 we have

$$m^*(A) \leq m^*(A \cap E) + m^*(A \cap E^c)$$

for any A and E. So our future task of verifying (2.6) has simplified: $E \in \mathcal{M}$ if and only if the following inequality holds:

$$m^*(A) \geq m^*(A \cap E) + m^*(A \cap E^c) \text{ for all } A \subseteq \mathbb{R}. \tag{2.7}$$

Now we give some examples of measurable sets.

Theorem 2.10

(i) Any null set is measurable.

(ii) Any interval is measurable.

Proof

(i) If N is a null set, then (Proposition 2.4) $m^*(N) = 0$. So for any $A \subset \mathbb{R}$ we have

$$m^*(A \cap N) \leq m^*(N) = 0 \quad \text{since } A \cap N \subseteq N$$
$$m^*(A \cap N^c) \leq m^*(A) \quad \text{since } A \cap N^c \subseteq A$$

and adding together we have proved (2.7).

(ii) Let $E = I$ be an interval. Suppose, for example, that $I = [a, b]$. Take any $A \subseteq \mathbb{R}$ and $\varepsilon > 0$. Find a covering of A with

$$m^*(A) \leq \sum_{n=1}^{\infty} l(I_n) \leq m^*(A) + \varepsilon.$$

Clearly the intervals $I_n' = I_n \cap [a, b]$ cover $A \cap [a, b]$ hence

$$m^*(A \cap [a, b]) \leq \sum_{n=1}^{\infty} l(I_n').$$

The intervals $I_n'' = I_n \cap (-\infty, a)$, $I_n''' = I_n \cap (b, +\infty)$ cover $A \cap [a, b]^c$ so

$$m^*(A \cap [a, b]^c) \leq \sum_{n=1}^{\infty} l(I_n'') + \sum_{n=1}^{\infty} l(I_n''').$$

Putting the above three inequalities together we obtain (2.7).

If I is unbounded, $I = [a, \infty)$ say, then the proof is even simpler since it is sufficient to consider $I_n' = I_n \cap [a, \infty)$ and $I_n'' = I_n \cap (-\infty, a)$. □

The fundamental properties of the class $\mathcal{M}$ of all Lebesgue-measurable sub-sets of $\mathbb{R}$ can now be proved. They fall into two categories: first we show that certain set operations on sets in $\mathcal{M}$ again produce sets in $\mathcal{M}$ (these are what we call 'closure properties') and second we prove that for sets in $\mathcal{M}$ the outer measure m^* has the property of countable additivity announced above.

Theorem 2.11

(i) $\mathbb{R} \in \mathcal{M}$,

(ii) if $E \in \mathcal{M}$ then $E^c \in \mathcal{M}$,

(iii) if $E_n \in \mathcal{M}$ for all $n = 1, 2, \ldots$ then $\bigcup_{n=1}^{\infty} E_n \in \mathcal{M}$.

Moreover, if $E_n \in \mathcal{M}$, $n = 1, 2, \ldots$ and $E_j \cap E_k = \emptyset$ for $j \neq k$, then

$$m^*\left(\bigcup_{n=1}^{\infty} E_n \right) = \sum_{n=1}^{\infty} m^*(E_n). \tag{2.8}$$

Remark 2.12

This result is the most important theorem in this chapter and provides the basis for all that follows. It also allows us to give names to the quantities under discussion.

Conditions (i)–(iii) mean that $\mathcal{M}$ is a σ-*field*. In other words, we say that a family of sets is a σ-field if it contains the base set and is closed under complements and countable unions. A $[0,\infty]$-valued function defined on a σ-field is called a *measure* if it satisfies (2.8) for pairwise disjoint sets, i.e. it is *countably additive*.

An alternative, rather more abstract and general, approach to measure theory is to begin with the above properties as *axioms*, i.e. to call a triple $(\Omega, \mathcal{F}, \mu)$ a *measure space* if Ω is an abstractly given set, $\mathcal{F}$ is a σ-field of subsets of Ω, and $\mu : \mathcal{F} \mapsto [0,\infty]$ is a function satisfying (2.8) (with μ instead of m^*). The task of defining Lebesgue measure on $\mathbb{R}$ then becomes that of verifying, with $\mathcal{M}$ and $m = m^*$ on $\mathcal{M}$ defined as above, that the triple $(\mathbb{R}, \mathcal{M}, m)$ satisfies these axioms, i.e. becomes a measure space.

Although the requirements of probability theory will mean that we have to consider such general measure spaces in due course, we have chosen our more concrete approach to the fundamental example of Lebesgue measure in order to demonstrate how this important measure space arises quite naturally from considerations of the 'lengths' of sets in $\mathbb{R}$, and leads to a theory of integration which greatly extends that of Riemann. It is also sufficient to allow us to develop most of the important examples of probability distributions.

Proof of the theorem

(i) Let $A \subseteq \mathbb{R}$. Note that $A \cap \mathbb{R} = A$, $\mathbb{R}^c = \emptyset$, so that $A \cap \mathbb{R}^c = \emptyset$. Now (2.6) reads $m^*(A) = m^*(A) + m^*(\emptyset)$ and is of course true since $m^*(\emptyset) = 0$.

(ii) Suppose $E \in \mathcal{M}$ and take any $A \subseteq \mathbb{R}$. We have to show (2.6) for E^c, i.e.

$$m^*(A) = m^*(A \cap E^c) + m^*(A \cap (E^c)^c),$$

but since $(E^c)^c = E$ this reduces to the condition for E which holds by hypothesis.

We split the proof of (iii) into several steps. But first:

A warm-up Suppose that $E_1 \cap E_2 = \emptyset$, $E_1, E_2 \in \mathcal{M}$. We shall show that $E_1 \cup E_2 \in \mathcal{M}$ and $m^*(E_1 \cup E_2) = m^*(E_1) + m^*(E_2)$.

Let $A \subseteq \mathbb{R}$. We have the condition for E_1:

$$m^*(A) = m^*(A \cap E_1) + m^*(A \cap E_1^c). \tag{2.9}$$

Now, apply (2.6) for E_2 with $A \cap E_1^c$ in place of A:

$$m^*(A \cap E_1^c) = m^*((A \cap E_1^c) \cap E_2) + m^*((A \cap E_1^c) \cap E_2^c).$$
$$= m^*(A \cap (E_1^c \cap E_2)) + m^*(A \cap (E_1^c \cap E_2^c))$$

(the situation is depicted in Figure 2.2).

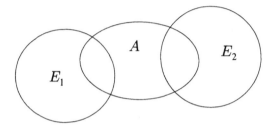

Figure 2.2 The sets A, E_1, E_2

Since E_1 and E_2 are disjoint, $E_1^c \cap E_2 = E_2$. By de Morgan's law $E_1^c \cap E_2^c = (E_1 \cup E_2)^c$. We substitute and we have

$$m^*(A \cap E_1^c) = m^*(A \cap E_2) + m^*(A \cap (E_1 \cup E_2)^c).$$

Inserting this into (2.9) we get

$$m^*(A) = m^*(A \cap E_1) + m^*(A \cap E_2) + m^*(A \cap (E_1 \cup E_2)^c). \tag{2.10}$$

Now by the subadditivity property of m^* we have

$$m^*(A \cap E_1) + m^*(A \cap E_2) \geq m^*((A \cap E_1) \cup (A \cap E_2))$$
$$= m^*(A \cap (E_1 \cup E_2)),$$

so (2.10) gives

$$m^*(A) \geq m^*(A \cap (E_1 \cup E_2)) + m^*(A \cap (E_1 \cup E_2)^c),$$

which is sufficient for $E_1 \cup E_2$ to belong to $\mathcal{M}$ (the inverse inequality is always true, as observed before (2.7)).

Finally, put $A = E_1 \cup E_2$ in (2.10) to get $m^*(E_1 \cup E_2) = m^*(E_1) + m^*(E_2)$, which completes the argument. $\square$

We return to the proof of the theorem.

Step 1. If pairwise disjoint E_k, $k = 1, 2, \ldots$, are in $\mathcal{M}$ then their union is in $\mathcal{M}$ and (2.8) holds.

We begin as in the proof of the warm-up and we have

$$m^*(A) = m^*(A \cap E_1) + m^*(A \cap E_1^c)$$

$$m^*(A) = m^*(A \cap E_1) + m^*(A \cap E_2) + m^*(A \cap (E_1 \cup E_2)^c)$$

(see (2.10)) and after n steps we expect

$$m^*(A) = \sum_{k=1}^{n} m^*(A \cap E_k) + m^*\left(A \cap \left(\bigcup_{k=1}^{n} E_k\right)^c\right). \tag{2.11}$$

Let us demonstrate this by induction. The case $n = 1$ is the first line above. Suppose that

$$m^*(A) = \sum_{k=1}^{n-1} m^*(A \cap E_k) + m^*\left(A \cap \left(\bigcup_{k=1}^{n-1} E_k\right)^c\right). \tag{2.12}$$

Since $E_n \in \mathcal{M}$, we may apply (2.6) with $A \cap \left(\bigcup_{k=1}^{n-1} E_k\right)^c$ in place of A:

$$m^*\left(A \cap \left(\bigcup_{k=1}^{n-1} E_k\right)^c\right) = m^*\left(A \cap \left(\bigcup_{k=1}^{n-1} E_k\right)^c \cap E_n\right) + m^*\left(A \cap \left(\bigcup_{k=1}^{n-1} E_k\right)^c \cap E_n^c\right). \tag{2.13}$$

Now we make the same observations as in the warm-up:

$$\left(\bigcup_{k=1}^{n-1} E_k\right)^c \cap E_n = E_n \quad (E_i \text{ are pairwise disjoint}),$$

$$\left(\bigcup_{k=1}^{n-1} E_k\right)^c \cap E_n^c = \left(\bigcup_{k=1}^{n} E_k\right)^c \quad (\text{by de Morgan's law}).$$

Inserting these into (2.13) we get

$$m^*\left(A \cap \left(\bigcup_{k=1}^{n-1} E_k\right)^c\right) = m^*(A \cap E_n) + m^*\left(A \cap \left(\bigcup_{k=1}^{n} E_k\right)^c\right),$$

and inserting this into the induction hypothesis (2.12) we get

$$m^*(A) = \sum_{k=1}^{n-1} m^*(A \cap E_k) + m^*(A \cap E_n) + m^*\left(A \cap \left(\bigcup_{k=1}^{n} E_k\right)^c\right)$$

as required to complete the induction step. Thus (2.11) holds for all n by induction.

As will be seen at the next step the fact that E_k are pairwise disjoint is not necessary in order to ensure that their union belongs to $\mathcal{M}$. However, with this assumption we have equality in (2.11) which does not hold otherwise. This equality will allow us to prove countable additivity (2.8).

Since

$$\left(\bigcup_{k=1}^{n} E_k\right)^c \supseteq \left(\bigcup_{k=1}^{\infty} E_k\right)^c,$$

from (2.11) by monotonicity (Proposition 2.5) we get

$$m^*(A) \geq \sum_{k=1}^{n} m^*(A \cap E_k) + m^*\left(A \cap \left(\bigcup_{k=1}^{\infty} E_k\right)^c\right).$$

The inequality remains true after we pass to the limit $n \to \infty$:

$$m^*(A) \geq \sum_{k=1}^{\infty} m^*(A \cap E_k) + m^*\left(A \cap \left(\bigcup_{k=1}^{\infty} E_k\right)^c\right). \tag{2.14}$$

By countable subadditivity (Theorem 2.7)

$$\sum_{k=1}^{\infty} m^*(A \cap E_k) \geq m^*\left(A \cap \bigcup_{k=1}^{\infty} E_k\right)$$

and so

$$m^*(A) \geq m^*\left(A \cap \bigcup_{k=1}^{\infty} E_k\right) + m^*\left(A \cap \left(\bigcup_{k=1}^{\infty} E_k\right)^c\right) \tag{2.15}$$

as required. So we have shown that $\bigcup_{k=1}^{\infty} E_k \in \mathcal{M}$ and hence the two sides of (2.15) are equal. The right-hand side of (2.14) is squeezed between the left and right of (2.15), which yields

$$m^*(A) = \sum_{k=1}^{\infty} m^*(A \cap E_k) + m^*\left(A \cap \left(\bigcup_{k=1}^{\infty} E_k\right)^c\right). \tag{2.16}$$

The equality here is a consequence of the assumption that E_k are pairwise disjoint. It holds for any set A, so we may insert $A = \bigcup_{j=1}^{\infty} E_j$. The last term on the right is zero because we have $m^*(\emptyset)$. Next $\left(\bigcup_{j=1}^{\infty} E_j\right) \cap E_n = E_n$ and so we have (2.8).

Step 2. If E_1, $E_2 \in \mathcal{M}$, then $E_1 \cup E_2 \in \mathcal{M}$ (not necessarily disjoint).

Again we begin as in the warm-up:

$$m^*(A) = m^*(A \cap E_1) + m^*(A \cap E_1^c). \tag{2.17}$$

Next, applying (2.6) to E_2 with $A \cap E_1^c$ in place of A we get

$$m^*(A \cap E_1^c) = m^*(A \cap E_1^c \cap E_2) + m^*(A \cap E_1^c \cap E_2^c).$$

We insert this into (2.17) to get

$$m^*(A) = m^*(A \cap E_1) + m^*(A \cap E_1^c \cap E_2) + m^*(A \cap E_1^c \cap E_2^c). \qquad (2.18)$$

By de Morgan's law, $E_1^c \cap E_2^c = (E_1 \cup E_2)^c$, so (as before)

$$m^*(A \cap E_1^c \cap E_2^c) = m^*(A \cap (E_1 \cup E_2)^c). \qquad (2.19)$$

By subadditivity of m^* we have

$$m^*(A \cap E_1) + m^*(A \cap E_1^c \cap E_2) \geq m^*(A \cap (E_1 \cup E_2)). \qquad (2.20)$$

Inserting (2.19) and (2.20) into (2.18) we get

$$m^*(A) \geq m^*(A \cap (E_1 \cup E_2)) + m^*(A \cap (E_1 \cup E_2)^c)$$

as required.

Step 3. If $E_k \in \mathcal{M}$, $k = 1, \dots, n$, then $E_1 \cup \cdots \cup E_n \in \mathcal{M}$ (not necessarily disjoint).

We argue by induction. There is nothing to prove for $n = 1$. Suppose the claim is true for $n - 1$. Then

$$E_1 \cup \cdots \cup E_n = (E_1 \cup \cdots \cup E_{n-1}) \cup E_n$$

so that the result follows from Step 2.

Step 4. If $E_1, E_2 \in \mathcal{M}$, then $E_1 \cap E_2 \in \mathcal{M}$.

We have $E_1^c, E_2^c \in \mathcal{M}$ by (ii), $E_1^c \cup E_2^c \in \mathcal{M}$ by Step 2, $(E_1^c \cup E_2^c)^c \in \mathcal{M}$ by (ii) again, but by de Morgan's law the last set is equal to $E_1 \cap E_2$.

Step 5. The general case: if $E_1, E_2, \dots$ are in $\mathcal{M}$, then so is $\bigcup_{k=1}^{\infty} E_k$.

Let $E_k \in \mathcal{M}$, $k = 1, 2, \dots$. We define an auxiliary sequence of pairwise disjoint sets F_k with the same union as E_k:

$$
\begin{aligned}
F_1 &= E_1 \\
F_2 &= E_2 \backslash E_1 = E_2 \cap E_1^c \\
F_3 &= E_3 \backslash (E_1 \cup E_2) = E_3 \cap (E_1 \cup E_2)^c \\
&\quad \cdots \\
F_k &= E_k \backslash (E_1 \cup \cdots \cup E_{k-1}) = E_k \cap (E_1 \cup \cdots \cup E_{k-1})^c;
\end{aligned}
$$

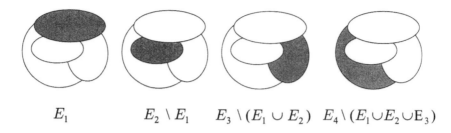

$$E_1 \qquad\qquad E_2 \setminus E_1 \qquad E_3 \setminus (E_1 \cup E_2) \quad E_4 \setminus (E_1 \cup E_2 \cup E_3)$$

Figure 2.3 The sets F_k

(see Figure 2.3).

By Steps 3 and 4 we know that all F_k are in $\mathcal{M}$. By the very construction they are pairwise disjoint, so by Step 1 their union is in $\mathcal{M}$. We shall show that

$$\bigcup_{k=1}^{\infty} F_k = \bigcup_{k=1}^{\infty} E_k.$$

This will complete the proof since the latter is now in $\mathcal{M}$. The inclusion

$$\bigcup_{k=1}^{\infty} F_k \subseteq \bigcup_{k=1}^{\infty} E_k$$

is obvious since for each k, $F_k \subseteq E_k$ by definition. For the inverse let $a \in \bigcup_{k=1}^{\infty} E_k$. Put $S = \{n \in \mathbb{N} : a \in E_n\}$ which is non-empty since a belongs to the union. Let $n_0 = \min S \in S$. If $n_0 = 1$, then $a \in E_1 = F_1$. Suppose $n_0 > 1$. So $a \in E_{n_0}$ and, by the definition of n_0, $a \notin E_1, \ldots, a \notin E_{n_0-1}$. By the definition of F_{n_0} this means that $a \in F_{n_0}$, so a is in $\bigcup_{k=1}^{\infty} F_k$. $\qquad\square$

Using de Morgan's laws you should easily verify an additional property of $\mathcal{M}$.

Proposition 2.13

If $E_k \in \mathcal{M}$, $k = 1, 2, \ldots$, then

$$E = \bigcap_{k=1}^{\infty} E_k \in \mathcal{M}.$$

We can therefore summarise the properties of the family $\mathcal{M}$ of Lebesgue-measurable sets as follows:

$\mathcal{M}$ is closed under countable unions, countable intersections, and complements. It contains intervals and all null sets.

Definition 2.14

We shall write $m(E)$ instead of $m^*(E)$ for any E in $\mathcal{M}$ and call $m(E)$ the *Lebesgue measure* of the set E.

Thus Theorems 2.11 and 2.6 now read as follows, and describe the construction which we have laboured so hard to establish:

> *Lebesgue measure* $m : \mathcal{M} \to [0, \infty]$ is a countably additive set function defined on the σ-field $\mathcal{M}$ of measurable sets. Lebesgue measure of an interval is equal to its length. Lebesgue measure of a null set is zero.

2.4 Basic properties of Lebesgue measure

Since Lebesgue measure is nothing else than the outer measure restricted to a special class of sets, some properties of the outer measure are automatically inherited by Lebesgue measure:

Proposition 2.15

Suppose that $A, B \in \mathcal{M}$.

(i) If $A \subset B$ then $m(A) \leq m(B)$.

(ii) If $A \subset B$ and $m(A)$ is finite then $m(B \setminus A) = m(B) - m(A)$.

(iii) m is translation-invariant.

Since $\emptyset \in \mathcal{M}$ we can take $E_i = \emptyset$ for all $i > n$ in (2.8) to conclude that Lebesgue measure is additive: if $E_i \in \mathcal{M}$ are pairwise disjoint, then

$$m(\bigcup_{i=1}^{n} E_i) = \sum_{i=1}^{n} m(E_i).$$

Exercise 2.6

Find a formula describing $m(A \cup B)$ and $m(A \cup B \cup C)$ in terms of measures of the individual sets and their intersections (we do not assume that the sets are pairwise disjoint).

Recalling that the *symmetric difference* $A \triangle B$ of two sets is defined by $A \triangle B = (A \setminus B) \cup (B \setminus A)$ the following result is also easy to check:

Proposition 2.16

If $A \in \mathcal{M}$, $m(A \triangle B) = 0$, then $B \in \mathcal{M}$ and $m(A) = m(B)$.

Hint Recall that null sets belong to $\mathcal{M}$ and that subsets of null sets are null.

As we noted in Chapter 1, every open set in $\mathbb{R}$ can be expressed as the union of a countable number of open intervals. This ensures that open sets in $\mathbb{R}$ are Lebesgue-measurable, since $\mathcal{M}$ contains intervals and is closed under countable unions. We can approximate the Lebesgue measure of any $A \in \mathcal{M}$ from above by the measures of a sequence of open sets containing A. This is clear from the following result:

Theorem 2.17

(i) For any $\varepsilon > 0$, $A \subset \mathbb{R}$ we can find an open set O such that

$$A \subset O, \qquad m(O) \le m^*(A) + \varepsilon.$$

Consequently, for any $E \in \mathcal{M}$ we can find an open set O containing E such that $m(O \setminus E) < \varepsilon$.

(ii) For any $A \subset \mathbb{R}$ we can find a sequence of open sets O_n such that

$$A \subset \bigcap_n O_n, \qquad m(\bigcap_n O_n) = m^*(A).$$

Proof

(i) By definition of $m^*(A)$ we can find a sequence (I_n) of intervals with $A \subset \bigcup_n I_n$ and

$$\sum_{n=1}^{\infty} l(I_n) - \frac{\varepsilon}{2} \le m^*(A).$$

Each I_n is contained in an open interval whose length is very close to that of I_n; if the left and right endpoints of I_n are a_n and b_n respectively let $J_n = (a_n - \frac{\varepsilon}{2^{n+2}}, b_n + \frac{\varepsilon}{2^{n+2}})$. Set $O = \bigcup_n J_n$, which is open. Then $A \subset O$ and

$$m(O) \le \sum_{n=1}^{\infty} l(J_n) \le \sum_{n=1}^{\infty} l(I_n) + \frac{\varepsilon}{2} \le m^*(A) + \varepsilon.$$

When $m(E) < \infty$ the final statement follows at once from (ii) in Proposition 2.15, since then $m(O \setminus E) = m(O) - m(E) \leq \varepsilon$. When $m(E) = \infty$ we first write $\mathbb{R}$ as a countable union of finite intervals: $\mathbb{R} = \bigcup_n (-n, n)$. Now $E_n = E \cap (-n, n)$ has finite measure, so we can find an open $O_n \supset E_n$ with $m(O_n \setminus E_n) \leq \frac{\varepsilon}{2^n}$. The set $O = \bigcup_n O_n$ is open and contains E. Now

$$O \setminus E = (\bigcup_n O_n) \setminus (\bigcup_n E_n) \subset \bigcup_n (O_n \setminus E_n),$$

so that $m(O \setminus E) \leq \sum_n m(O_n \setminus E_n) \leq \varepsilon$ as required.

(ii) In (i) use $\varepsilon = \frac{1}{n}$ and let O_n be the open set so obtained. With $E = \bigcap_n O_n$ we obtain a measurable set containing A such that $m(E) < m(O_n) \leq m^*(A) + \frac{1}{n}$ for each n, hence the result follows.   □

Remark 2.18

Theorem 2.17 shows how the freedom of movement allowed by the closure properties of the σ-field $\mathcal{M}$ can be exploited by producing, for any set $A \subset \mathbb{R}$, a measurable set $O \supset A$ which is obtained from open intervals with two operations (countable unions followed by countable intersections) and whose measure equals the outer measure of A.

Finally we show that monotone sequences of measurable sets behave as one would expect with respect to m.

Theorem 2.19

Suppose that $A_n \in \mathcal{M}$ for all $n \geq 1$. Then we have:

(i) if $A_n \subset A_{n+1}$ for all n, then

$$m(\bigcup_n A_n) = \lim_{n \to \infty} m(A_n),$$

(ii) if $A_n \supset A_{n+1}$ for all n and $m(A_1) < \infty$, then

$$m(\bigcap_n A_n) = \lim_{n \to \infty} m(A_n).$$

Proof

(i) Let $B_1 = A_1$, $B_i = A_i - A_{i-1}$ for $i > 1$. Then $\bigcup_{i=1}^{\infty} B_i = \bigcup_{i=1}^{\infty} A_i$ and the $B_i \in \mathcal{M}$ are pairwise disjoint, so that

$$m(\bigcup_i A_i) = m(\bigcup_i B_i)$$

$$= \sum_{i=1}^{\infty} m(B_i) \quad \text{(by countable additivity)}$$

$$= \lim_{n \to \infty} \sum_{i=1}^{n} m(B_i)$$

$$= \lim_{n \to \infty} m(\bigcup_{n=1}^{n} B_i) \quad \text{(by additivity)}$$

$$= \lim_{n \to \infty} m(A_n),$$

since $A_n = \bigcup_{i=1}^{n} B_i$ by construction – see Figure 2.4.

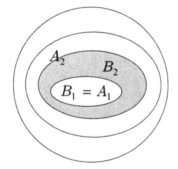

Figure 2.4 Sets A_n, B_n

(ii) $A_1 \setminus A_1 = \emptyset \subset A_1 \setminus A_2 \subset \cdots \subset A_1 \setminus A_n \subset \cdots$ for all n, so that by (i)

$$m(\bigcup_n (A_1 \setminus A_n)) = \lim_{n \to \infty} m(A_1 \setminus A_n)$$

and since $m(A_1)$ is finite, $m(A_1 \setminus A_n) = m(A_1) - m(A_n)$. On the other hand, $\bigcup_n (A_1 \setminus A_n) = A_1 \setminus \bigcap_n A_n$, so that

$$m(\bigcup_n (A_1 \setminus A_n)) = m(A_1) - m(\bigcap_n A_n) = m(A_1) - \lim_{n \to \infty} m(A_n).$$

The result follows.   □

Remark 2.20

The proof of Theorem 2.19 simply relies on the countable additivity of m and on the definition of the sum of a series in $[0, \infty]$, i.e.

$$\sum_{i=1}^{\infty} m(A_i) = \lim_{n \to \infty} \sum_{i=1}^{n} m(A_i).$$

Consequently the result is true, not only for the set function m we have constructed on $\mathcal{M}$, but for any countably additive set function defined on a σ-field. It also leads us to the following claim, which, though we consider it here only for m, actually characterizes countably additive set functions.

Theorem 2.21

The set function m satisfies:

(i) m is finitely additive, i.e. for pairwise disjoint sets (A_i) we have

$$m(\bigcup_{i=1}^{n} A_i) = \sum_{i=1}^{n} m(A_i)$$

for each n;

(ii) m is continuous at $\varnothing$, i.e. if (B_n) decreases to $\varnothing$, then $m(B_n)$ decreases to 0.

Proof

To prove this claim, recall that $m : \mathcal{M} \mapsto [0, \infty]$ is countably additive. This implies (i), as we have already seen. To prove (ii), consider a sequence (B_n) in $\mathcal{M}$ which decreases to $\varnothing$. Then $A_n = B_n \setminus B_{n+1}$ defines a disjoint sequence in $\mathcal{M}$, and $\bigcup_n A_n = B_1$. We may assume that B_1 is bounded, so that $m(B_n)$ is finite for all n, so that, by Proposition 2.15 (ii), $m(A_n) = m(B_n) - m(B_{n+1}) \geq 0$ and hence we have

$$m(B_1) = \sum_{n=1}^{\infty} m(A_n)$$
$$= \lim_{k \to \infty} \sum_{n=1}^{k} [m(B_n) - m(B_{n+1})]$$
$$= m(B_1) - \lim_{n \to \infty} m(B_n)$$

which shows that $m(B_n) \to 0$, as required. $\qquad\square$

2.5 Borel sets

The definition of $\mathcal{M}$ does not easily lend itself to verification that a particular set belongs to $\mathcal{M}$; in our proofs we have had to work quite hard to show that $\mathcal{M}$ is closed under various operations. It is therefore useful to add another construction to our armoury, one which shows more directly how open sets (and indeed open intervals) and the structure of σ-fields lie at the heart of many of the concepts we have developed.

We begin with an auxiliary construction enabling us to produce new σ-fields.

Theorem 2.22

The intersection of a family of σ-fields is a σ-field.

Proof

Let $\mathcal{F}_\alpha$ be σ-fields for $\alpha \in \Lambda$ (the index set Λ can be arbitrary). Put

$$\mathcal{F} = \bigcap_{\alpha \in \Lambda} \mathcal{F}_\alpha.$$

We verify the conditions of the definition.

(i) $\mathbb{R} \in \mathcal{F}_\alpha$ for all $\alpha \in \Lambda$, so $\mathbb{R} \in \mathcal{F}$.

(ii) If $E \in \mathcal{F}$, then $E \in \mathcal{F}_\alpha$ for all $\alpha \in \Lambda$. Since the $\mathcal{F}_\alpha$ are σ-fields, $E^c \in \mathcal{F}_\alpha$ and so $E^c \in \mathcal{F}$.

(iii) If $E_k \in \mathcal{F}$ for $k = 1, 2, \ldots$, then $E_k \in \mathcal{F}_\alpha$, all α, k, hence $\bigcup_{k=1}^\infty E_k \in \mathcal{F}_\alpha$, all α, and so $\bigcup_{k=1}^\infty E_k \in \mathcal{F}$. $\qquad\qquad\square$

Definition 2.23

Put

$$\mathcal{B} = \bigcap \{\mathcal{F} : \mathcal{F} \text{ is a } \sigma\text{-field containing all intervals}\}.$$

We say that $\mathcal{B}$ is the σ-field generated by all intervals and we call the elements of $\mathcal{B}$ *Borel* sets (after Emile Borel, 1871–1956). It is obviously the smallest σ-field containing all intervals. In general, we say that $\mathcal{G}$ is the σ-field generated by a family of sets $\mathcal{A}$ if $\mathcal{G} = \bigcap \{\mathcal{F} : \mathcal{F} \text{ is a } \sigma\text{-field such that } \mathcal{F} \supset \mathcal{A}\}$.

Example 2.24

(Borel sets) The following examples illustrate how the closure properties of the σ-field $\mathcal{B}$ may be used to verify that most familiar sets in $\mathbb{R}$ belong to $\mathcal{B}$.

(i) By construction, all intervals belong to $\mathcal{B}$, and since $\mathcal{B}$ is a σ-field, all open sets must belong to $\mathcal{B}$, as any open set is a countable union of (open) intervals.

(ii) Countable sets are Borel sets, since each is a countable union of closed intervals of the form $[a, a]$; in particular $\mathbb{N}$ and $\mathbb{Q}$ are Borel sets. Hence, as the complement of a Borel set, the set of irrational numbers is also Borel. Similarly, finite and cofinite sets are Borel sets.

The definition of $\mathcal{B}$ is also very flexible – as long as we start with *all* intervals of a particular type, these collections generate the same Borel σ-field:

Theorem 2.25

If instead of the family of all intervals we take all open intervals, all closed intervals, all intervals of the form (a, ∞) (or of the form $[a, \infty)$, $(-\infty, b)$, or $(-\infty, b]$), all open sets, or all closed sets, then the σ-field generated by them is the same as $\mathcal{B}$.

Proof

Consider for example the σ-field generated by the family of open intervals $\mathcal{J}$ and denote it by $\mathcal{C}$:

$$\mathcal{C} = \bigcap \{\mathcal{F} \supset \mathcal{J}, \ \mathcal{F} \text{ is a } \sigma\text{-field}\}.$$

We have to show that $\mathcal{B} = \mathcal{C}$. Since open intervals are intervals, $\mathcal{J} \subset I$ (the family of all intervals), then

$$\{\mathcal{F} \supset I\} \subset \{\mathcal{F} \supset \mathcal{J}\}$$

i.e. the collection of all σ-fields $\mathcal{F}$ which contain I is smaller than the collection of all σ-fields which contain the smaller family $\mathcal{J}$, since it is a more demanding requirement to contain a bigger family, so there are fewer such objects. The inclusion is reversed after we take the intersection on both sides, thus $\mathcal{C} \subset \mathcal{B}$ (the intersection of a smaller family is bigger, as the requirement of belonging to each of its members is a less stringent one).

We shall show that $\mathcal{C}$ contains all intervals. This will be sufficient, since $\mathcal{B}$ is the intersection of such σ-fields, so it is contained in each, so $\mathcal{B} \subset \mathcal{C}$.

To this end consider intervals $[a, b)$, $[a, b]$, $(a, b]$ (the intervals of the form (a, b) are in $\mathcal{C}$ by definition):

$$[a, b) = \bigcap_{n=1}^{\infty} (a - \frac{1}{n}, b),$$

$$[a, b] = \bigcap_{n=1}^{\infty} (a - \frac{1}{n}, b + \frac{1}{n}),$$

$$(a, b] = \bigcap_{n=1}^{\infty} (a, b + \frac{1}{n}).$$

$\mathcal{C}$ as a σ-field is closed with respect to countable intersection, so it contains the sets on the right. The argument for unbounded intervals is similar. The proof is complete. □

Exercise 2.7

Show that the family of intervals of the form $(a, b]$ also generates the σ-field of Borel sets. Show that the same is true for the family of all intervals $[a, b)$.

Remark 2.26

Since $\mathcal{M}$ is a σ-field containing all intervals, and $\mathcal{B}$ is the smallest such σ-field, we have the inclusion $\mathcal{B} \subset \mathcal{M}$, i.e. every Borel set in $\mathbb{R}$ is Lebesgue-measurable. The question therefore arises whether these σ-fields might be the same. In fact the inclusion is proper. It is not altogether straightforward to construct a set in $\mathcal{M} \setminus \mathcal{B}$, and we shall not attempt this here (but see the Appendix). However, by Theorem 2.17 (ii), given any $E \in \mathcal{M}$ we can find a Borel set $B \supset E$ of the form $B = \bigcap_n O_n$, where the (O_n) are open sets, and such that $m(E) = m(B)$. In particular,

$$m(B \Delta E) = m(B \setminus E) = 0.$$

Hence m cannot distinguish between the measurable set E and the Borel set B we have constructed.

Thus, given a Lebesgue-measurable set E we can find a Borel set B such that their symmetric difference $E \Delta B$ is a null set. Now we know that $E \Delta B \in \mathcal{M}$, and it is obvious that subsets of null sets are also null, and hence in $\mathcal{M}$. However, we cannot conclude that every null set will be a Borel set (if $\mathcal{B}$ did contain all null sets then by Theorem 2.17 (ii) we would have $\mathcal{B} = \mathcal{M}$), and this points to an 'incompleteness' in $\mathcal{B}$ which explains why, even if we begin by defining m on intervals and then extend the definition to Borel sets, we would also need to extend it further in order to be able to identify precisely which sets are 'negligible' for our purposes. On the other hand, extension of the measure m to the σ-field $\mathcal{M}$ will suffice, since $\mathcal{M}$ does contain all m-null sets and all subsets of null sets also belong to $\mathcal{M}$.

We show that $\mathcal{M}$ is the smallest σ-field on $\mathbb{R}$ with this property, and we say that $\mathcal{M}$ is the *completion* of $\mathcal{B}$ relative to m and $(\mathbb{R}, \mathcal{M}, m)$ is complete (whereas the measure space $(\mathbb{R}, \mathcal{B}, m)$ is not complete). More precisely, a measure space $(X, \mathcal{F}, \mu)$ is *complete* if for all $F \in \mathcal{F}$ with $\mu(F) = 0$, for all $N \subset F$ we have $N \in \mathcal{F}$ (and so $\mu(N) = 0$).

The *completion* of a σ-field $\mathcal{G}$, relative to a given measure μ, is defined as the smallest σ-field $\mathcal{F}$ containing $\mathcal{G}$ such that, if $N \subset G \in \mathcal{G}$ and $\mu(G) = 0$, then $N \in \mathcal{F}$.

Proposition 2.27

The completion of $\mathcal{G}$ is of the form $\{G \cup N : G \in \mathcal{F}, N \subset F \in \mathcal{F} \text{ with } \mu(F) = 0\}$.

This allows us to extend the measure μ uniquely to a measure $\bar{\mu}$ on $\mathcal{F}$ by setting $\bar{\mu}(G \cup N) = \mu(G)$ for $G \in \mathcal{G}$.

Theorem 2.28

$\mathcal{M}$ is the completion of $\mathcal{B}$.

Proof

We show first that $\mathcal{M}$ contains all subsets of null sets in $\mathcal{B}$: so let $N \subset B \in \mathcal{B}$, B null, and suppose $A \subset \mathbb{R}$. To show that $N \in \mathcal{M}$ we need to show that

$$m^*(A) \geq m^*(A \cap N) + m^*(A \cap N^c).$$

First note that $m^*(A \cap N) \leq m^*(N) \leq m^*(B) = 0$. So it remains to show that $m^*(A) \geq m^*(A \cap N^c)$ but this follows at once from monotonicity of m^*.

Thus we have shown that $N \in \mathcal{M}$. Since $\mathcal{M}$ is a complete σ-field containing $\mathcal{B}$, this means that $\mathcal{M}$ also contains the completion $\mathcal{C}$ of $\mathcal{B}$.

Finally, we show that $\mathcal{M}$ is the minimal such σ-field, i.e. that $\mathcal{M} \subset \mathcal{C}$: first consider $E \in \mathcal{M}$ with $m^*(E) < \infty$, and choose $B = \bigcap_n O_n \in \mathcal{B}$ as described above such that $B \supset E$, $m(B) = m^*(E)$. (We reserve the use of m for sets in $\mathcal{B}$ throughout this argument.)

Consider $N = B \setminus E$, which is in $\mathcal{M}$ and has $m^*(N) = 0$, since m^* is additive on $\mathcal{M}$. By Theorem 2.17 (ii) we can find $L \supset N$, $L \in \mathcal{B}$ and $m(L) = 0$. In other words, N is a subset of a null set in $\mathcal{B}$, and therefore $E = B \setminus N$ belongs to the completion $\mathcal{C}$ of $\mathcal{B}$. For $E \in \mathcal{M}$ with $m^*(E) = \infty$, apply the above to $E_n = E \cap [-n, n]$ for each $n \in \mathbb{N}$. Each $m^*(E_n)$ is finite, so the E_n all belong to $\mathcal{C}$ and hence so does their countable union E. Thus $\mathcal{M} \subset \mathcal{C}$ and so they are equal. $\square$

Despite these technical differences, measurable sets are never far from 'nice' sets, and, in addition to approximations from above by open sets, as observed in Theorem 2.17, we can approximate the measure of any $E \in \mathcal{M}$ from below by those of closed subsets.

Theorem 2.29

If $E \in \mathcal{M}$ then for given $\varepsilon > 0$ there exists a closed set $F \subset E$ such that $m(E \setminus F) < \varepsilon$. Hence there exists $B \subset E$ in the form $B = \bigcup_n F_n$, where all the F_n are closed sets, and $m(E \setminus B) = 0$.

Proof

The complement E^c is measurable and by Theorem 2.17 we can find an open set O containing E^c such that $m(O \setminus E^c) \leq \varepsilon$. But $O \setminus E^c = O \cap E = E \setminus O^c$, and $F = O^c$ is closed and contained in E. Hence this F is what we need. The final part is similar to Theorem 2.17 (ii), and the proof is left to the reader. $\square$

Exercise 2.8

Show that each of the following two statements is equivalent to saying that $E \in \mathcal{M}$:

(a) given $\varepsilon > 0$ there is an open set $O \supset E$ with $m^*(O \setminus E) < \varepsilon$,

(b) given $\varepsilon > 0$ there is a closed set $F \subset E$ with $m^*(E \setminus F) < \varepsilon$.

Remark 2.30

The two statements in the above Exercise are the key to a considerable generalization, linking the ideas of measure theory to those of topology:

A non-negative countably additive set function μ defined on $\mathcal{B}$ is called a *regular Borel measure* if for every Borel set B we have:

$$\mu(B) = \inf\{\mu(O) : O \text{ open}, O \supset B\},$$
$$\mu(B) = \sup\{\mu(F) : F \text{ closed}, F \subset B\}.$$

In Theorems 2.17 and 2.29 we have verified these relations for Lebesgue measure. We shall consider other concrete examples of regular Borel measures later.

2.6 Probability

The ideas which led to Lebesgue measure may be adapted to construct measures generally on arbitrary sets: any set Ω carrying an outer measure (i.e. a mapping from $P(\Omega)$ to $[0, \infty]$, monotone and countably subadditive) can be equipped with a measure μ defined on an appropriate σ-field $\mathcal{F}$ of its subsets. The resulting triple $(\Omega, \mathcal{F}, \mu)$ is then called a measure space, as observed in Remark 2.12. Note that in the construction of Lebesgue measure we only used the properties, not the particular form of the outer measure.

For the present, however, we shall be content with noting simply how to restrict Lebesgue measure to any Lebesgue-measurable subset B of $\mathbb{R}$ with $m(B) > 0$:

Given Lebesgue measure m on the Lebesgue σ-field $\mathcal{M}$, let

$$\mathcal{M}_B = \{A \cap B : A \in \mathcal{M}\}$$

and for $A \in \mathcal{M}_B$ write

$$m_B(A) = m(A).$$

Proposition 2.31

$(B, \mathcal{M}_B, m_B)$ is a complete measure space.

Hint $\bigcup_i (A_i \cap B) = (\bigcup_i A_i) \cap B$ and $(A_1 \cap B) \setminus (A_2 \cap B) = (A_1 \setminus A_2) \cap B$.

We can finally state precisely what we mean by 'selecting a number from [0,1] at random': restrict Lebesgue measure m to the interval $B = [0, 1]$ and consider the σ-field of $\mathcal{M}_{[0,1]}$ of measurable subsets of $[0, 1]$. Then $m_{[0,1]}$ is a measure on $\mathcal{M}_{[0,1]}$ with 'total mass' 1. Since all subintervals of [0,1] with the same length have equal measure, the 'mass' of $m_{[0,1]}$ is spread uniformly over [0,1], so that, for example, the 'probability' of choosing a number from $[0, \frac{1}{10})$ is the same as that of choosing a number from $[\frac{6}{10}, \frac{7}{10})$, namely $\frac{1}{10}$. Thus all numerals are equally likely to appear as first digits of the decimal expansion of the chosen number. On the other hand, with this measure, the probability that the chosen number will be rational is 0, as is the probability of drawing an element of the Cantor set C.

We now have the basis for some probability theory, although a general development still requires the extension of the concept of measure from $\mathbb{R}$ to abstract sets. Nonetheless the building blocks are already evident in the detailed development of the example of Lebesgue measure. The main idea in providing a mathematical foundation for probability theory is to use the concept of measure to provide the mathematical model of the intuitive notion of probability. The distinguishing feature of probability is the concept of *independence*, which we introduce below. We begin by defining the general framework.

2.6.1 Probability space

Definition 2.32

A *probability space* is a triple $(\Omega, \mathcal{F}, P)$ where Ω is an arbitrary set, $\mathcal{F}$ is a σ-field of subsets of Ω, and P is a measure on $\mathcal{F}$ such that

$$P(\Omega) = 1,$$

called *probability measure* or, briefly, *probability*.

Remark 2.33

The original definition, given by Kolmogorov in 1932, is a variant of the above (see Theorem 2.21): $(\Omega, \mathcal{F}, P)$ is a probability space if $(\Omega, \mathcal{F})$ are given as above, and P is a finitely additive set function with $P(\varnothing) = 0$ and $P(\Omega) = 1$ such that $P(B_n) \searrow 0$ whenever (B_n) in $\mathcal{F}$ decreases to $\varnothing$.

Example 2.34

We see at once that Lebesgue measure restricted to $[0, 1]$ is a probability measure. More generally: suppose we are given an arbitrary Lebesgue-measurable set $\Omega \subset \mathbb{R}$, with $m(\Omega) > 0$. Then $P = c \cdot m_\Omega$, where $c = \frac{1}{m(\Omega)}$, and $m = m_\Omega$ denotes the restriction of Lebesgue measure to measurable subsets of Ω, provides a probability measure on Ω, since P is complete and $P(\Omega) = 1$.

For example, if $\Omega = [a, b]$, we obtain $c = \frac{1}{b-a}$, and P becomes the 'uniform distribution' over $[a, b]$. However, we can also use less familiar sets for our base space; for example, $\Omega = [a, b] \cap (\mathbb{R} \setminus \mathbb{Q})$, $c = \frac{1}{b-a}$ gives the same distribution over the irrationals in $[a, b]$.

2.6.2 Events: conditioning and independence

The word 'event' is used to indicate that something is happening. In probability a typical event is to draw elements from a set and then the event is concerned with the outcome belonging to a particular subset. So, as described above, if $\Omega = [0, 1]$ we may be interested in the fact that a number drawn at random from $[0,1]$ belongs to some $A \subset [0,1]$. We want to estimate the probability of this happening, and in the mathematical setup this is the number $P(A)$, here $m_{[0,1]}(A)$. So it is natural to require that A should belong to $\mathcal{M}_{[0,1]}$, since these are the sets we may measure. By a slight abuse of the language, probabilists tend to identify the actual 'event' with the set A which features in the event. The next definition simply confirms this abuse of language.

Definition 2.35

Given a probability space $(\Omega, \mathcal{F}, P)$ we say that the elements of $\mathcal{F}$ are *events*.

Suppose next that a number has been drawn from $[0,1]$ but has not been revealed yet. We would like to bet on it being in $[0, \frac{1}{4}]$ and we get a tip that it certainly belongs to $[0, \frac{1}{2}]$. Clearly, given this 'inside information', the probability of success is now $\frac{1}{2}$ rather than $\frac{1}{4}$. This motivates the following general definition.

Definition 2.36

Suppose that $P(B) > 0$. Then the number

$$P(A|B) = \frac{P(A \cap B)}{P(B)}$$

is called the *conditional probability of A given B*.

Proposition 2.37

The mapping $A \mapsto P(A|B)$ is countably additive on the σ-field $\mathcal{F}_B$.

Hint Use the fact that $A \mapsto P(A \cap B)$ is countably additive on $\mathcal{F}$.

A classical application of the conditional probability is the total probability formula which enables the computation of the probability of an event by means of conditional probabilities given some disjoint hypotheses:

Exercise 2.9

Prove that if H_i are pairwise disjoint events such that $\bigcup_{i=1}^{\infty} H_i = \Omega$, $P(H_i) \neq 0$, then

$$P(A) = \sum_{i=1}^{\infty} P(A|H_i)P(H_i).$$

It is natural to say that the event A is *independent of* B if the fact that B takes place has no influence on the chances of A, i.e.

$$P(A|B) = P(A).$$

By definition of $P(A|B)$ this immediately implies the relation

$$P(A \cap B) = P(A)P(B),$$

which is usually taken as the definition of independence. The advantage of this practice is that we may dispose of the assumption $P(B) > 0$.

Definition 2.38

The events A, B are *independent* if

$$P(A \cap B) = P(A) \cdot P(B).$$

Exercise 2.10

Suppose that A and B are independent events. Show that A^c and B are also independent.

The exercise indicates that if A and B are independent events, then all elements of the σ-fields they generate are mutually independent, since these σ-fields are simply the collections $\mathcal{F}_A = \{\emptyset, A, A^c, \Omega\}$ and $\mathcal{F}_B = \{\emptyset, B, B^c, \Omega\}$ respectively. This leads us to a natural extension of the definition: two σ-fields $\mathcal{F}_1$ and $\mathcal{F}_2$ are *independent* if for any choice of sets $A_1 \in \mathcal{F}_1$ and $A_2 \in \mathcal{F}_2$ we have $P(A_1 \cap A_2) = P(A_1)P(A_2)$.

However, the extension of these definitions to three or more events (or several σ-fields) needs a little care, as the following simple examples show:

Example 2.39

Let $\Omega = [0,1]$, $A = [0, \frac{1}{4}]$ as before; then A is independent of $B = [\frac{1}{8}, \frac{5}{8}]$ and of $C = [\frac{1}{8}, \frac{3}{8}] \cup [\frac{3}{4}, 1]$. In addition, B and C are independent. However,

$$P(A \cap B \cap C) \neq P(A)P(B)P(C).$$

Thus, given three events, the pairwise independence of each of the three possible pairs does *not* suffice for the extension of 'independence' to all three events.

On the other hand, with $A = [0, \frac{1}{4}]$, $B = C = [0, \frac{1}{16}] \cup [\frac{1}{4}, \frac{11}{16}]$ (or alternatively with $C = [0, \frac{1}{16}] \cup [\frac{9}{16}, 1]$),

$$P(A \cap B \cap C) = P(A)P(B)P(C) \qquad (2.21)$$

but none of the pairs make independent events.

This confirms further that we need to demand rather more if we wish to extend the above definition – pairwise independence is not enough, nor is (2.21); therefore we need to require both conditions to be satisfied together. Extending this to n events leads to:

Definition 2.40

The events $A_1, \ldots, A_n$ are *independent* if for all $k \leq n$ for each choice of k events, the probability of their intersection is the product of the probabilities.

Again there is a powerful counterpart for σ-fields (which can be extended to sequences, and even arbitrary families):

Definition 2.41

The σ-fields $\mathcal{F}_1, \mathcal{F}_2, \ldots, \mathcal{F}_n$ defined on a given probability space $(\Omega, \mathcal{F}, P)$ are *independent* if, for all choices of distinct indices $i_1, i_2, \ldots, i_k$ from $\{1, 2, \ldots, n\}$ and all choices of sets $F_{i_n} \in \mathcal{F}_{i_n}$, we have

$$P(F_{i_1} \cap F_{i_2} \cap \cdots \cap F_{i_k}) = P(F_{i_1})P(F_{i_2}) \cdots P(F_{i_k}).$$

The issue of independence will be revisited in the subsequent chapters where we develop some more tools to calculate probabilities.

2.6.3 Applications to mathematical finance

As indicated in the Preface, we will explore briefly how the ideas developed in each chapter can be applied in the rapidly growing field of mathematical finance. This is not intended as an introduction to this subject, but hopefully it will demonstrate how a consistent mathematical formulation can help to clarify ideas central to many disciplines. Readers who are unfamiliar with mathematical finance should consult texts such as [4], [5], [7] for definitions and a discussion of the main ideas of the subject.

Probabilistic modelling in finance centres on the analysis of models for the evolution of the value of traded assets, such as *stocks* or *bonds*, and seeks to identify trends in their future behaviour. Much of the modern theory is concerned with evaluating *derivative securities* such as *options*, whose value is determined by the (random) future values of some underlying security, such as a stock.

We illustrate the above probability ideas on a classical model of stock prices, namely the *binomial tree*. This model is based on finitely many time instants at which the prices may change, and the changes are of a very simple nature. Suppose that the number of steps is N, and denote the price at the k-th step by $S(k)$, $0 \le k \le N$. At each step the stock price changes in the following way: the price at a given step is the price at the previous step multiplied by U with probability p or D with probability $q = 1 - p$, where $0 < D < U$. Therefore the final price depends on the sequence $\omega = (\omega_1, \omega_2, \ldots, \omega_N)$ where $\omega_i = 1$ indicates the application of the factor U or $\omega_i = 0$, which indicates application of the factor D. Such a sequence is called a *path* and we take Ω to consist of all possible paths. In other words,

$$S(k) = S(0) \prod_{i=1}^{k} \eta(i),$$

where

$$\eta(k) = \begin{cases} U \text{ with probability } p \\ D \text{ with probability } q. \end{cases}$$

Exercise 2.11

Suppose $N = 5$, $U = 1.2$, $D = 0.9$, and $S(0) = 500$. Find the number of all paths. How many paths lead to the price $S(5) = 524.88$? What is the probability that $S(5) > 900$ if the probability of going up in a single step is 0.5?

In general, the total number of paths is clearly 2^N and at step k there are $k + 1$ possible prices.

We construct a probability space by equipping Ω with the σ-field 2^Ω of all subsets of Ω, and the probability defined on single-element sets by $P(\{\omega\}) = p^k q^{N-k}$, where $k = \sum_{i=1}^{N} \omega_i$.

As time progresses we gather information about stock prices, or, what amounts to the same, about paths. This means that having observed some prices, the range of possible future developments is restricted. Our information increases with time and this idea can be captured by the following family of σ-fields.

Fix $m < n$ and define a σ-field $\mathcal{F}_m = \{A : \omega, \omega' \in A \implies \omega_1 = \omega'_1, \omega_2 = \omega'_2, \ldots, \omega_m = \omega'_m\}$. So all paths from a particular set A in this σ-field have identical initial segments while the remaining coordinates are arbitrary. Note that

$\mathcal{F}_0 = \{\Omega, \emptyset\},$

$\mathcal{F}_1 = \{A_1, A_1^c, \Omega, \varnothing\}$, where $A_1 = \{\omega : \omega_1 = 1\}$, i.e. $S(1) = S(0)U$, and $A_1^c = \{\omega : \omega_1 = 0\}$ i.e. $S(1) = S(0)D$.

Exercise 2.12

Prove that $\mathcal{F}_m$ has 2^{2^m} elements.

Exercise 2.13

Prove that the sequence $\mathcal{F}_m$ is increasing.

This sequence is an example of a *filtration* (the identifying features are that the σ-fields should be contained in $\mathcal{F}$ and form an increasing chain), a concept which we shall revisit later on.

The consecutive choices of stock prices are closely related to coin-tossing. Intuition tells us that the latter are independent. This can be formally seen by introducing another σ-field describing the fact that at a particular step we have a particular outcome. Suppose ω is such that $\omega_k = 1$. Then we can identify the set of all paths with this property $A_k = \{\omega : \omega_k = 1\}$ and extend to a σ-field: $\mathcal{G}_k = \{A_k, A_k^c, \Omega, \varnothing\}$. In fact, $A_k^c = \{\omega : \omega_k = 0\}$.

Exercise 2.14

Prove that $\mathcal{G}_m$ and $\mathcal{G}_k$ are independent if $m \neq k$.

2.7 Proofs of propositions

Proof (of Proposition 2.5)

If the intervals I_n cover B, then they also cover A: $A \subset B \subset \bigcup_n I_n$, hence $Z_B \subset Z_A$. The infimum of a larger set cannot be greater than the infimum of a smaller set (trivial illustration: $\inf\{0, 1, 2\} < \inf\{1, 2\}$, $\inf\{0, 1, 2\} = \inf\{0, 2\}$), hence the result. $\qquad\square$

Proof (of Proposition 2.8)

If a system I_n of intervals covers A then the intervals $I_n + t$ cover $A + t$. Conversely, if J_n cover $A + t$ then $J_n - t$ cover A. Moreover, the total length of a family of intervals does not change when we shift each by a number. So we have a one–one correspondence between the interval coverings of A and $A + t$ and this correspondence preserves the total length of the covering. This implies that the sets Z_A and Z_{A+t} are the same, so their infima are equal. $\square$

Proof (of Proposition 2.13)

By de Morgan's law

$$\bigcap_{k=1}^{\infty} E_k = \Big(\bigcup_{k=1}^{\infty} E_k^c \Big)^c.$$

By Theorem 2.11 (ii) all E_k^c are in $\mathcal{M}$, hence by (iii) the same can be said about the union $\bigcup_{k=1}^{\infty} E_k^c$. Finally, by (ii) again, the complement of this union is in $\mathcal{M}$, and so the intersection $\bigcap_{k=1}^{\infty} E_k$ is in $\mathcal{M}$. $\square$

Proof (of Proposition 2.15)

(i) Proposition 2.5 tells us that the outer measure is monotone, but since m is just the restriction of m^* to $\mathcal{M}$, then the same is true for m: $A \subset B$ implies $m(A) = m^*(A) \leq m^*(B) = m(B)$.

(ii) We write B as a disjoint union $B = A \cup (B \setminus A)$ and then by additivity of m we have $m(B) = m(A) + m(B \setminus A)$. Subtracting $m(A)$ (here it is important that $m(A)$ is finite) we get the result.

(iii) Translation invariance of m follows at once from translation invariance of the outer measure in the same way as in (i) above. $\square$

Proof (of Proposition 2.16)

The set $A \triangle B$ is null, hence so are its subsets $A \setminus B$ and $B \setminus A$. Thus these sets are measurable, and so is $A \cap B = A \setminus (A \setminus B)$, and therefore also $B = (A \cap B) \cup (B \setminus A) \in \mathcal{M}$. Now $m(B) = m(A \cap B) + m(B \setminus A)$ as the sets on the right are disjoint. But $m(B \setminus A) = 0 = m(A \setminus B)$, so $m(B) = m(A \cap B) = m(A \cap B) + m(A \setminus B) = m((A \cap B) \cup (A \setminus B)) = m(A)$. $\square$

Proof (of Proposition 2.27)

The family $\mathcal{G} = \{ G \cup N : G \in \mathcal{F}, N \subset F \in \mathcal{F} \text{ with } \mu(F) = 0 \}$ contains the set X since $X \in \mathcal{F}$. If $G_i \cup N_i \in \mathcal{G}$, $N_i \subset F_i$, $\mu(F_i) = 0$, then $\bigcup G_i \cup N_i = \bigcup G_i \cup \bigcup N_i$ is in $\mathcal{G}$ since the first set on the right is in $\mathcal{F}$ and the second is a subset of a null set $\bigcup F_i \in \mathcal{F}$. If $G \cup N \in \mathcal{G}$, $N \subset F$, then $(G \cup N)^c = (G \cup F)^c \cup ((F \setminus N) \cap G^c)$, which is also in $\mathcal{G}$ Thus $\mathcal{G}$ is a σ-field. Consider any other σ-field $\mathcal{H}$ containing

$\mathcal{F}$ and all subsets of null sets. Since $\mathcal{H}$ is closed with respect to the unions, it contains $\mathcal{G}$ and so $\mathcal{G}$ is the smallest σ-field with this property.   □

Proof (of Proposition 2.31)

It follows at once from the definitions and the hint that $\mathcal{M}_B$ is a σ-field. To see that m_B is a measure we check countable additivity: with $C_i = A_i \cap B$ pairwise disjoint in $\mathcal{M}_B$, we have

$$m_B\left(\bigcup_i C_i\right) = m\left(\bigcup_i (A_i \cap B)\right) = \sum_i m(A_i \cap B) = \sum_i m(C_i) = \sum_i m_B(C_i).$$

Therefore $(B, \mathcal{M}_B, m_B)$ is a measure space. It is complete, since subsets of null sets contained in B are by definition m_B-measurable.   □

Proof (of Proposition 2.37)

Assume that A_n are measurable and pairwise disjoint. By the definition of conditional probability

$$P\left(\bigcup_{n=1}^{\infty} A_n \middle| B\right) = \frac{1}{P(B)} P\left(\left(\bigcup_{n=1}^{\infty} A_n\right) \cap B\right)$$

$$= \frac{1}{P(B)} P\left(\bigcup_{n=1}^{\infty} (A_n \cap B)\right)$$

$$= \frac{1}{P(B)} \sum_{n=1}^{\infty} P(A_n \cap B)$$

$$= \sum_{n=1}^{\infty} P(A_n | B)$$

since $A_n \cap B$ are also pairwise disjoint and P is countably additive.   □

3
Measurable functions

3.1 The extended real line

The length of $\mathbb{R}$ is unbounded above, i.e. 'infinite'. To deal with this we defined Lebesgue measure for sets of infinite as well as finite measure. In order to handle functions between such sets comprehensively, it is convenient to allow functions which take infinite values: we take their range to be (part of) the 'extended real line' $\overline{\mathbb{R}} = [-\infty, \infty]$, obtained by adding the 'points at infinity' $-\infty$ and $+\infty$ to $\mathbb{R}$. Arithmetic in this set needs a little care, as already observed in Section 2.2: we assume that $a + \infty = \infty$ for all real a, $a \times \infty = \infty$ for $a > 0$, $a \times \infty = -\infty$ for $a < 0$, $\infty \times \infty = \infty$ and $0 \times \infty = 0$, with similar definitions for $-\infty$. These are all 'obvious' intuitively (except possibly $0 \times \infty$), and (as for measures) we avoid ever forming 'sums' of the form $\infty + (-\infty)$. With these assumptions 'arithmetic works as before'.

3.2 Lebesgue-measurable functions

The domain of the functions we shall be considering is usually $\mathbb{R}$. Now we have the freedom of defining f only 'up to null sets': once we have shown two functions f and g to be equal on $\mathbb{R} \setminus E$ where E is some null set, then $f = g$ for all practical purposes. To formalise this, we say that f has a property (P) *almost everywhere* (a.e.) if f has this property at all points of its domain, except

55

possibly on some null set.

For example, the function

$$f(x) = \begin{cases} 1 & \text{for } x \neq 0 \\ 0 & \text{for } x = 0 \end{cases}$$

is almost everywhere continuous, since it is continuous on $\mathbb{R} \setminus \{0\}$, and the exceptional set $\{0\}$ is null. (Note that probabilists tend to say 'almost surely' (a.s.) instead of 'almost everywhere' (a.e.) and we shall follow their lead in the sections devoted to probability.)

The next definition will introduce the class of Lebesgue-measurable functions. The condition imposed on $f : \mathbb{R} \to \mathbb{R}$ will be necessary (though not sufficient) to give meaning to the (Lebesgue) integral $\int f \, dm$. Let us first give some motivation.

Integration is always concerned with the process of approximation. In the Riemann integral we split the interval $I = [a, b]$, over which we integrate into small pieces I_n – again intervals. The simplest method of doing this is to divide the interval into N equal parts. Then we construct approximating sums by multiplying the lengths of the small intervals by certain numbers c_n (related to the values of the function in question; for example $c_n = \inf_{I_n} f$, $c_n = \sup_{I_n} f$, or $c_n = f(x)$ for some $x \in I_n$):

$$\sum_{n=1}^{N} c_n l(I_n).$$

For large N this sum is close to the Riemann integral $\int_a^b f(x) \, dx$ (given some regularity of f).

The approach to the Lebesgue integral is similar but there is a crucial difference. Instead of splitting the integration domain into small parts, we decompose the range of the function (see Figure 3.1).

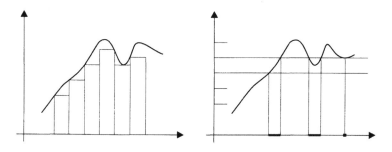

Figure 3.1 Riemann vs. Lebesgue

Again, a simple way is to introduce short intervals J_n of equal length. To build the approximating sums we first take the inverse images of J_n by f, i.e. $f^{-1}(J_n)$. These may be complicated sets, not necessarily intervals. Here the theory of measure developed previously comes into its own. We are able to measure sets provided they are measurable, i.e. they are in $\mathcal{M}$. Given that, we compute

$$\sum_{n=1}^{N} c_n m(f^{-1}(J_n)),$$

where $c_n \in J_n$ or $c_n = \inf J_n$, for example.

The following definition guarantees that this procedure makes sense (though some extra care may be needed to arrive at a finite number as $N \to \infty$).

Definition 3.1

Suppose that E is a measurable set. We say that a function $f : E \longrightarrow \mathbb{R}$ is (*Lebesgue-*)*measurable* if for any interval $I \subseteq \mathbb{R}$

$$f^{-1}(I) = \{x \in \mathbb{R} : f(x) \in I\} \in \mathcal{M}.$$

In what follows, the term *measurable* (without qualification) will refer to Lebesgue-measurable functions.

If all the sets $f^{-1}(I) \in \mathcal{B}$, i.e. if they are Borel sets, we call f *Borel-measurable*, or simply a *Borel* function.

The underlying philosophy is one which is common for various mathematical notions: the inverse image of a 'nice' set is 'nice'. Remember continuous functions, for example, where the inverse image of any open set is required to be open. The actual meaning of the word 'nice' depends on the particular branch of mathematics. In the above definitions, note that since $\mathcal{B} \subset \mathcal{M}$, every Borel function is (Lebesgue-)measurable.

Remark 3.2

The terminology is somewhat unfortunate. 'Measurable' objects should be measured (as with measurable sets). However, measurable functions will be integrated. This confusion stems from the fact that the word *integrable*, which would probably fit best here, carries a more restricted meaning, as we shall see later. This terminology is widely accepted and we are not going to try to fight the whole world here.

We give some equivalent formulations:

Theorem 3.3

The following conditions are equivalent:

(i) f is measurable,

(ii) for all a, $f^{-1}((a, \infty))$ is measurable,

(iii) for all a, $f^{-1}([a, \infty))$ is measurable,

(iv) for all a, $f^{-1}((-\infty, a))$ is measurable,

(v) for all a, $f^{-1}((-\infty, a])$ is measurable.

Proof

Of course (i) implies any of the other conditions. We show that (ii) implies (i). The proofs of the other implications are similar, and are left as exercises (which you should attempt).

We have to show that for any interval I, $f^{-1}(I) \in \mathcal{M}$. By (ii) we have that for the particular case $I = (a, \infty)$. Suppose $I = (-\infty, a]$. Then

$$f^{-1}((-\infty, a]) = f^{-1}(\mathbb{R} \setminus (a, \infty)) = E \setminus f^{-1}((a, \infty)) \in \mathcal{M} \qquad (3.1)$$

since both E and $f^{-1}((a, \infty))$ are in $\mathcal{M}$ (we use the closure properties of $\mathcal{M}$ established before). Next

$$f^{-1}((-\infty, b)) = f^{-1}\left(\bigcup_{n=1}^{\infty} (-\infty, b - \frac{1}{n}] \right)$$

$$= \bigcup_{n=1}^{\infty} f^{-1}\left((-\infty, b - \frac{1}{n}] \right).$$

By (3.1), $f^{-1}\left((-\infty, b - \frac{1}{n}] \right) \in \mathcal{M}$ and the same is true for the countable union. From this we can easily deduce that

$$f^{-1}([b, \infty)) \in \mathcal{M}.$$

Now let $I = (a, b)$, and

$$f^{-1}((a, b)) = f^{-1}((-\infty, b) \cap (a, \infty))$$

$$= f^{-1}((-\infty, b)) \cap f^{-1}((a, \infty))$$

is in $\mathcal{M}$ as the intersection of two elements of $\mathcal{M}$. By the same reasoning $\mathcal{M}$ contains

$$f^{-1}([a, b]) = f^{-1}((-\infty, b] \cap [a, \infty))$$

$$= f^{-1}((-\infty, b]) \cap f^{-1}([a, \infty))$$

and half-open intervals are handled similarly. $\qquad \square$

3.3 Examples

The following simple results show that most of the functions encountered 'in practice' are measurable.

1. Constant functions are measurable. Let $f(x) \equiv c$. Then

$$f^{-1}((a, \infty)) = \begin{cases} \mathbb{R} & \text{if } a < c \\ \varnothing & \text{otherwise} \end{cases}$$

   and in both cases we have measurable sets.

2. Continuous functions are measurable. For we note that (a, ∞) is an open set and so is $f^{-1}((a, \infty))$. As we know, all open sets are measurable.

3. Define the indicator function of a set A by

$$\mathbf{1}_A(x) = \begin{cases} 1 & \text{if } x \in A \\ 0 & \text{otherwise.} \end{cases}$$

Then

$$A \in \mathcal{M} \quad \Leftrightarrow \quad \mathbf{1}_A \text{ is measurable}$$

since

$$\mathbf{1}_A^{-1}((a, \infty)) = \begin{cases} \mathbb{R} & \text{if } a < 0 \\ A & \text{if } 0 \le a < 1 \\ \varnothing & \text{if } a \ge 1. \end{cases}$$

Exercise 3.1

Prove that every monotone function is measurable.

Exercise 3.2

Prove that if f is a measurable function, then the level set $\{x : f(x) = a\}$ is measurable for every $a \in \mathbb{R}$.

Remark 3.4

In the Appendix, assuming the validity of the axiom of choice, we show that there are subsets of $\mathbb{R}$ which fail to be Lebesgue-measurable, and that there are Lebesgue-measurable sets which are not Borel sets. Thus, if $\mathcal{P}(\mathbb{R})$ denotes the σ-field of all subsets of $\mathbb{R}$, the following inclusions are strict:

$$\mathcal{B} \subset \mathcal{M} \subset \mathcal{P}(\mathbb{R}).$$

These (rather esoteric) facts can be used, by considering the indicator functions of these sets, to construct examples of non-measurable functions and of measurable functions which are not Borel functions. While it is important to be aware of these distinctions in order to understand why these different concepts are introduced at all, such examples will not feature in the applications of the theory which we have in mind.

3.4 Properties

The class of measurable functions is very rich, as the following results show.

Theorem 3.5

The set of real-valued measurable functions defined on $E \in \mathcal{M}$ is a vector space and closed under multiplication, i.e. if f and g are measurable functions then $f + g$, and fg are also measurable (in particular, if g is a constant function $g \equiv c$, cf is measurable for all real c).

Proof

Fix measurable functions $f, g : E \to \mathbb{R}$. First consider $f + g$. Our goal is to show that for each $a \in \mathbb{R}$,

$$B = (f + g)^{-1}(-\infty, a) = \{t : f(t) + g(t) < a\} \in \mathcal{M}.$$

Suppose that all the rationals are arranged in a sequence $\{q_n\}$. Now

$$B = \bigcup_{n=1}^{\infty} \{t : f(t) < q_n, g(t) < a - q_n\}$$

– we decompose the half-plane below the line $x + y = a$ into a countable union of unbounded 'boxes': $\{(x, y) : x < q_n, y < a - q_n\}$ (see Figure 3.2). Clearly

$$\{t : f(t) < q_n, g(t) < a - q_n\} = \{t : f(t) < q_n\} \cap \{t : g(t) < a - q_n\}$$

is measurable as an intersection of measurable sets. Hence $B \in \mathcal{M}$ as a countable union of elements of $\mathcal{M}$.

To deal with fg we adopt a slightly indirect approach in order to remain 'one-dimensional': first note that if g is measurable, then so is $-g$. Hence $f - g = f + (-g)$ is measurable. Since $fg = \frac{1}{4}\{(f + g)^2 - (f - g)^2\}$, it will suffice to prove that the square of a measurable function is measurable. So take a measurable

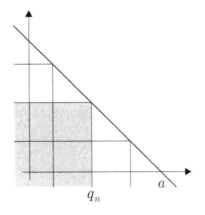

Figure 3.2 Boxes

$h : E \to \mathbb{R}$ and consider $\{x \in E : h^2(x) > a\}$. For $a < 0$ this set is $E \in \mathcal{M}$, and for $a \geq 0$

$$\{x : h^2(x) > a\} = \{x : h(x) > \sqrt{a}\} \cup \{x : h(x) < -\sqrt{a}\}.$$

Both sets on the right are measurable, hence we have shown that h^2 is measurable. Apply this with $h = f + g$ and $h = f - g$ respectively, to conclude that fg is measurable. It follows that cf is measurable for constant c, hence the class of real-valued measurable functions forms a vector space under addition. $\quad\square$

Remark 3.6

An elegant proof of the theorem is based on the following lemma, which will also be useful later. Its proof makes use of the simple topological fact that every open set in $\mathbb{R}^2$ decomposes into a countable union of rectangles, in precise analogy with open sets in $\mathbb{R}$ and intervals.

Lemma 3.7

Suppose that $F : \mathbb{R} \times \mathbb{R} \to \mathbb{R}$ is a continuous function. If f and g are measurable, then $h(x) = F(f(x), g(x))$ is also measurable.

It now suffices to take $F(u, v) = u + v$, $F(u, v) = uv$ to obtain a second proof of Theorem 3.5.

Proof of the lemma

For any real a,

$$\{x : h(x) > a\} = \{x : (f(x), g(x)) \in G_a\}$$

where $G_a = \{(u, v) : F(u, v) > a\} = F^{-1}((a, \infty))$. Suppose for the moment that we have been lucky and G_a is a rectangle: $G_a = (a_1, b_1) \times (c_1, d_1)$.

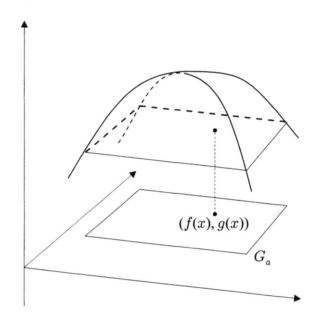

Figure 3.3 The sets G_a

It is clear from Figure 3.3 that

$$\{x : h(x) > a\} = \{x : f(x) \in (a_1, b_1) \text{ and } g(x) \in (c_1, d_1)\}$$
$$= \{x : f(x) \in (a_1, b_1)\} \cap \{x : g(x) \in (c_1, d_1)\}.$$

In general, we have to decompose the set G_a into a union of rectangles. The set G_a is an open subset of $\mathbb{R} \times \mathbb{R}$ since F is continuous. Hence it can be written as

$$G_a = \bigcup_{n=1}^{\infty} R_n$$

where R_n are open rectangles $R_n = (a_n, b_n) \times (c_n, d_n)$. So

$$\{x : h(x) > a\} = \bigcup_{n=1}^{\infty} \{x : f(x) \in (a_n, b_n)\} \cap \{x : g(x) \in (c_n, d_n)\}$$

is measurable due to the stability properties of $\mathcal{M}$.   □

A simple application of Theorem 3.5 is to consider the product $f\mathbf{1}_A$. If f is a measurable function and A is a measurable set, then $f\mathbf{1}_A$ is measurable. This function is simply f on A and 0 outside A. Applying this to the set $A = \{x \in E : f(x) > 0\}$ we see that the positive part f^+ of a measurable function is measurable: we have

$$f^+(x) = \begin{cases} f(x) & \text{if } f(x) > 0 \\ 0 & \text{if } f(x) \leq 0. \end{cases}$$

Similarly the negative part f^- of f is measurable, since

$$f^-(x) = \begin{cases} 0 & \text{if } f(x) > 0 \\ -f(x) & \text{if } f(x) \leq 0. \end{cases}$$

Proposition 3.8

Let E be a measurable subset of $\mathbb{R}$.

(i) $f : E \to \mathbb{R}$ is measurable if and only if both f^+ and f^- are measurable.

(ii) If f is measurable, then so is $|f|$; but the converse is false.

Hint Part (ii) requires the existence of non-measurable sets (as proved in the Appendix), not their particular form.

Exercise 3.3

Show that if f is measurable, then the truncation of f:

$$f^a(x) = \begin{cases} a & \text{if } f(x) > a \\ f(x) & \text{if } f(x) \leq a, \end{cases}$$

is also measurable.

Exercise 3.4

Find a non-measurable f such that f^2 is measurable.

Passage to the limit does not destroy measurability – all the work needed was done when we established the stability properties of $\mathcal{M}$!

Theorem 3.9

If $\{f_n\}$ is a sequence of measurable functions defined on the set E in $\mathbb{R}$, then the following are measurable functions also:

$$\max_{n \leq k} f_n, \quad \min_{n \leq k} f_n, \quad \sup_{n \in \mathbb{N}} f_n, \quad \inf_{n \in \mathbb{N}} f_n, \quad \limsup_{n \to \infty} f_n, \quad \liminf_{n \to \infty} f_n.$$

Proof

It is sufficient to note that the following are measurable sets:

$$\{x : (\max_{n \leq k} f_n)(x) > a\} = \bigcup_{n=1}^{k} \{x : f_n(x) > a\},$$

$$\{x : (\min_{n \leq k} f_n)(x) > a\} = \bigcap_{n=1}^{k} \{x : f_n(x) > a\},$$

$$\{x : (\sup_{n \geq k} f_n)(x) > a\} = \bigcup_{n=k}^{\infty} \{x : f_n(x) > a\},$$

$$\{x : (\inf_{n \geq k} f_n)(x) \geq a\} = \bigcap_{n=k}^{\infty} \{x : f_n(x) \geq a\}.$$

For the upper limit, by definition

$$\limsup_{n \to \infty} f_n = \inf_{n \geq 1} \{\sup_{m \geq n} f_m\}$$

and the above relations show that $h_n = \sup_{m \geq n} f_m$ is measurable, hence $\inf_{n \geq 1} h_n(x)$ is measurable. The lower limit is done similarly.   □

Corollary 3.10

If a sequence f_n of measurable functions converges (pointwise) then the limit is a measurable function.

Proof

This is immediate since $\lim_{n \to \infty} f_n = \limsup_{n \to \infty} f_n$, which is measurable.   □

Remark 3.11

Note that Theorems 3.5 and 3.9 have counterparts for Borel functions, i.e. they remain valid upon replacing 'measurable' by 'Borel' throughout.

Things are slightly more complicated when we consider the role of null sets. On the one hand, changing a function on a null set cannot destroy its measurability, i.e. any measurable function which is altered on a null set remains measurable. However, as not all null sets are Borel sets, we cannot conclude similarly for Borel sets, and thus the following results have no natural 'Borel' counterparts.

Theorem 3.12

If $f : E \to \mathbb{R}$ is measurable, $E \in \mathcal{M}$, $g : E \to \mathbb{R}$ is such that the set $\{x : f(x) = g(x)\}$ is null, then g is measurable.

Proof

Consider the difference $d(x) = g(x) - f(x)$. It is zero except on a null set, so

$$\{x : d(x) > a\} = \begin{cases} \text{a null set} & \text{if } a \geq 0 \\ \text{a full set} & \text{if } a < 0 \end{cases}$$

where a full set is the complement of a null set. Both null and full sets are measurable, hence d is a measurable function. Thus $g = f + d$ is measurable. $\square$

Corollary 3.13

If (f_n) is a sequence of measurable functions and $f_n(x) \to f(x)$ almost everywhere for x in E, then f is measurable.

Proof

Let A be the null set such that $f_n(x)$ converges for all $x \in E \setminus A$. Then $\mathbf{1}_{A^c} f_n$ converge everywhere to $g = \mathbf{1}_{A^c} f$ which is therefore measurable. But $f = g$ almost everywhere, so f is also measurable. $\square$

Exercise 3.5

Let f_n be a sequence of measurable functions. Show that the set $E = \{x : f_n(x) \text{ converges}\}$ is measurable.

Since we are able to adjust a function f at will on a null set without altering its measurability properties, the following definition is a useful means of concentrating on the values of f that 'really matter' for integration theory, by identifying its bounds 'outside null sets':

Definition 3.14

Suppose $f : E \to \overline{\mathbb{R}}$ is measurable. The *essential supremum* ess sup f is defined as $\inf\{z : f \leq z \text{ a.e.}\}$ and the *essential infimum* ess inf f is $\sup\{z : f \geq z \text{ a.e.}\}$.

Note that ess sup f can be $+\infty$. If ess sup $f = -\infty$, then $f = -\infty$ a.e. since by definition of ess sup, $f \leq -n$ a.e. for all $n \geq 1$. Now if ess sup f is finite, and $A = \{x : \text{ess sup } f < f(x)\}$, define A_n for $n \geq 1$ by

$$A_n = \{x : \text{ess sup } f < f(x) - \frac{1}{n}\}.$$

These are null sets, hence so is $A = \bigcup_n A_n$, and thus we have verified:

$$f \leq \text{ess sup } f \text{ a.e.}$$

The following is now straightforward to prove.

Proposition 3.15

If f, g are measurable functions, then

$$\text{ess sup } (f + g) \leq \text{ess sup } f + \text{ess sup } g.$$

Exercise 3.6

Show that for measurable f, ess sup $f \leq \sup f$. Show that these quantities coincide when f is continuous.

3.5 Probability

3.5.1 Random variables

In the special case of probability spaces we use the phrase *random variable* to mean a measurable function. That is, if $(\Omega, \mathcal{F}, P)$ is a probability space, then $X : \Omega \to \mathbb{R}$ is a random variable if for all $a \in \mathbb{R}$ the set $X^{-1}([a, \infty))$ is in $\mathcal{F}$:

$$\{\omega \in \Omega : X(\omega) \geq a\} \in \mathcal{F}.$$

In the case where $\Omega \subset \mathbb{R}$ is a measurable set and $\mathcal{F} = \mathcal{B}$ is the σ-field of Borel subsets of Ω, random variables are just Borel functions $\mathbb{R} \to \mathbb{R}$.

In applied probability, the set Ω represents the outcomes of a random experiment that can be observed by means of various measurements. These measurements assign numbers to outcomes and thus we arrive at the notion of random variable in a natural way. The condition imposed guarantees that questions of the following sort make sense: What is the probability that the value of the random variable lies within given limits?

3.5.2 σ-fields generated by random variables

As indicated before, the random variables we encounter will in fact be Borel-measurable functions. The values of the random variable X will not lead us to non-Borel sets; in fact, they are likely to lead us to discuss much coarser distinctions between sets than are already available within the complexity of the Borel σ-field $\mathcal{B}$. We should therefore be ready to consider different σ-fields contained within $\mathcal{F}$. To be precise, the family of sets

$$X^{-1}(\mathcal{B}) = \{S \subset \mathcal{F} : S = X^{-1}(B) \text{ for some } B \in \mathcal{B}\}$$

is a σ-field. If X is a random variable, $X^{-1}(\mathcal{B}) \subset \mathcal{F}$ but it may be a much smaller subset depending on the degree of sophistication of X. We denote this σ-field by $\mathcal{F}_X$ and call it the σ-field *generated* by X.

The simplest possible case is where X is constant, $X \equiv a$. The $X^{-1}(B)$ is either Ω or $\emptyset$, depending on whether $a \in B$ or not and the σ-field generated is trivial: $\mathcal{F} = \{\emptyset, \Omega\}$.

If X takes two values $a \neq b$, then $\mathcal{F}_X$ contains four elements: $\mathcal{F}_X = \{\emptyset, \Omega, X^{-1}(\{a\}), X^{-1}(\{b\})\}$. If X takes finitely many values, $\mathcal{F}_X$ is finite. If X takes denumerably many values, $\mathcal{F}_X$ is uncountable (it may be identified with the σ-field of all subsets of a countable set). We can see that the size of $\mathcal{F}_X$ grows together with the level of complication of X.

Exercise 3.7

Show that $\mathcal{F}_X$ is the smallest σ-field containing the inverse images $X^{-1}(B)$ of all Borel sets B.

Exercise 3.8

Is the family of sets $\{X(A) : A \in \mathcal{F}\}$ a σ-field?

The notion of $\mathcal{F}_X$ has the following interpretation. The values of the measurement X are all we can observe. From these we deduce some information on the level of complexity of the random experiment, that is the size of Ω and $\mathcal{F}_X$, and we can estimate the probabilities of the sets in $\mathcal{F}_X$ by statistical methods. The σ-field generated represents the amount of information produced by the random variable. For example, suppose that a die is thrown and only 0 and 1 are reported depending on the number shown being odd or even. We will never distinguish this experiment from coin-tossing. The information provided by the measurement is insufficient to explore the complexity of the experiment (which has six possible outcomes, here grouped together into two sets).

3.5.3 Probability distributions

For any random variable X we can introduce a measure on the σ-field of Borel sets B by setting

$$P_X(B) = P(X^{-1}(B)).$$

We call P_X the *probability distribution* of the random variable X.

Theorem 3.16

The set function P_X is countably additive.

Proof

Given pairwise disjoint Borel sets B_i their inverse images $X^{-1}(B_i)$ are pairwise disjoint and $X^{-1}(\bigcup_i B_i) = \bigcup_i X^{-1}(B_i)$, so

$$P_X(\bigcup_i B_i) = P(X^{-1}(\bigcup_i B_i)) = P(\bigcup_i X^{-1}(B_i)) = \sum_i P(X^{-1}(B_i))$$
$$= \sum_i P_X(B_i)$$

as required. $\square$

Thus $(\mathbb{R}, \mathcal{B}, P_X)$ is a probability space. For this it is sufficient to note that $P_X(\mathbb{R}) = P(\Omega) = 1$.

We consider some simple examples. Suppose that X is constant, i.e. $X \equiv a$. Then we call P_X the *Dirac measure* concentrated at a and denote it by δ_a. Clearly

$$\delta_a(B) = \begin{cases} 1 & \text{if } a \in B \\ 0 & \text{if } a \notin B. \end{cases}$$

In particular, $\delta_a(\{a\}) = 1$.

If X takes two values,

$$X(\omega) = \begin{cases} a & \text{with probability } p \\ b & \text{with probability } 1 - p, \end{cases}$$

then

$$P_X(B) = \begin{cases} 1 & \text{if } a, b \in B \\ p & \text{if } a \in B, b \notin B \\ 1-p & \text{if } b \in B, a \notin B \\ 0 & \text{otherwise,} \end{cases}$$

and so

$$P_X(B) = p\delta_a(B) + (1-p)\delta_b(B).$$

The distribution of a general discrete random variable (i.e. one which takes only finitely many different values, except possibly on some null set) is of the form: if the values of X are a_i taken with probabilities $p_i > 0$, $i = 1, 2, \dots$ $\sum p_i = 1$, then

$$P_X(B) = \sum_{i=1}^{\infty} p_i \delta_{a_i}(B).$$

Classical examples are:

(1) the geometric distribution, where $p_i = (1-q)q^i$ for some $q \in (0,1)$,

(2) the Poisson distribution where $p_i = \frac{\lambda^i}{i!}e^{-\lambda}$.

We shall not discuss the discrete case further since this is not our primary goal in this text, and it is covered in many elementary texts on probability theory (such as [10]).

Now consider the classical probability space with $\Omega = [0,1]$, $\mathcal{F} = \mathcal{B}$, $P = m|_{[0,1]}$ – Lebesgue measure restricted to [0,1]. We can give examples of random variables given by explicit formulae.

For instance, let $X(\omega) = a\omega + b$. Then the image of $[0,1]$ is the interval $[b, a+b]$ and $P_X = \frac{1}{a}m|_{[b,a+b]}$, i.e. for Borel B

$$P_X(B) = \frac{m(B \cap [b, a+b])}{a}.$$

Example 3.17

Suppose a car leaves city A at random between 12 am and 1 pm. It travels at 50 mph towards B which is 25 miles from A. What is the probability distribution of the distance between the car and B at 1 pm?

Clearly, this distance is 0 with probability $\frac{1}{2}$, i.e. if the car departs before 12.30. As a function of the starting time (represented as $\omega \in [0,1]$) the distance has the form

$$X(\omega) = \begin{cases} 0 & \text{if } \omega \in [0, \frac{1}{2}] \\ 50\omega - 25 & \text{if } \omega \in (\frac{1}{2}, 1] \end{cases}$$

and $P_X = \frac{1}{2}P_1 + \frac{1}{2}P_2$ where $P_1 = \delta_0$, $P_2 = \frac{1}{25}m_{[0,25]}$. In this example, therefore, P_X is a combination of Dirac and Lebesgue measures.

In later chapters we shall explore more complicated forms of X and the corresponding distributions after developing further machinery needed to handle the computations.

3.5.4 Independence of random variables

Definition 3.18

X, Y are *independent* if the σ-fields generated by them are independent.

In other words, for any Borel sets B, C in $\mathbb{R}$,

$$P(X^{-1}(B) \cap Y^{-1}(C)) = P(X^{-1}(B))P(Y^{-1}(C)).$$

Example 3.19

Let $(\Omega = [0, 1], \mathcal{M})$ be equipped with Lebesgue measure. Consider $X = \mathbf{1}_{[0, \frac{1}{2}]}$, $Y = \mathbf{1}_{[\frac{1}{4}, \frac{3}{4}]}$. Then $\mathcal{F}_X = \{\varnothing, [0, 1], [0, \frac{1}{2}], (\frac{1}{2}, 1]\}$, $\mathcal{F}_Y = \{\varnothing, [0, 1], [\frac{1}{4}, \frac{3}{4}], [0, \frac{1}{4}) \cup (\frac{3}{4}, 1]\}$ are clearly independent.

Example 3.20

Let Ω be as above and let $X(\omega) = \omega$, $Y(\omega) = 1 - \omega$. Then $\mathcal{F}_X = \mathcal{F}_Y = \mathcal{M}$. A σ-field cannot be independent with itself (unless it is trivial): Take $A \in \mathcal{M}$ and then independence requires $P(A \cap A) = P(A) \times P(A)$ (the set A belongs to 'both' σ-fields), i.e. $P(A) = P(A)^2$, which can happen only if either $P(A) = 0$ or $P(A) = 1$. So a σ-field independent with itself consists of sets of measure zero or one.

3.5.5 Applications to mathematical finance

Consider a model of stock prices, discrete in time, i.e. assume that the stock prices are given by a sequence $S(n)$ of random variables, $n = 1, 2, \ldots, N$. If the length of one step is h, then we have the time horizon $T = Nh$ and we shall often write $S(T)$ instead of $S(N)$. An example of such a model is the binomial tree considered in the previous chapter. Recall that a European call option is the random variable of the form $(S(N) - K)^+$ (N is the exercise time, K is the strike price, S is the underlying asset). A natural generalisation of this is a random variable of the form $f(S(N))$ for some measurable function

$f : \mathbb{R} \to \mathbb{R}$. This random variable is of course measurable with respect to the σ-field generated by $S(N)$. This allows us to formulate a general definition:

Definition 3.21

A European derivative security (contingent claim) with the underlying asset represented by a sequence $S(n)$ and exercise time N is a random variable X measurable with respect to the σ-field $\mathcal{F}$ generated by $S(N)$.

Proposition 3.22

A European derivative security X must be of the form $X = f(S(N))$ for some measurable real function f.

The above definition is not sufficient for applications. For example, it does not cover one of the basic derivative instruments, namely futures. Recall that a holder of the *futures* contract has the right to receive (or an obligation to pay in case of negative values) a certain sequence $(X(1), \ldots, X(N))$ of cash payments, depending on the values of the underlying security. To be specific, if for example the length of one step is one year and r is the risk-free interest rate for annual compounding, then

$$X(n) = S(n)(1 + r)^{N-n} - X(n-1)(1 + r)^{N-n+1}.$$

In order to introduce a general notion of derivative security which would cover futures, we first consider a natural generalisation,

$$X(n) = f_n(S(0), S(1), \ldots, S(n)),$$

and then we push the level of generality ever further:

Definition 3.23

A derivative security (contingent claim) with the underlying asset represented by a sequence $(S(n))$ and the expiry time N is a sequence $(X(1), \ldots, X(N))$ of random variables such that $X(n)$ is measurable with respect to the σ-field $\mathcal{F}_n$ generated by $(S(0), S(1), \ldots, S(n))$, for each $n = 1, \ldots, N$.

Proposition 3.24

A derivative security X must be of the form $X = f(S(0), S(1), \ldots, S(N))$ for some measurable $f : \mathbb{R}^{N+1} \to \mathbb{R}$.

We could make one more step and dispose of the underlying random variables. The role of the underlying object would be played by an increasing sequence of σ-fields $\mathcal{F}_n$ and we would say that a contingent claim (avoiding here the other term) is a sequence of random variables $X(n)$ such that $X(n)$ is $\mathcal{F}_n$-measurable, but there is little need for such a generality in practical applications. The only case where that formulation would be relevant is the situation where there are no numerical observations but only some flow of information modelled by events and σ-fields.

Example 3.25

Payoffs of exotic options depend on the whole paths of consecutive stock prices. For example, the payoff of a European lookback option with exercise time N is determined by

$$f(x_0, x_1, \ldots, x_N) = \max\{x_0, x_1, \ldots, x_N\} - x_N.$$

Exercise 3.9

Find the function f for a down-and-out call (which is a European call except that it ceases to exist if the stock price at any time before the exercise date goes below the barrier $L < S(0)$).

Example 3.26

Consider an American put option in a binomial model. We shall see that it fits the above abstract scheme. Recall that *American* options can be exercised at any time before expiry and the payoff of a put exercised at time n is $(K-S(n))^+$, written $g(S(n))$ for brevity, $g(x) = (K-x)^+$. This option offers to the holder cash flow of the same nature as the stock. The latter is determined by the stock price and stock can be sold at any time, of course only once. The American option can be sold or exercised also only once. The value of this option will be denoted by $P^A(n)$ and we shall show that it is a derivative security in the sense of Definition 3.23.

We shall demonstrate that it is possible to write

$$P^A(n) = f_n(S(n))$$

for some functions f_n. Consider an option expiring at $N = 2$. Clearly

$$f_2(x) = g(x).$$

At time $n = 1$ the holder of the option can exercise or wait till $n = 2$. The value of waiting is the same as the value of European put issued at $n = 1$ with exercise time $N = 2$ (which is well known and will be seen in Section 7.4.4 in some detail) can be computed as the expectation with respect to some probability p of the discounted payoff. The value of the American put is the greater of the two, so

$$f_1(x) = \max \left\{ g(x), \frac{1}{1+r} \left[p f_2(xU) + (1-p) f_2(xD) \right] \right\}.$$

The same argument gives

$$f_0(x) = \max \left\{ g(x), \frac{1}{1+r} \left[p f_1(xU) + (1-p_*) f_1(xD) \right] \right\}.$$

In general, for an American option expiring at time N we have the following chain of recursive formulae:

$$f_N(x) = g(x),$$
$$f_{n-1}(x) = \max \left\{ g(x), \frac{1}{1+r} \left[p f_n(xU) + (1-p) f_n(xD) \right] \right\}.$$

3.6 Proofs of propositions

Proof (of Proposition 3.8)

(i) We have proved that if f is measurable then so are f^+, f^-. Conversely, note that $f(x) = f^+(x) - f^-(x)$, so Theorem 3.5 gives the result.

(ii) The function $u \mapsto |u|$ is continuous, so Lemma 3.7 with $F(u, v) = |u|$ gives measurability of $|f|$ (an alternative is to use $|f| = f^+ + f^-$). To see that the converse is not true take a non-measurable set A and let $f = \mathbf{1}_A - \mathbf{1}_{A^c}$. It is non-measurable since $\{x : f(x) > 0\} = A$ is non-measurable. But $|f| = 1$ is clearly measurable. $\square$

Proof (of Proposition 3.15)

Since $f \le \mathrm{ess\,sup} f$ and $g \le \mathrm{ess\,sup} g$ a.e., by adding we have $f + g \le \mathrm{ess\,sup} f + \mathrm{ess\,sup} g$ a.e. So the number $\mathrm{ess\,sup} f + \mathrm{ess\,sup} g$ belongs to the set $\{z : f + g \le z \text{ a.e.}\}$ and hence the infimum of this set is smaller than this number. $\square$

Proof (of Proposition 3.22)

First note that the σ-field generated by $S(N)$ is of the form $\mathcal{F} = \{S(N)^{-1}(B) : B\text{-Borel}\}$ since these sets form a σ-field and any other σ-field such that $S(N)$ is measurable with respect to it has to contain all inverse images of Borel sets. Next we proceed in three steps:

1. Suppose $X = \mathbf{1}_A$ for $A \in \mathcal{F}$. Then $A = S(N)^{-1}(B)$ for a Borel subset of $\mathbb{R}$. Put $f = \mathbf{1}_B$ and clearly $X = f \circ S(N)$.

2. If X is a step function, $X = \sum c_i \mathbf{1}_{A_i}$ then take $f = \sum c_i \mathbf{1}_{B_i}$ where $A_i = S(N)^{-1}(B_i)$.

3. In general, a measurable function X can be approximated by step functions $X_n = \sum_{k=0}^{2^{2n}} \frac{k}{2^n} \cdot \mathbf{1}_{Y^{-1}([\frac{k}{2^n}, \frac{k+1}{2^n}))}$ (see Proposition 4.15 for more details) and we take $f = \limsup f_n$, where f_n corresponds to Y_n as in step 2) and the sequence clearly converges on the range of $S(N)$. $\square$

Proof (of Proposition 3.24)

We follow the lines of the previous proof.

1. Suppose $X = \mathbf{1}_A$ for $A \in \mathcal{F}$. Then $A = (S(1), \ldots, S(N))^{-1}(B)$ for Borel $B \subset \mathbb{R}^N$, and $f = \mathbf{1}_B$ satisfies the claim.

Steps 2. and 3. are the same as in the proof of the previous proposition. $\square$

4
Integral

The theory developed in this chapter deals with Lebesgue measure for the sake of simplicity. However, all we need (except for the section where we discuss the Riemann integration) is the property of m being a measure, i.e. a countably additive (extended-) real-valued function μ defined on a σ-field $\mathcal{F}$ of subsets of a fixed set Ω. Therefore, the theory developed for the measure space $(\mathbb{R}, \mathcal{M}, m)$ in the following sections can be extended virtually without change to an abstractly given measure space $(\Omega, \mathcal{F}, \mu)$.

We encourage the reader to bear in mind the possibility of such a generalisation. We will need it in the probability section at the end of the chapter, and in the following chapters.

4.1 Definition of the integral

We are now able to resolve one of the problems we identified earlier: how to integrate functions like $\mathbf{1}_{\mathbb{Q}}$, which take only finitely many values, but where the sets on which these values are taken are not at all 'like intervals'.

Definition 4.1

A non-negative function $\varphi : \mathbb{R} \to \mathbb{R}$ which takes only finitely many values, i.e. the range of φ is a *finite* set of distinct non-negative reals $\{a_1, a_2, \ldots, a_n\}$, is a *simple function* if all the sets

$$A_i = \varphi^{-1}(\{a_i\}) = \{x : \varphi(x) = a_i\}, \qquad i = 1, 2, \ldots, n,$$

are measurable sets. Note that the sets $A_i \in \mathcal{M}$ are pairwise disjoint and their union is $\mathbb{R}$.

Clearly we can write

$$\varphi(x) = \sum_{i=1}^{n} a_i \mathbf{1}_{A_i}(x),$$

so that (by Theorem 3.5) each simple function is measurable.

Definition 4.2

The (Lebesgue) *integral* over $E \in \mathcal{M}$ of the simple function φ is given by

$$\int_E \varphi \, dm = \sum_{i=1}^{n} a_i m(A_i \cap E)$$

(see Figure 4.1). Note that since we shall allow $m(A_i) = +\infty$, we use the convention $0 \times \infty = 0$ here.

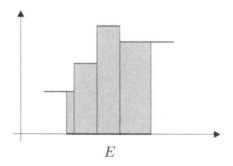

E

Figure 4.1   Integral of a simple function

Example 4.3

Consider the simple function $\mathbf{1}_\mathbb{Q}$ which takes the value 1 on $\mathbb{Q}$ and 0 on $\mathbb{R} \setminus \mathbb{Q}$. By the above definition we have

$$\int_\mathbb{R} \mathbf{1}_\mathbb{Q} \, dm = 1 \times m(\mathbb{Q}) + 0 \times m(\mathbb{R} \setminus \mathbb{Q}) = 0$$

since $\mathbb{Q}$ is a null set. Recall that this function is *not* Riemann-integrable. Similarly, $\mathbf{1}_C$ has integral 0, where C is the Cantor set.

Exercise 4.1

Find the integral of φ over E where

(a) $\varphi(x) = \mathrm{Int}(x)$, $E = [0, 10]$,

(b) $\varphi(x) = \mathrm{Int}(x^2)$, $E = [0, 2]$,

(c) $\varphi(x) = \mathrm{Int}(\sin x)$, $E = [0, 2\pi]$

and Int denotes the integer part of a real number. (Note that many texts use the symbol $[x]$ to denote $\mathrm{Int}(x)$. We prefer to use Int for increased clarity.)

In order to extend the integral to more general functions, Henri Lebesgue (in 1902) adopted an apparently obvious, but subtle device: instead of partitioning the *domain* of a bounded function f into many small intervals, he partitioned its *range* into a finite number of small intervals of the form $A_i = [a_{i-1}, a_i)$, and approximated the 'area' under the graph of f by the *upper sum*

$$S(n) = \sum_{i=1}^{n} a_i m(f^{-1}(A_i))$$

and the *lower sum*

$$s(n) = \sum_{i=1}^{n} a_{i-1} m(f^{-1}(A_i))$$

respectively; then integrable functions had the property that the infimum of all upper sums equals the supremum of all lower sums – mirroring Riemann's construction (see also Figure 3.1).

A century of experience with the Lebesgue integral has led to many equivalent definitions, some of them technically (if not always conceptually) simpler. We shall follow a version which, while very similar to Lebesgue's original construction, allows us to make full use of the measure theory developed already. First we stay with non-negative functions:

Definition 4.4

For any non-negative measurable function f and $E \in \mathcal{M}$ the *integral* $\int_E f \, dm$ is defined as

$$\int_E f \, dm = \sup Y(E, f)$$

where

$$Y(E, f) = \left\{ \int_E \varphi \, dm : 0 \leq \varphi \leq f, \varphi \text{ is simple} \right\}.$$

Note that the integral can be $+\infty$, and is always non-negative. Clearly, the set $Y(E, f)$ is always of the form $[0, x]$ or $[0, x)$, where the value $x = +\infty$ is allowed.

If $E = [a, b]$ we write the integral as

$$\int_a^b f \, dm, \quad \int_a^b f(x) \, dm(x),$$

or even as $\int_a^b f(x) \, dx$, when no confusion is possible (and we set $\int_a^b f \, dm = -\int_b^a f \, dm$ if $a > b$). The notation $\int f \, dm$ means $\int_{\mathbb{R}} f \, dm$.

Clearly, if for some $A \in \mathcal{M}$ and a non-negative measurable function g we have $g = 0$ on A^c, then any non-negative simple function that lies below g must be zero on A^c. Applying this to $g = f.\mathbf{1}_A$ we obtain the important identity

$$\int_A f \, dm = \int f \mathbf{1}_A \, dm.$$

Exercise 4.2

Suppose that $f : [0, 1] \to \mathbb{R}$ is defined by letting $f(x) = 0$ on the Cantor set and $f(x) = k$ for all x in each interval of length 3^{-k} which has been removed from $[0, 1]$. Calculate $\int_0^1 f \, dm$.

Hint Recall that $\sum_{k=1}^{\infty} kx^{k-1} = \frac{d}{dx} \left(\sum_{k=0}^{\infty} x^k \right) = \frac{1}{(1-x)^2}$ when $|x| < 1$.

If f is a simple function, we now have two definitions of the integral; thus for consistency you should check carefully that the above definitions coincide.

Proposition 4.5

For simple functions, Definitions 4.2 and 4.4 are equivalent.

Furthermore, we can prove the following basic properties of integrals of simple functions:

Theorem 4.6

Let φ, ψ be simple functions. Then:

(i) if $\varphi \le \psi$ then $\int_E \varphi \, dm \le \int_E \psi \, dm$,

(ii) if A, B are disjoint sets in $\mathcal{M}$, then

$$\int_{A \cup B} \varphi \, dm = \int_A \varphi \, dm + \int_B \varphi \, dm,$$

(iii) for all constants $a > 0$

$$\int_E a\varphi \, dm = a \int_E \varphi \, dm.$$

Proof

(i) Notice that $Y(E, \varphi) \subseteq Y(E, \psi)$ (we use Definition 4.4).

(ii) Employing the properties of m we have ($\varphi = \sum c_i \mathbf{1}_{D_i}$)

$$\int_{A \cup B} \varphi \, dm = \sum c_i m(D_i \cap (A \cup B))$$
$$= \sum c_i \big(m(D_i \cap A) + m(D_i \cap B) \big)$$
$$= \sum c_i m(D_i \cap A) + \sum c_i (D_i \cap B)$$
$$= \int_A \varphi \, dm + \int_B \varphi \, dm.$$

(iii) If $\varphi = \sum c_i \mathbf{1}_{A_i}$ then $a\varphi = \sum a c_i \mathbf{1}_{A_i}$ and

$$\int_E a\varphi \, dm = \sum a c_i m(E \cap A_i) = a \sum c_i m(E \cap A_i) = a \int_E \varphi \, dm$$

as required. $\qquad\qquad\qquad\qquad\qquad\qquad\qquad\qquad\qquad\qquad\qquad$ $\square$

Next we show that the properties of the integrals of simple functions extend to the integrals of non-negative measurable functions:

Theorem 4.7

Suppose f and g are non-negative measurable functions.

(i) If $A \in \mathcal{M}$, and $f \le g$ on A, then

$$\int_A f \, dm \le \int_A g \, dm.$$

(ii) If $B \subseteq A$, A, $B \in \mathcal{M}$, then

$$\int_B f \, dm \le \int_A f \, dm.$$

(iii) For $a \geq 0$,

$$\int_A af \, dm = a \int_A f \, dm.$$

(iv) If A is null then

$$\int_A f \, dm = 0.$$

(v) If $A, B \in \mathcal{M}, A \cap B = \emptyset$, then

$$\int_{A \cup B} f \, dm = \int_A f \, dm + \int_B f \, dm.$$

Proof

(i) Notice that $Y(A, f) \subseteq Y(A, g)$ (there is more room to squeeze simple functions under g than under f) and the sup of a bigger set is larger.

(ii) If φ is a simple function lying below f on B, then extending it by zero outside B we obtain a simple function which is below f on A. The integrals of these simple functions are the same, so $Y(B, f) \subseteq Y(A, f)$ and we conclude as in (i).

(iii) The elements of the set $Y(A, af)$ are of the form $a \times x$ where $x \in Y(A, f)$, so the same relation holds between their suprema.

(iv) For any simple function φ, $\int_A \varphi \, dm = 0$. To see this, take $\varphi = \sum c_i \mathbf{1}_{E_i}$, say, then $m(A \cap E_i) = 0$ for each i, so $Y(A, f) = \{0\}$.

(v) The elements of $Y(A \cup B, f)$ are of the form $\int_{A \cup B} \varphi \, dm$, so by Theorem 4.6 (ii) they are of the form $\int_A \varphi \, dm + \int_B \varphi \, dm$. So $Y(A \cup B, f) = Y(A, f) + Y(B, f)$ and taking suprema this yields $\int_{A \cup B} f \, dm \leq \int_A f \, dm + \int_B f \, dm$. For the opposite inequality, suppose that the simple functions φ and ψ satisfy: $\varphi \leq f$ on A and $\varphi = 0$ off A, while $\psi \leq f$ on B and $\psi = 0$ off B. Since $A \cap B = \emptyset$, we can construct a new simple function $\gamma \leq f$ by setting $\gamma = \varphi$ on A, $\gamma = \psi$ on B and $\gamma = 0$ outside $A \cup B$. Then

$$\int_A \varphi \, dm + \int_B \psi \, dm = \int_A \gamma \, dm + \int_B \gamma \, dm$$

$$= \int_{A \cup B} \gamma \, dm$$

$$\leq \int_{A \cup B} f \, dm.$$

On the right we have an upper bound which remains valid for all simple functions that lie below f on $A \cup B$. Thus taking suprema over φ and ψ separately on the left gives $\int_A f \, dm + \int_B f \, dm \leq \int_{A \cup B} f \, dm$. $\qquad \square$

Exercise 4.3

Prove the following mean value theorem for the integral: if $a \leq f(x) \leq b$ for $x \in A$, then $am(A) \leq \int_A f \, dm \leq bm(A)$.

We now confirm that null sets are precisely the 'negligible sets' for integration theory:

Theorem 4.8

Suppose f is a non-negative measurable function. Then $f = 0$ a.e. if and only if $\int_{\mathbb{R}} f \, dm = 0$.

Proof

First, note that if $f = 0$ a.e. and $0 \leq \varphi \leq f$ is a simple function, then $\varphi = 0$ a.e. since neither f nor φ takes a negative value. Thus $\int_{\mathbb{R}} \varphi \, dm = 0$ for all such φ and so $\int_{\mathbb{R}} f \, dm = 0$ also.

Conversely, given $\int_{\mathbb{R}} f \, dm = 0$, let $E = \{x : f(x) > 0\}$. Our goal is to show that $m(E) = 0$. Put

$$E_n = f^{-1}([\frac{1}{n}, \infty)) \quad \text{for} \quad n \geq 1.$$

Clearly, (E_n) increases to E with

$$E = \bigcup_{n=1}^{\infty} E_n.$$

To show that $m(E) = 0$ it is sufficient to prove that $m(E_n) = 0$ for all n. (See Theorem 2.19.) The function $\varphi = \frac{1}{n}\mathbf{1}_{E_n}$ is simple and $\varphi \leq f$ by the definition of E_n. So

$$\int_{\mathbb{R}} \varphi \, dm = \frac{1}{n}m(E_n) \leq \int_{\mathbb{R}} f \, dm = 0,$$

hence $m(E_n) = 0$ for all n.   □

Using the results proved so far, the following 'a.e.' version of the monotonicity of the integral is not difficult to prove:

Proposition 4.9

If f and g are measurable then $f \leq g$ a.e. implies $\int f \, dm \leq \int g \, dm$.

Hint Let $A = \{x : f(x) \leq g(x)\}$; then $B = A^c$ is null and $f1_A \leq g1_A$. Now use Theorems 4.7 and 4.8.

Using Theorems 3.5 and 3.9 you should now provide a second proof of a result we already noted in Proposition 3.8 but repeat here for emphasis:

Proposition 4.10

The function $f : \mathbb{R} \to \mathbb{R}$ is measurable iff both f^+ and f^- are measurable.

4.2 Monotone convergence theorems

The crux of Lebesgue integration is its convergence theory. We can make a start on that by giving the following famous result:

Theorem 4.11 (Fatou's lemma)

If $\{f_n\}$ is a sequence of non-negative measurable functions then

$$\liminf_{n \to \infty} \int_E f_n \, dm \geq \int_E \left(\liminf_{n \to \infty} f_n \right) dm.$$

Proof

Write

$$f = \liminf_{n \to \infty} f_n$$

and recall that

$$f = \lim_{n \to \infty} g_n$$

where $g_n = \inf_{k \geq n} f_k$ (the sequence g_n is non-decreasing). Let φ be a simple function, $\varphi \leq f$. To show that

$$\int_E f \, dm \leq \liminf_{n \to \infty} \int_E f_n \, dm$$

it is sufficient to see that

$$\int_E \varphi \, dm \leq \liminf_{n \to \infty} \int_E f_n \, dm$$

for any such φ.

The set where $f = 0$ is irrelevant since it does not contribute to $\int_E f \, dm$, so we can assume, without loss of generality, that $f > 0$ on E. Put

$$\overline{\varphi}(x) = \begin{cases} \varphi(x) - \varepsilon > 0 & \text{if } \varphi(x) > 0 \\ 0 & \text{if } \varphi(x) = 0 \text{ or } x \notin E \end{cases}$$

where ε is sufficiently small to ensure $\overline{\varphi} \geq 0$.

Now $\overline{\varphi} < f$, $g_n \nearrow f$, so 'eventually' $g_n \geq \overline{\varphi}$. We make the last statement more precise: put

$$A_k = \{x : g_k(x) \geq \overline{\varphi}(x)\}$$

and we have

$$A_k \subseteq A_{k+1}, \qquad \bigcup_{k=1}^{\infty} A_k = \mathbb{R}.$$

Next,

$$\int_{A_n \cap E} \overline{\varphi} \, dm \leq \int_{A_n \cap E} g_n \, dm \quad (\text{as } g_n \text{ dominates } \overline{\varphi} \text{ on } A_n)$$

$$\leq \int_{A_n \cap E} f_k \, dm \quad \text{for } k \geq n \text{ (by the definition of } g_n)$$

$$\leq \int_E f_k \, dm \quad (\text{as } E \text{ is the larger set})$$

for $k \geq n$. Hence

$$\int_{A_n \cap E} \overline{\varphi} \, dm \leq \liminf_{k \to \infty} \int_E f_k \, dm. \qquad (4.1)$$

Now we let $n \to \infty$: writing $\overline{\varphi} = \sum_{i=1}^{l} c_i \mathbf{1}_{B_i}$ for some $c_i \geq 0$, $B_i \in \mathcal{M}$, $i \leq l$,

$$\int_{A_n \cap E} \overline{\varphi} \, m = \sum_{i=1}^{l} c_i m(A_n \cap E \cap B_i) \longrightarrow \sum_{i=1}^{l} c_i m(E \cap B_i) = \int_E \overline{\varphi} \, dm$$

and the inequality (4.1) remains true in the limit:

$$\int_E \overline{\varphi} \, dm \leq \liminf_{k \to \infty} \int_E f_k \, dm.$$

We are close – all we need is to replace $\overline{\varphi}$ by φ in the last relation. This will be done by letting $\varepsilon \to 0$ but some care will be needed.

Suppose that $m(\{x : \varphi(x) > 0\}) < \infty$. Then

$$\int_E \overline{\varphi} \, dm = \int_E \varphi \, dm - \varepsilon m(\{x : \varphi(x) > 0\})$$

and we get the result by letting $\varepsilon \to 0$.

The case $m(\{x : \varphi(x) > 0\}) = \infty$ has to be treated separately. Here $\int_E \varphi\, dm = \infty$, so $\int_E f\, dm = \infty$. We have to show that

$$\liminf_{k\to\infty} \int_E f_k\, dm = \infty.$$

Let c_i be the values of φ and let $a = \frac{1}{2}\min\{c_i\}$ ($\{c_i\}$ is a finite set!). Similarly to above, put

$$D_n = \{x : g_n(x) > a\}$$

and

$$\int_{D_n \cap E} g_n\, dm \to \infty$$

since $D_n \nearrow \mathbb{R}$. As before,

$$\int_{D_n \cap E} g_n\, dm \le \int_{D_n \cap E} f_k\, dm \le \int_E f_k\, dm$$

for $k \ge n$, so $\liminf \int_E f_k\, dm$ has to be infinite. $\square$

Example 4.12

Let $f_n = \mathbf{1}_{[n,n+1]}$. Clearly $\int f_n\, dm = 1$ for all n, $\liminf f_n = 0\ (= \lim f_n)$, so the above inequality may be strict and we have

$$\int (\lim f_n)\, dm \ne \lim \int f_n\, dm.$$

Exercise 4.4

Construct an example of a sequence of functions with the strict inequality as above, such that all f_n are zero outside the interval $[0, 1]$.

It is now easy to prove one of the two main convergence theorems.

Theorem 4.13 (monotone convergence theorem)

If $\{f_n\}$ is a sequence of non-negative measurable functions, and $\{f_n(x) : n \ge 1\}$ increases monotonically to $f(x)$ for each x, i.e. $f_n \nearrow f$ pointwise, then

$$\lim_{n\to\infty} \int_E f_n(x)\, dm = \int_E f\, dm.$$

Proof

Since $f_n \leq f$, $\int_E f_n \, dm \leq \int_E f \, dm$ and so

$$\limsup_{n \to \infty} \int_E f_n \, dm \leq \int_E f \, dm.$$

Fatou's lemma gives

$$\int_E f \, dm \leq \liminf_{n \to \infty} \int_E f_n \, dm,$$

which together with the basic relation

$$\liminf_{n \to \infty} \int_E f_n \, dm \leq \limsup_{n \to \infty} \int_E f_n \, dm$$

gives

$$\int_E f \, dm = \liminf_{n \to \infty} \int_E f_n \, dm = \limsup_{n \to \infty} \int_E f_n \, dm,$$

hence the sequence $\int_E f_n \, dm$ converges to $\int_E f \, dm$.   □

Corollary 4.14

Suppose $\{f_n\}$ and f are non-negative and measurable. If $\{f_n\}$ increases to f almost everywhere, then we still have $\int_E f_n \, dm \nearrow \int_E f \, dm$ for all measurable E.

Proof

Suppose that $f_n \nearrow f$ a.e. and A is the set where the convergence holds, so that A^c is null. We can define

$$g_n = \begin{cases} f_n & \text{on } A \\ 0 & \text{on } A^c, \end{cases}$$

$$g = \begin{cases} f & \text{on } A \\ 0 & \text{on } A^c. \end{cases}$$

Then using $E = [E \cap A^c] \cup [E \cap A]$ we get

$$\int_E g_n \, dm = \int_{E \cap A} f_n \, dm + \int_{E \cap A^c} 0 \, dm$$

$$= \int_{E \cap A} f_n \, dm + \int_{E \cap A^c} f_n \, dm$$

$$= \int_E f_n \, dm$$

(since $E \cap A^c$ is null), and similarly $\int_E g \, dm = \int_E f \, dm$. The convergence $g_n \to g$ holds everywhere, so by Theorem 4.13, $\int_E g_n \, dm \to \int_E g \, dm$.   □

To apply the monotone convergence theorem it is convenient to approximate non-negative measurable functions by increasing sequences of simple functions.

Proposition 4.15

For any non-negative measurable f there is a sequence s_n of non-negative simple functions such that $s_n \nearrow f$.

Hint Put

$$s_n = \sum_{k=0}^{2^{2n}} \frac{k}{2^n} \cdot \mathbf{1}_{f^{-1}([\frac{k}{2^n}, \frac{k+1}{2^n}))}$$

(see Figure 4.2).

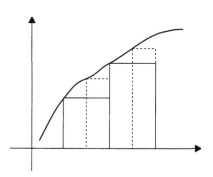

Figure 4.2 Approximation by simple functions

4.3 Integrable functions

All the hard work is done: we can extend the integral very easily to general real functions, using the positive part $f^+ = \max(f, 0)$, and the negative part $f^- = \max(-f, 0)$, of any measurable function $f : \mathbb{R} \to \mathbb{R}$. We will not use the non-negative measurable function $|f|$ alone: as we saw in Proposition 3.8, $|f|$ can be measurable without f being measurable!

Definition 4.16

If $E \in \mathcal{M}$ and the measurable function f has both $\int_E f^+ \, dm$ and $\int_E f^- \, dm$ finite, then we say that f is *integrable*, and define

$$\int_E f \, dm = \int_E f^+ \, dm - \int_E f^- \, dm.$$

The set of all functions that are integrable over E is denoted by $\mathcal{L}^1(E)$. In what follows E will be fixed and we often simply write $\mathcal{L}^1$ for $\mathcal{L}^1(E)$.

Exercise 4.5

For which α is $f(x) = x^\alpha$ in $\mathcal{L}^1(E)$ where (a) $E = (0,1)$; (b) $E = (1,\infty)$?

Hint See the remark following Example 4.27 below.

Note that f is integrable iff $|f|$ is integrable, and that

$$\int_E |f|\,\mathrm{d}m = \int_E f^+\,\mathrm{d}m + \int_E f^-\,\mathrm{d}m.$$

Thus the Lebesgue integral is an 'absolute' integral: we cannot 'make' a function integrable by cancellation of large positive and negative parts. This has the consequence that some functions which have improper Riemann integrals fail to be Lebesgue-integrable (see Section 4.5).

The properties of the integral of non-negative functions extend to any, not necessarily non-negative, integrable functions.

Proposition 4.17

If f and g are integrable, $f \le g$, then $\int f\,\mathrm{d}m \le \int g\,\mathrm{d}m$.

Hint If $f \le g$, then $f^+ \le g^+$ but $f^- \ge g^-$.

Remark 4.18

We observe (following [13], 5.12) that many proofs of results concerning integrable functions follow a standard pattern, utilising linearity and monotone convergence properties. To prove that a 'linear' result holds for all functions in a space such as $\mathcal{L}^1(E)$ we proceed in four steps:

1. verify that the required property holds for indicator functions – this is usually so by definition,

2. use linearity to extend the property to non-negative simple functions,

3. then use monotone convergence to show that the property is shared by all non-negative measurable functions,

4. finally, extend to the whole class of functions by writing $f = f^+ - f^-$ and using linearity again.

The next result gives a good illustration of the technique.

We wish to show that the mapping $f \mapsto \int_A f \, dm$ is linear. This fact is interesting on its own, but will also allow us to show that $\mathcal{L}^1$ is a vector space.

Theorem 4.19

For any integrable functions f, g their sum $f + g$ is also integrable and

$$\int_E (f + g) \, dm = \int_E f \, dm + \int_E g \, dm.$$

Proof

We apply the technique described in Remark 4.18. In this instance, we can conveniently begin with simple functions thus bypassing Step 1.

Step 2. Suppose first that f and g are non-negative simple functions. The result is a matter of routine calculation: let $f = \sum a_i \mathbf{1}_{A_i}$, $g = \sum b_j \mathbf{1}_{B_j}$. The sum $f + g$ is also a simple function which can be written in the form

$$f + g = \sum_{i,j} (a_i + b_j) \mathbf{1}_{A_i \cap B_j}.$$

Therefore

$$
\begin{aligned}
\int_E (f + g) \, dm &= \sum_{i,j} (a_i + b_j) m(A_i \cap B_j \cap E) \\
&= \sum_i \sum_j a_i m(A_i \cap B_j \cap E) + \sum_j \sum_i b_j m(A_i \cap B_j \cap E) \\
&= \sum_i a_i \sum_j m(A_i \cap B_j \cap E) + \sum_j b_j \sum_i m(A_i \cap B_j \cap E) \\
&= \sum_i a_i m\Big(\bigcup_j (A_i \cap B_j \cap E)\Big) + \sum_j b_j m\Big(\bigcup_i (A_i \cap B_j \cap E)\Big) \\
&= \sum_i a_i m\Big(A_i \cap \bigcup_j B_j \cap E\Big) + \sum_j b_j m\Big(B_j \cap \bigcup_i A_i \cap E\Big) \\
&= \sum_i a_i m(A_i \cap E) + \sum_j b_j m(B_j \cap E) \\
&= \int_E f \, dm + \int_E g \, dm
\end{aligned}
$$

where we have used the additivity of m and the facts that A_i cover $\mathbb{R}$ and the same is true for B_j.

Step 3. Now suppose that f, g are non-negative measurable (not necessarily simple) functions. By Proposition 4.15 we can find sequences s_n, t_n of simple functions such that $s_n \nearrow f$ and $t_n \nearrow g$. Clearly $s_n + t_n \nearrow f+g$, hence using the monotone convergence theorem and the additivity property for simple functions we obtain

$$\int_E (f+g)\,\mathrm{d}m = \lim_{n\to\infty} \int_E (s_n + t_n)\,\mathrm{d}m$$

$$= \lim_{n\to\infty} \int_E s_n\,\mathrm{d}m + \lim_{n\to\infty} \int_E t_n\,\mathrm{d}m$$

$$= \int_E f\,\mathrm{d}m + \int_E g\,\mathrm{d}m.$$

This, in particular, implies that the integral of $f+g$ is finite if the integrals of f and g are finite.

Step 4. Finally, let f, g be arbitrary integrable functions. Since

$$\int_E |f+g|\,\mathrm{d}m \le \int_E (|f| + |g|)\,\mathrm{d}m,$$

we can use Step 2 to deduce that the left-hand side is finite.

We have

$$f + g = (f+g)^+ - (f+g)^-$$
$$f + g = (f^+ - f^-) + (g^+ - g^-),$$

so

$$(f+g)^+ - (f+g)^- = f^+ - f^- + g^+ - g^-.$$

We rearrange the equality to have only additions on both sides:

$$(f+g)^+ + f^- + g^- = f^+ + g^+ + (f+g)^-.$$

We have non-negative functions on both sides, so by what we have proved so far,

$$\int_E (f+g)^+\,\mathrm{d}m + \int_E f^-\,\mathrm{d}m + \int_E g^-\,\mathrm{d}m = \int_E f^+\,\mathrm{d}m + \int_E g^+\,\mathrm{d}m + \int_E (f+g)^-\,\mathrm{d}m,$$

hence

$$\int_E (f+g)^+\,\mathrm{d}m - \int_E (f+g)^-\,\mathrm{d}m = \int_E f^+\,\mathrm{d}m - \int_E f^-\,\mathrm{d}m + \int_E g^+\,\mathrm{d}m - \int_E g^-\,\mathrm{d}m.$$

By definition of the integral the last relation implies the claim of the theorem.

$\square$

The following result is a routine application of monotone convergence:

Proposition 4.20

If f is integrable and $c \in \mathbb{R}$, then

$$\int_E (cf)\, \mathrm{d}m = c \int_E f\, \mathrm{d}m.$$

Hint Approximate f by a sequence of simple functions.

We complete the proof that $\mathcal{L}^1$ is a vector space:

Theorem 4.21

For any measurable E, $\mathcal{L}^1(E)$ is a vector space.

Proof

Let $f, g \in \mathcal{L}^1$. To show that $f + g \in \mathcal{L}^1$ we have to prove that $|f+g|$ is integrable:

$$\int_E |f + g|\, \mathrm{d}m \le \int_E (|f| + |g|)\, \mathrm{d}m = \int_E |f|\, \mathrm{d}m + \int_E |g|\, \mathrm{d}m < \infty.$$

Now let c be a constant:

$$\int_E |cf|\, \mathrm{d}m = \int_E |c|\,|f|\, \mathrm{d}m = |c| \int_E |f|\, \mathrm{d}m < \infty,$$

so that $cf \in \mathcal{L}^1(E)$.   □

We can now answer an important question on the extent to which the integral determines the integrand.

Theorem 4.22

If $\int_A f\, \mathrm{d}m \le \int_A g\, \mathrm{d}m$ for all $A \in \mathcal{M}$, then $f \le g$ almost everywhere. In particular, if $\int_A f\, \mathrm{d}m = \int_A g\, \mathrm{d}m$ for all $A \in \mathcal{M}$, then $f = g$ almost everywhere.

Proof

By additivity of the integral (and Proposition 4.19 below) it is sufficient to show that $\int_A h\, \mathrm{d}m \ge 0$ for all $A \in \mathcal{M}$ implies $h \ge 0$ (and then take $h = g - f$).

Write $A = \{x : h(x) < 0\}$; then $A = \bigcup A_n$ where $A_n = \{x : h(x) \le -\frac{1}{n}\}$. By monotonicity of the integral,

$$\int_{A_n} h \, dm \le \int_{A_n} \left(-\frac{1}{n}\right) dm = -\frac{1}{n} m(A_n),$$

which is non-negative but this can only happen if $m(A_n) = 0$. The sequence of sets A_n increases with n, hence $m(A) = 0$, and so $h(x) \ge 0$ almost everywhere.

A similar argument shows that if $\int_A h \, dm \le 0$ for all A, then $h \le 0$ a.e. This implies the second claim of the theorem: put $h = g - f$ and $\int_A h \, dm$ is both non-negative and non-positive, hence $h \ge 0$ and $h \le 0$ a.e., thus $h = 0$ a.e. $\qquad\qquad\square$

The next proposition lists further important properties of integrable functions, whose straightforward proofs are typical applications of the results proved so far.

Proposition 4.23

(i) An integrable function is a.e. finite.

(ii) For measurable f and A,

$$m(A) \inf_A f \le \int_A f \, dm \le m(A) \sup_A f.$$

(iii) $|\int f \, dm| \le \int |f| \, dm$.

(iv) Assume that $f \ge 0$ and $\int f \, dm = 0$. Then $f = 0$ a.e.

The following theorem gives us the possibility of constructing many interesting measures, and is essential for the development of probability distributions.

Theorem 4.24

Let $f \ge 0$. Then $A \mapsto \int_A f \, dm$ is a measure.

Proof

Denote $\mu(A) = \int_A f \, dm$. The goal is to show

$$\mu(\bigcup_i E_i) = \sum \mu(E_i)$$

for pairwise disjoint E_i. To this end consider the sequence $g_n = f\mathbf{1}_{\bigcup_{i=1}^n E_i}$ and note that $g_n \nearrow g$, where $g = f\mathbf{1}_{\bigcup_{i=1}^\infty E_i}$. Now

$$\int g\,dm = \mu(\bigcup_{i=1}^\infty E_i),$$

$$\int g_n\,dm = \int_{\bigcup_{i=1}^n E_i} f\,dm = \sum_{i=1}^n \int_{E_i} f\,dm = \sum_{i=1}^n \mu(E_i)$$

and the monotone convergence theorem completes the proof. $\qquad\square$

4.4 The dominated convergence theorem

Many questions in analysis centre on conditions under which the order of two limit processes, applied to certain functions, can be interchanged. Since integration is a limit process applied to measurable functions, it is natural to ask under what conditions on a pointwise (or pointwise a.e.) convergent sequence (f_n), the limit of the integrals is the integral of the pointwise limit function f, i.e. when can we state that $\lim \int f_n\,dm = \int (\lim f_n)\,dm$? The monotone convergence theorem (Theorem 4.13) provided the answer that this conclusion is valid for monotone increasing sequences of non-negative measurable functions, though in that case, of course, the limits may equal $+\infty$. The following example shows that for general sequences of integrable functions the conclusion will not hold without some further conditions:

Example 4.25

Let $f_n(x) = n\mathbf{1}_{[0,\frac{1}{n}]}(x)$. Clearly $f_n(x) \to 0$ for all x but $\int f_n(x)\,dx = 1$.

The limit theorem which turns out to be the most useful in practice states that convergence holds for an a.e. convergent sequence which is *dominated* by an integrable function. Again Fatou's lemma holds the key to the proof.

Theorem 4.26 (dominated convergence theorem)

Suppose $E \in \mathcal{M}$. Let (f_n) be a sequence of measurable functions such that $|f_n| \le g$ a.e. on E for all $n \ge 1$, where g is integrable over E. If $f = \lim_{n\to\infty} f_n$ a.e. then f is integrable over E and

$$\lim_{n\to\infty} \int_E f_n(x)\,dm = \int_E f\,dm.$$

Proof

Suppose for the moment that $f_n \geq 0$. Fatou's lemma gives

$$\int_E f \, dm \leq \liminf_{n \to \infty} \int_E f_n \, dm.$$

It is therefore sufficient to show that

$$\limsup_{n \to \infty} \int_E f_n \, dm \leq \int_E f \, dm. \qquad (4.2)$$

Fatou's lemma applied to $g - f_n$ gives

$$\int_E \lim_{n \to \infty} (g - f_n) \, dm \leq \liminf_{n \to \infty} \int_E (g - f_n) \, dm.$$

On the left we have

$$\int_E (g - f) \, dm = \int_E g \, dm - \int_E f \, dm.$$

On the right

$$\liminf_{n \to \infty} \int_E (g - f_n) \, dm = \liminf_{n \to \infty} \left(\int_E g \, dm - \int_E f_n \, dm \right)$$
$$= \int_E g \, dm - \limsup_{n \to \infty} \int_E f_n \, dm,$$

where we have used the elementary fact that

$$\liminf_{n \to \infty} (-a_n) = -\limsup_{n \to \infty} a_n.$$

Putting this together we get

$$\int_E g \, dm - \int_E f \, dm \leq \int_E g \, dm - \limsup_{n \to \infty} \int_E f_n \, dm.$$

Finally, subtract $\int_E g \, dm$ (which is finite) and multiply by -1 to arrive at (4.2).

Now consider a general, not necessarily non-negative sequence (f_n). Since by the hypothesis

$$-g(x) \leq f_n(x) \leq g(x)$$

we have

$$0 \leq f_n(x) + g(x) \leq 2g(x)$$

and we can apply the result proved for non-negative functions to the sequence $f_n(x) + g(x)$ (the function $2g$ is of course integrable). $\qquad \square$

Example 4.27

Going back to the example preceding the theorem, $f_n = n\mathbf{1}_{[0,\frac{1}{n}]}$, we can see that an integrable g to dominate f_n cannot be found. The least upper bound is $g(x) = \sup_n f_n(x)$, $g(x) = k$ on $(\frac{1}{k+1}, \frac{1}{k}]$, so

$$\int g(x)\,dx = \sum_{k=1}^{\infty} k\left(\frac{1}{k} - \frac{1}{k+1}\right) = \sum_{k=1}^{\infty} \frac{1}{k+1} = +\infty.$$

For a typical positive example consider

$$f_n(x) = \frac{n\sin x}{1 + n^2 x^{1/2}}$$

for $x \in (0,1)$. Clearly $f_n(x) \to 0$. To conclude that $\lim_n \int f_n\,dm = 0$ we need an integrable dominating function. This is usually where some ingenuity is needed; however in the present example the most straightforward estimate will suffice:

$$\left|\frac{n\sin x}{1 + n^2 x^{1/2}}\right| \leq \frac{n}{1 + n^2 x^{1/2}} \leq \frac{n}{n^2 x^{1/2}} = \frac{1}{nx^{1/2}} \leq \frac{1}{x^{1/2}}.$$

(To see from first principles that the dominating function $g : x \mapsto \frac{1}{\sqrt{x}}$ is integrable over $[0,1]$ can be rather tedious – cf. the worked example in Chapter 1 for the Riemann integral of $x \mapsto \sqrt{x}$. However, we shall show shortly that the Lebesgue and Riemann integrals of a bounded function coincide if the latter exists, and hence we can apply the fundamental theorem of the calculus to confirm the integrability of g.)

The following facts will be useful later.

Proposition 4.28

Suppose f is integrable and define $g_n = f\mathbf{1}_{[-n,n]}$, $h_n = \min(f,n)$ (both truncate f in some way: the g_n vanish outside a bounded interval, the h_n are bounded). Then $\int |f - g_n|\,dm \to 0$, $\int |f - h_n|\,dm \to 0$.

Hint Use the dominated convergence theorem.

Exercise 4.6

Use the dominated convergence theorem to find

$$\lim_{n\to\infty} \int_1^{\infty} f_n(x)\,dx$$

where

$$f_n(x) = \frac{\sqrt{x}}{1 + nx^3}.$$

Exercise 4.7

Investigate the convergence of

$$\int_a^\infty \frac{n^2 x e^{-n^2 x^2}}{1 + x^2}\,\mathrm{d}x$$

for $a > 0$, and for $a = 0$.

Exercise 4.8

Investigate the convergence of

$$\int_0^\infty \frac{1}{(1 + \frac{x}{n})^n \sqrt[n]{x}}\,\mathrm{d}x.$$

We will need the following extension of Theorem 4.19:

Proposition 4.29

For a sequence of non-negative measurable functions f_n we have

$$\int \sum_{n=1}^\infty f_n\,\mathrm{d}m = \sum_{n=1}^\infty \int f_n\,\mathrm{d}m.$$

Hint The sequence $g_k = \sum_{n=1}^k f_n$ is increasing and converges to $\sum_{n=1}^\infty f_n$.

We cannot yet conclude that the sum of the series on the right-hand side is a.e. finite, so $\sum_{n=1}^\infty f_n$ need not be integrable. However:

Theorem 4.30 (Beppo–Levi)

Suppose that

$$\sum_{k=1}^\infty \int |f_k|\,\mathrm{d}m \quad \text{is finite.}$$

Then the series $\sum_{k=1}^\infty f_k(x)$ converges for almost all x, its sum is integrable, and

$$\int \sum_{k=1}^\infty f_k\,\mathrm{d}m = \sum_{k=1}^\infty \int f_k\,\mathrm{d}m.$$

Proof

The function $\varphi(x) = \sum_{k=1}^{\infty} |f_k(x)|$ is non-negative, measurable, and by Proposition 4.29

$$\int \varphi \, dm = \sum_{k=1}^{\infty} \int |f_k| \, dm.$$

This is finite, so φ is integrable. Therefore φ is finite a.e. Hence the series $\sum_{k=1}^{\infty} |f_k(x)|$ converges a.e. and so the series $\sum_{k=1}^{\infty} f_k(x)$ converges (since it converges absolutely) for almost all x. Let $f(x) = \sum_{k=1}^{\infty} f_k(x)$ (put $f(x) = 0$ for x for which the series diverges – the value we choose is irrelevant since the set of such x is null). For all partial sums we have

$$|\sum_{k=1}^{n} f_k(x)| \le \varphi(x),$$

so we can apply the dominated convergence theorem to find

$$\int f \, dm = \int \lim_{n \to \infty} \sum_{k=1}^{n} f_k \, dm = \lim_{n \to \infty} \int \sum_{k=1}^{n} f_k \, dm$$

$$= \lim_{n \to \infty} \sum_{k=1}^{n} \int f_k \, dm = \sum_{k=1}^{\infty} \int f_k \, dm$$

as required.   □

Example 4.31

Recalling that

$$\sum_{k=1}^{\infty} k x^{k-1} = \frac{1}{(1-x)^2},$$

we can use the Beppo–Levi theorem to evaluate the integral $\int_0^1 (\frac{\log x}{1-x})^2 \, dx$: first let $f_n(x) = n x^{n-1} (\log x)^2$ for $n \ge 1$, $x \in (0,1)$, so that $f_n \ge 0$, f_n is continuous, hence measurable, and

$$\sum_{n=1}^{\infty} f_n(x) = (\frac{\log x}{1-x})^2 = f(x)$$

is finite for $x \in (0,1)$. By Beppo–Levi the sum is integrable and

$$\int_0^1 f(x) \, dx = \sum_{n=1}^{\infty} \int_0^1 f_n(x) \, dx.$$

To calculate $\int_0^1 f_n(x)\,\mathrm{d}x$ we first use integration by parts to obtain

$$\int_0^1 x^{n-1}(\log x)^2\,\mathrm{d}x = \frac{2}{n^3}.$$

Thus $\int_0^1 f(x)\,\mathrm{d}x = 2\sum_{n=1}^\infty \frac{1}{n^2} = \frac{\pi^2}{3}.$

Exercise 4.9

The following are variations on the above theme:

(a) For which values of $a \in \mathbb{R}$ does the power series $\sum_{n\geq 0} n^a x^n$ define an integrable function on $[-1,1]$?

(b) Show that $\int_0^\infty \frac{x}{e^x - 1}\,\mathrm{d}x = \frac{\pi^2}{6}.$

4.5 Relation to the Riemann integral

Our prime motivation for introducing the Lebesgue integral has been to provide a sound theoretical foundation for the twin concepts of measure and integral, and to serve as the model upon which an abstract theory of measure spaces can be built. Such a general theory has many applications, a principal one being the mathematical foundations of the theory of probability. At the same time, Lebesgue integration has greater scope and more flexibility in dealing with limit operations than does its Riemann counterpart.

However, just as with the Riemann integral, the computation of specific integrals from first principles is laborious, and we have, as yet, no simple 'recipes' for handling particular functions. To link the theory with the convenient techniques of elementary calculus we therefore need to take two further steps: to prove the fundamental theorem of the calculus as stated in Chapter 1 and to show that the Lebesgue and Riemann integrals coincide whenever the latter exists. In the process we shall find necessary and sufficient conditions for the existence of the Riemann integral.

In fact, given Proposition 4.23 the proof of the fundamental theorem becomes a simple application of the intermediate value theorem for continuous functions, and is left to the reader:

Proposition 4.32

If $f : [a,b] \to \mathbb{R}$ is continuous then f is integrable and the function F given by $F(x) = \int_a^x f\,\mathrm{d}m$ is differentiable for $x \in (a,b)$, with derivative $F' = f$.

Hint Note that if $f \in \mathcal{L}^1$ and $A, B \in \mathcal{M}$ are disjoint, then $\int_{A \cup B} f \, dm = \int_A f \, dm + \int_B f \, dm$. Thus show that we can write $F(x+h) - F(x) = \int_x^{x+h} f \, dm$ for fixed $[x, x+h] \subset (a, b)$.

We turn to showing that Lebesgue's theory extends that of Riemann:

Theorem 4.33

Let $f : [a, b] \mapsto \mathbb{R}$ be bounded.

(i) f is Riemann-integrable if and only if f is a.e. continuous with respect to Lebesgue measure on $[a, b]$.

(ii) Riemann-integrable functions on $[a, b]$ are integrable with respect to Lebesgue measure on $[a, b]$ and the integrals are the same.

Proof

We need to prepare a little for the proof by recalling notation and some basic facts. Recall from Chapter 1 that any partition

$$\mathcal{P} = \{a_i : a = a_0 < a_1 < \cdots < a_n = b\}$$

of the interval $[a, b]$, with $\Delta_i = a_i - a_{i-1}$ $(i = 1, 2, \ldots, n)$ and with M_i (resp. m_i) the sup (resp. inf) of f on $I_i = [a_{i-1}, a_i]$, induces upper and lower Riemann sums $U_{\mathcal{P}} = \sum_{i=1}^n M_i \Delta_i$ and $L_{\mathcal{P}} = \sum_{i=1}^n m_i \Delta_i$. But these are just the Lebesgue integrals of the simple functions $u_{\mathcal{P}} = \sum_{i=1}^n M_i \mathbf{1}_{I_i}$ and $l_{\mathcal{P}} = \sum_{i=1}^n m_i \mathbf{1}_{I_i}$, by definition of the integral for such functions.

Choose a sequence of partitions $(\mathcal{P}_n)$ such that each $\mathcal{P}_{n+1}$ refines $\mathcal{P}_n$ and the length of the largest subinterval in $\mathcal{P}_n$ goes to 0; writing u_n for $u_{\mathcal{P}_n}$ and l_n for $l_{\mathcal{P}_n}$ we have $l_n \leq f \leq u_n$ for all n. Apply this on the measure space $([a, b], \mathcal{M}_{[a,b]}, m)$ where $m = m_{[a,b]}$ denotes Lebesgue measure restricted to $[a, b]$. Then $u = \inf_n u_n$ and $l = \sup_n l_n$ are measurable functions, and both sequences are monotone, since

$$l_1 \leq l_2 \leq \cdots \leq f \leq \cdots \leq u_2 \leq u_1. \tag{4.3}$$

Thus $u = \lim_n u_n$ and $l = \lim_n l_n$ (pointwise) and all functions in (4.3) are bounded on $[a, b]$ by $M = \sup\{f(x) : x \in [a, b]\}$, which is integrable on $[a, b]$. By dominated convergence we conclude that

$$\lim_n U_n = \lim_n \int_a^b u_n \, dm = \int_a^b u \, dm, \qquad \lim_n L_n = \lim_n \int_a^b l_n \, dm = \int_a^b l \, dm$$

and the limit functions u and l are (Lebesgue-)integrable.

Now suppose that x is not an endpoint of any of the intervals in the partitions $(\mathcal{P}_n)$ – which excludes only countably many points of $[a, b]$. Then we have:

$$f \text{ is continuous at } x \text{ iff } u(x) = f(x) = l(x).$$

This follows at once from the definition of continuity, since the length of each subinterval approaches 0 and so the variation of f over the intervals containing x approaches 0 iff f is continuous at x.

The Riemann integral $\int_a^b f(x)\,dx$ was defined as the common value of $\lim_n U_n = \int_a^b u\,dm$ and $\lim_n L_n = \int_a^b l\,dm$ whenever these limits are equal.

To prove (i), assume first that f is Riemann-integrable, so that the upper and lower integrals coincide: $\int_a^b u\,dm = \int_a^b l\,dm$. But $l \leq f \leq u$, hence $\int_a^b (u - l)\,dm = 0$ means that $u = l = f$ a.e. by Theorem 4.22. Hence f is continuous a.e. by the above characterization of continuity of f at x, which only excludes a further null set of partition points.

Conversely, if f is a.e. continuous, then $u = f = l$ a.e. and u and l are Lebesgue-measurable, hence so is f (note that this uses the completeness of Lebesgue measure!). But f is also bounded by hypothesis, so it is Lebesgue-integrable over $[a, b]$, and as the integrals are a.e. equal, the integrals coincide (but note that $\int_a^b f\,dm$ denotes the *Lebesgue* integral of f!):

$$\int_a^b l\,dm = \int_a^b f\,dm = \int_a^b u\,dm. \tag{4.4}$$

Since the outer integrals are the same, f is by definition also Riemann-integrable, which proves (i).

To prove (ii), note simply that if f is Riemann-integrable, (i) shows that f is a.e. continuous, hence measurable, and then (4.4) shows that its Lebesgue integral coincides with the two outer integrals, hence with its Riemann integral. □

Example 4.34

Recall the following example from Section 1.2: Dirichlet's function defined on $[0, 1]$ by

$$f(x) = \begin{cases} \frac{1}{n} & \text{if } x = \frac{m}{n} \in \mathbb{Q} \\ 0 & \text{if } x \notin \mathbb{Q} \end{cases}$$

is a.e. continuous, hence Riemann-integrable, and its Riemann integral equals its Lebesgue integral, which is 0, since f is zero outside the null set $\mathbb{Q}$.

We have now justified the unproven claims made in earlier examples when evaluating integrals, since, at least for any continuous functions on bounded intervals, the techniques of elementary calculus also give the Lebesgue integrals of the functions concerned. Since the integral is additive over disjoint domains use of these techniques also extends to piecewise continuous functions.

Example 4.35 (improper Riemann integrals)

Dealing with improper Riemann integrals involves an additional limit operation; we define such an integral by:

$$\int_{-\infty}^{\infty} f(x)\,\mathrm{d}x := \lim_{a\to-\infty, b\to\infty} \int_{a}^{b} f(x)\,\mathrm{d}x$$

whenever the double limit exists. (Other cases of 'improper integrals' are discussed in Remark 4.37.)

Now suppose for the function $f : \mathbb{R} \mapsto \mathbb{R}$ this improper Riemann integral exists. Then the Riemann integral $\int_a^b f(x)\,\mathrm{d}x$ exists for each bounded interval $[a, b]$, so that f is a.e. continuous on each $[a, b]$, and thus on $\mathbb{R}$. The converse is false, however: the function f which takes the value 1 on $[n, n + 1)$ when n is even, and -1 when n is odd, is a.e. continuous (and thus Lebesgue-measurable on $\mathbb{R}$), but clearly the above limits fail to exist.

More generally, it is not hard to show that if $f \in \mathcal{L}^1(\mathbb{R})$ then the above double limits will always exist. On the other hand, the existence of the double limit does not by itself guarantee that $f \in \mathcal{L}^1$ without further conditions: consider the function (see Figure 4.3)

$$f(x) = \begin{cases} \frac{(-1)^n}{n+1} & \text{if } x \in [n, n+1),\ n \geq 0 \\ 0 & \text{if } x < 0. \end{cases}$$

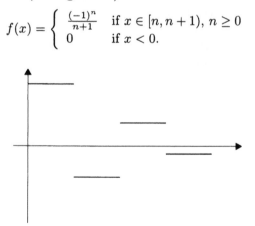

Figure 4.3 Graph of f

Clearly the improper Riemann integral exists,

$$\int_{-\infty}^{\infty} f(x)\,\mathrm{d}x = \sum_{n=0}^{\infty} \frac{(-1)^n}{n+1}$$

and the series converges. However, $f \notin \mathcal{L}^1$, since $\int_{\mathbb{R}} |f| \, dm = \sum_{n=0}^{\infty} \frac{1}{n+1}$, which diverges.

This yields another illustration of the 'absolute' nature of the Lebesgue integral: $f \in \mathcal{L}^1$ iff $|f| \in \mathcal{L}^1$, so we cannot expect a finite sum for an integral whose 'pieces' make up a conditionally convergent series. For non-negative functions these problems do not arise; we have:

Theorem 4.36

If $f \geq 0$ and the above improper Riemann integral of f exists, then the Lebesgue integral $\int_{\mathbb{R}} f \, dm$ always exists and equals the improper integral.

Proof

To see this, simply note that the sequence (f_n) with $f_n = f\mathbf{1}_{[-n,n]}$ increases monotonically to f, hence f is Lebesgue-measurable. Since f_n is Riemann-integrable on $[-n, n]$, the integrals coincide there, i.e.

$$\int_{\mathbb{R}} f_n \, dm = \int_{-n}^{n} f(x) \, dx$$

for each n, so that $f_n \in \mathcal{L}^1(\mathbb{R})$ for all n. By hypothesis the double limit

$$\lim_n \int_{-n}^{n} f(x) \, dx = \int_{-\infty}^{\infty} f(x) \, dx$$

exists. On the other hand,

$$\lim_n \int_{\mathbb{R}} f_n \, dm = \int_{\mathbb{R}} f \, dm$$

by monotone convergence, and so $f \in \mathcal{L}^1(\mathbb{R})$ and

$$\int_{\mathbb{R}} f \, dm = \int_{-\infty}^{\infty} f(x) \, dx$$

as required.   □

Exercise 4.10

Show that the function f given by $f(x) = \frac{\sin x}{x}$ $(x \neq 0)$ has an improper Riemann integral over $\mathbb{R}$, but is not in $\mathcal{L}^1$.

Remark 4.37

A second kind of improper Riemann integral is designed to handle functions which have asymptotes on a bounded interval, such as $f(x) = \frac{1}{x}$ on $(0,1)$. For such cases we can define

$$\int_a^b f(x)\,\mathrm{d}x = \lim_{\varepsilon \searrow 0} \int_{a+\varepsilon}^b f(x)\,\mathrm{d}x$$

when the limit exists. (Similar remarks apply to the upper limit of integration.)

4.6 Approximation of measurable functions

The previous section provided an indication of the extent of the additional 'freedom' gained by developing the Lebesgue integral: Riemann integration binds us to functions whose discontinuities form an m-null set, while we can still find the Lebesgue integral of functions that are *nowhere* continuous, such as $\mathbf{1}_{\mathbb{Q}}$. We may ask, however, how real this additional generality is: can we, for example, *approximate* an arbitrary $f \in \mathcal{L}^1$ by continuous functions? In fact, since continuity is a local property, can we do this for arbitrary measurable functions? And this, in turn, provides a link with simple functions, since every measurable function is a limit of simple functions. We can go further, and ask whether for a simple function g approximating a given measurable function f we can choose the inverse image $g^{-1}(\{a_i\})$ of each element of the range of g to be an interval (such a g is usually called a *step function*; $g = \sum_n c_n \mathbf{1}_{I_n}$, where I_n are intervals). We shall tackle this question first:

Theorem 4.38

If f is a bounded measurable function on $[a, b]$ and $\varepsilon > 0$ is given, then there exists a step function h such that $\int_a^b |f - h|\,\mathrm{d}m < \varepsilon$.

Proof

First assume additionally that $f \geq 0$. Then $\int_a^b f\,\mathrm{d}m$ is well defined as

$$\sup\{\int_a^b \varphi\,\mathrm{d}m : 0 \leq \varphi \leq f,\ \text{simple}\}.$$

Since $f \geq \varphi$ we have $|f - \varphi| = f - \varphi$, so we can find a simple function φ

satisfying

$$\int_a^b |f - \varphi|\, dm = \int_a^b f\, dm - \int_a^b \varphi\, dm < \frac{\varepsilon}{2}.$$

It then remains to approximate an arbitrary simple function φ which vanishes off $[a,b]$ by a step function h. The finite range $\{a_1, a_2, ..., a_n\}$ of the function φ partitions $[a,b]$, yielding disjoint measurable sets $E_i = \varphi^{-1}(\{a_i\})$ such that $\bigcup_{i=1}^n E_i = [a,b]$. We now approximate each E_i by intervals: note that since φ is simple, $M = \sup\{\varphi(x) : x \in [a,b]\} < \infty$. By Theorem 2.17 we can find open sets O_i such that $E_i \subset O_i$ and $m(O_i \setminus E_i) < \frac{\varepsilon}{2nM}$ for $i \leq n$. Since each E_i has finite measure, so do the O_i, hence each O_i can in turn be approximated by a *finite* union of disjoint open intervals: we know that $O_i = \bigcup_{j=1}^\infty I_{ij}$, where the open intervals can be chosen disjoint, so that $m(O_i) = \sum_{j=1}^\infty m(I_{ij}) < \infty$. As the series converges, we can find k_i such that $m(O_i) - m(\bigcup_{j=1}^{k_i} I_{ij}) < \frac{\varepsilon}{2nM}$. Thus with $G_i = \bigcup_{j=1}^{k_i} I_{ij}$ we have $\int_a^b |\mathbf{1}_{E_i} - \mathbf{1}_{G_i}|\, dm = m(E_i \Delta G_i) < \frac{\varepsilon}{nM}$ for each $i \leq n$. So set $h = \sum_{i=1}^n a_i \mathbf{1}_{G_i}$. This step function satisfies $\int_a^b |\varphi - h|\, dm < \frac{\varepsilon}{2}$ and hence $\int_a^b |f - h|\, dm < \varepsilon$.

The extension to general f is clear: f^+ and f^- can be approximated to within $\frac{\varepsilon}{2}$ by step functions h_1 and h_2 say, so with $h = h_1 - h_2$ we obtain

$$\int_a^b |f - h|\, dm \leq \int_a^b |f^+ - h_1|\, dm + \int_a^b |f^- - h_2|\, dm < \varepsilon,$$

which completes the proof. $\square$

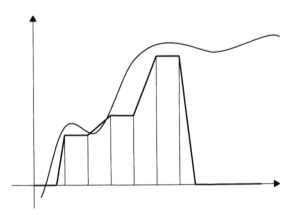

Figure 4.4 Approximation by continuous functions

The 'payoff' is now immediate: with f and h as above, we can re-order the intervals I_{ij} into a single finite sequence $(J_m)_{m \leq n}$ with $J_m = (c_m, d_m)$ and $h = \sum_{m=1}^n a_m \mathbf{1}_{J_m}$. We may assume that $l(J_m) = (d_m - c_m) > \frac{\varepsilon'}{2}$, and approximate

$\mathbf{1}_{J_m}$ by a continuous function g_m by setting $g_m = 1$ on the slightly smaller interval $(c_m + \frac{\varepsilon'}{4}, d - \frac{\varepsilon'}{4})$ and 0 outside J_m, while extending linearly in between (see Figure 4.4). It is obvious that g_m is continuous and $\int_a^b |\mathbf{1}_{J_m} - g_m| \, dm < \frac{\varepsilon'}{2}$. Repeating for each J_m and taking $\varepsilon' < \frac{\varepsilon}{nK}$, where $K = \max_{m \le n} |a_m|$, shows that the continuous function $g = \sum_{m=1}^n a_m g_m$ satisfies $\int_a^b |h - g| \, dm < \frac{\varepsilon}{2}$. Combining this inequality with Theorem 4.38 yields:

Theorem 4.39

Given $f \in \mathcal{L}^1$ and $\varepsilon > 0$, we can find a continuous function g, vanishing outside some finite interval, such that $\int |f - g| \, dm < \varepsilon$.

Proof

The preceding argument has verified this when f is a bounded measurable function vanishing off some interval $[a, b]$. For a given $f \in \mathcal{L}^1[a, b]$ we can again assume without loss that $f \ge 0$. Let $f_n = \min(f, n)$; then the f_n are bounded measurable functions dominated by f, $f_n \to f$, so that $\int_a^b |f - f_N| \, dm < \frac{\varepsilon}{2}$ for some N. We can now find a continuous g, vanishing outside a finite interval, such that $\int_a^b |f_N - g| \, dm < \frac{\varepsilon}{2}$. Thus $\int_a^b |f - g| dm < \varepsilon$.

Finally, let $f \in \mathcal{L}^1(\mathbb{R})$ and $f \ge 0$ be given. Choose n large enough to ensure that $\int_{\{|x| \ge n\}} f \, dm < \frac{\varepsilon}{3}$ (which we can do as $\int_{\mathbb{R}} |f| \, dm$ is finite; Proposition 4.28), and simultaneously choose a continuous g with $\int_{\{|x| \ge n\}} g \, dm < \frac{\varepsilon}{3}$ which satisfies $\int_{-n}^n |f - g| \, dm < \frac{\varepsilon}{3}$. Thus $\int_{\mathbb{R}} |f - g| \, dm < \varepsilon$. $\qquad \square$

The well-known *Riemann–Lebesgue lemma*, which is very useful in the discussion of Fourier series, can be easily deduced from the above approximation theorems:

Lemma 4.40 (Riemann–Lebesgue)

Suppose $f \in \mathcal{L}^1(\mathbb{R})$. Then the sequences $s_k = \int_{-\infty}^{\infty} f(x) \sin kx \, dx$ and $c_k = \int_{-\infty}^{\infty} f(x) \cos kx \, dx$ both converge to 0 as $k \to \infty$.

Proof

We prove this for (s_k) leaving the other, similar, case to the reader. For simplicity of notation write $\int$ for $\int_{-\infty}^{\infty}$. The transformation $x = y + \frac{\pi}{k}$ shows that

$$s_k = \int f(y + \frac{\pi}{k}) \sin(ky + \pi) \, dy = - \int f(y + \frac{\pi}{k}) \sin(ky) \, dy.$$

Since $|\sin x| \leq 1$,

$$\int |f(x) - f(x + \frac{\pi}{k})| \, dx \geq |\int (f(x) - f(x + \frac{\pi}{k})) \sin kx \, dx| = 2|s_k|.$$

It will therefore suffice to prove that $\int |f(x) - f(x+h)| \, dx \to 0$ when $h \to 0$. This is most easily done by approximating f by a continuous g which vanishes outside some finite interval $[a, b]$, and such that $\int |f - g| \, dm < \frac{\varepsilon}{3}$ for a given $\varepsilon > 0$. For $|h| < 1$, the continuous function $g_h(x) = g(x + h)$ then vanishes off $[a - 1, b + 1]$ and

$$\int |f(x + h) - f(x)| \, dm \leq \int |f(x + h) - g(x + h)| \, dm$$

$$+ \int |g(x + h) - g(x)| \, dm + \int |g(x) - f(x)| \, dm.$$

The first and last integrals on the right are less than $\frac{\varepsilon}{3}$, while the integrand of the second can be made less than $\frac{\varepsilon}{3(b-a+2)}$ whenever $|h| < \delta$, by an appropriate choice of $\delta > 0$, as g is continuous. As g vanishes outside $[a - 1, b + 1]$, the second integral is also less than $\frac{\varepsilon}{3}$. Thus if $|h| < \delta$, $\int |f(x + h) - f(x)| \, dm < \varepsilon$. This proves that $\lim_{k \to \infty} \int f(x) \sin kx \, dx = 0$. $\square$

4.7 Probability

4.7.1 Integration with respect to probability distributions

Let X be a random variable with probability distribution P_X. The following theorem shows how to perform a change of variable when integrating a function of X. In other words, it shows how to change the measure in an integral. This is fundamental in applying integration theory to probabilities. We emphasise again that only the closure properties of σ-fields and the countable additivity of measures are needed for the theorems we shall apply here, so that we can use an abstract formulation of a probability space $(\Omega, \mathcal{F}, P)$ in discussing their applications.

Theorem 4.41

Given a random variable $X : \Omega \to \mathbb{R}$,

$$\int_\Omega g(X(\omega)) \, dP(\omega) = \int_\mathbb{R} g(x) \, dP_X(x). \tag{4.5}$$

Proof

We employ the technique described in Remark 4.18. For the indicator function $g = \mathbf{1}_A$ we have $P(X \in A)$ on both sides. Then by linearity we have the result for simple functions. Approximation of non-negative measurable g by a monotone sequence of simple functions combined with the monotone convergence theorem gives the equality for such g. The case of general $g \in \mathcal{L}^1$ follows as before from the linearity of the integral, using $g = g^+ - g^-$. $\square$

The formula is useful in the case where the form of P_X is known and allows one to carry out explicit computations.

Before we proceed to these situations, consider a very simple case as an illustration of the formula. Suppose that X is constant, i.e. $X(\omega) \equiv a$. Then on the left in (4.5) we have the integral of a constant function, which equals $g(a)P(\Omega) = g(a)$ according to the general scheme of integrating indicator functions. On the right $P_X = \delta_a$ and thus we have a method of computing an integral with respect to Dirac measure: $\int g(x) \, \mathrm{d}\delta_a = g(a)$.

For discrete X taking values a_i with probabilities p_i we have

$$\int g(X) \, \mathrm{d}P = \sum_i g(a_i) p_i,$$

which is a well-known formula from elementary probability theory (see also Section 3.5.3). In this case we have $P_X = \sum_i p_i \delta_{a_i}$ and on the right, the integral with respect to the combination of measures is the combination of the integrals:

$$\int g(x) \, \mathrm{d}P_X = \sum_i p_i \int g(x) \, \mathrm{d}\delta_{a_i}(x).$$

In fact, this is a general property.

Theorem 4.42

If $P_X = \sum_i p_i P_i$, where the P_i are probability measures, $\sum p_i = 1$, $p_i \geq 0$, then

$$\int g(x) \, \mathrm{d}P_X(x) = \sum_i p_i \int g(x) \, \mathrm{d}P_i(x).$$

Proof

The method is the same as above: first consider indicator functions $\mathbf{1}_A$ and the claim is just the definition of P_X: on the left we have $P_X(A)$, on the right $\sum_i p_i P_i(A)$. Then by additivity we get the formula for simple functions, and finally, approximation and use of the monotone convergence theorem completes the proof as before.   □

4.7.2 Absolutely continuous measures: examples of densities

The measures P of the form

$$A \mapsto P(A) = \int_A f \, \mathrm{d}m$$

with non-negative integrable f will be called *absolutely continuous*, and the function f will be called a *density of P with respect to Lebesgue measure*, or simply a *density*. Clearly, for P to be a probability we have to impose the condition

$$\int f \, \mathrm{d}m = 1.$$

Students of probability often have an oversimplified mental picture of the world of random variables, believing that a random variable is either discrete or absolutely continuous. This image stems from the practical computational approach of many elementary textbooks, which present probability without the necessary background in measure theory. We have already provided a simple example which shows this to be a false dichotomy (Example 3.17).

The simplest example of a density is this: let $\Omega \subset \mathbb{R}$ be a Borel set with finite Lebesgue measure and put

$$f(x) = \begin{cases} \frac{1}{m(\Omega)} & \text{if } x \in \Omega \\ 0 & \text{otherwise.} \end{cases}$$

We have already come across this sort of measure in the previous chapter, that is, the probability distribution of a specific random variable. We say that in this case the measure (distribution) is *uniform*. It corresponds to the case where the values of the random variable are spread evenly across some set, typically an interval, such as in choosing a number at random (Example 2.34).

Slightly more complicated is the so-called *triangle* distribution with the density of the form shown in Figure 4.5.

The most famous is the *Gaussian* or *normal* density

$$n(x) = \frac{1}{\sqrt{2\pi}\sigma} e^{-\frac{(x-\mu)^2}{2\sigma^2}}. \tag{4.6}$$

This function is symmetric with respect to $x = \mu$, and vanishes at infinity, i.e. $\lim_{x \to -\infty} n(x) = 0 = \lim_{x \to \infty} n(x)$ (see Figure 4.6).

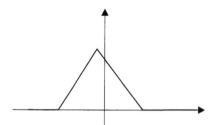

Figure 4.5 Triangle distribution

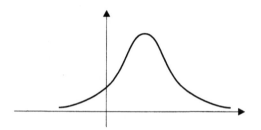

Figure 4.6 Gaussian distribution

Exercise 4.11

Show that $\int_{-\infty}^{\infty} n(x)\,dx = 1$.

Hint First consider the case $\mu = 0$, $\sigma = 1$ and then transform the general case to this.

The meaning of the number μ will become clear below and σ will be explained in the next chapter.

Another widely used example is the *Cauchy* density:

$$c(x) = \frac{1}{\pi}\frac{1}{1+x^2}.$$

This density gives rise to many counterexamples to 'theorems' which are too good to be true.

Exercise 4.12

Show that $\int_{-\infty}^{\infty} c(x)\,dx = 1$.

The *exponential* density is given by

$$f(x) = \begin{cases} ce^{-\lambda x} & \text{if } x \geq 0 \\ 0 & \text{otherwise.} \end{cases}$$

Exercise 4.13

Find the constant c for f to be a density of probability distribution.

The *gamma* distribution is really a large family of distributions, indexed by a parameter $t > 0$. It contains the exponential distribution as the special case where $t = 1$. Its density is defined as

$$f(x) = \begin{cases} \frac{1}{\Gamma(t)}\lambda^t x^{t-1}e^{-\lambda x} & \text{if } x \geq 0 \\ 0 & \text{otherwise} \end{cases}$$

where the gamma function $\Gamma(t) = \int_0^\infty x^{t-1}e^{-x}\,dx$.

The gamma distribution contains another widely used distribution as a special case: the distribution obtained from the density f when $\lambda = \frac{1}{2}$ and $t = \frac{d}{2}$ for some $d \in \mathbb{N}$ is denoted by $\chi^2(d)$ and called the *chi-squared distribution with d degrees of freedom*.

The (cumulative) *distribution function* corresponding to a density is given by

$$F(y) = \int_{-\infty}^y f(x)\,dx.$$

If f is continuous then F is differentiable and $F'(x) = f(x)$ by the fundamental theorem of calculus (see Proposition 4.32). We say that F is absolutely continuous if this relation holds with integrable f, and then f is the density of the probability measure induced by F. The following example due to Lebesgue shows that continuity of F is not sufficient for the existence of a density.

Example 4.43

Recall the Lebesgue function F defined on page 20. We have $F(y) = 0$ for $y \leq 0$, $F(y) = 1$ for $y \geq 1$, $F(y) = \frac{1}{2}$ for $y \in [\frac{1}{3}, \frac{2}{3})$, $F(y) = \frac{1}{4}$ for $y \in [\frac{1}{9}, \frac{2}{9})$, $F(y) = \frac{3}{4}$ for $y \in [\frac{7}{9}, \frac{8}{9})$ and so on. The function F is constant on the intervals removed in the process of constructing the Cantor set (see Figure 4.7).

It is differentiable almost everywhere and the derivative is zero. So F cannot be absolutely continuous since then f would be zero almost everywhere, but on the other hand its integral is 1.

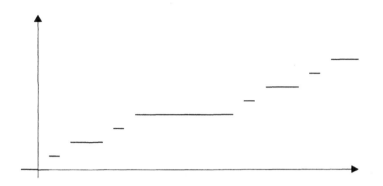

Figure 4.7 Lebesgue's function

We now define the (cumulative) distribution function of a random variable $X : \Omega \to \mathbb{R}$, where, as above, $(\Omega, \mathcal{F}, P)$ is a given probability space:

$$F_X(y) = P(\{\omega : X(\omega) \le y\}) = P_X((-\infty, y]).$$

Proposition 4.44

(i) F_X is non-decreasing ($y_1 \le y_2$ implies $F_X(y_1) \le F_X(y_2)$),

(ii) $\lim_{y \to \infty} F_X(y) = 1$, $\lim_{y \to -\infty} F_X(y) = 0$,

(iii) F_X is right-continuous (if $y \to y_0$, $y \ge y_0$, then $F_X(y) \to F(y_0)$).

Exercise 4.14

Show that F_X is continuous if and only if $P_X(\{y\}) = 0$ for all y.

Exercise 4.15

Find F_X for

(a) a constant random variable X, $X(\omega) = a$ for all ω,

(b) $X : [0, 1] \to \mathbb{R}$ given by $X(\omega) = \min\{\omega, 1 - \omega\}$ (the distance to the nearest endpoint of the interval $[0, 1]$),

(c) $X : [0, 1]^2 \to \mathbb{R}$, the distance to the nearest edge of the square $[0, 1]^2$.

The fact that we are doing probability on subsets of $\mathbb{R}^n$ as sample spaces turns out to be not restrictive. In fact, the interval $[0, 1]$ is sufficient, as the following Skorokhod representation theorem shows.

Theorem 4.45

If a function $F : \mathbb{R} \to [0,1]$ satisfies conditions (i)–(iii) of Proposition 4.44, then there is a random variable defined on the probability space $([0;1], \mathcal{B}, m_{[0,1]})$, $X : [0,1] \to \mathbb{R}$, such that $F = F_X$.

Proof

We write, for $\omega \in [0,1]$,

$$X^+(\omega) = \inf\{x : F(x) > \omega\}, \qquad X^-(\omega) = \sup\{x : F(x) < \omega\}.$$

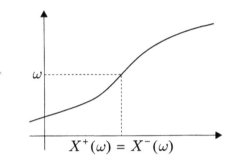

Figure 4.8 Construction of X^-; continuity point

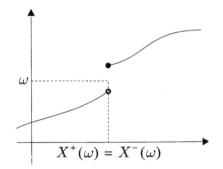

Figure 4.9 Construction of X^-; discontinuity point

Three possible cases are illustrated in Figures 4.8, 4.9 and 4.10. We show that $F_{X^-} = F$, and for that we have to show that $F(y) = m(\{\omega : X^-(\omega) \le y\})$. The set $\{\omega : X^-(\omega) \le y\}$ is an interval with left endpoint 0. We are done if we show that its right endpoint is $F(y)$, i.e. if $X^-(\omega) \le y$ is equivalent to $\omega \le F(y)$.

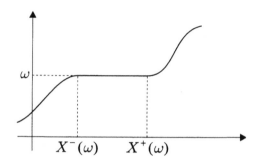

Figure 4.10 Construction of X^-; 'flat' piece

Suppose that $\omega \leq F(y)$. Then

$$\{x : F(x) < \omega\} \subset \{x : F(x) < F(y)\} \subset \{x : x \leq y\}$$

(the last inclusion by the monotonicity of F), hence $X^-(\omega) = \sup\{x : F(x) < \omega\} \leq y$.

Suppose that $X^-(\omega) \leq y$. By monotonicity $F(X^-(\omega)) \leq F(y)$. By the right-continuity of F, $\omega \leq F(X^-(\omega))$ (if $\omega > F(X^-(\omega))$, then there is $x_0 > X^-(\omega)$ such that $F(X^-(\omega)) < F(x_0) < \omega$, which is impossible since x_0 is in the set whose supremum is taken to get $X^-(\omega)$), so $\omega \leq F(y)$.

For future use we also show that $F_{X+} = F$. It is sufficient to see that $m(\{\omega : X^-(\omega) < X^+(\omega)\}) = 0$ (which is intuitively clear as this may happen only when the graph of F is 'flat', and there are countably many values corresponding to the 'flat' pieces, their Lebesgue measure being zero). More rigorously,

$$\{\omega : X^-(\omega) < X^+(\omega)\} = \bigcup_{q \in \mathbb{Q}} \{\omega : X^-(\omega) \leq q < X^+(\omega)\}$$

and $m(\{\omega : X^-(\omega) \leq q < X^+(\omega)\}) = m(\{\omega : X^-(\omega) \leq q\} \setminus \{\omega : X^+(\omega) \leq q\}) = F(q) - F(q) = 0$. $\qquad \square$

The following theorem provides a powerful method for calculating integrals relative to absolutely continuous distributions. The result holds for general measures but we formulate it for a probability distribution of a random variable in order not to overload or confuse the notation.

Theorem 4.46

If P_X defined on $\mathbb{R}^n$ is absolutely continuous with density f_X, $g : \mathbb{R}^n \to \mathbb{R}$ is integrable with respect to P_X, then

$$\int_{\mathbb{R}^n} g(x) \, \mathrm{d}P_X(x) = \int_{\mathbb{R}^n} f_X(x) g(x) \, \mathrm{d}x.$$

Proof

For an indicator function $g(x) = \mathbf{1}_A(x)$ we have $P_X(A)$ on the left which equals $\int_A f_X(x) \, \mathrm{d}x$ by the form of P, and consequently is equal to $\int_{\mathbb{R}^n} \mathbf{1}_A(x) f_X(x) \, \mathrm{d}x$, i.e. the right-hand side. Extension to simple functions by linearity and to general integrable g by limit passage is routine. $\square$

Corollary 4.47

In the situation of the previous theorem we have

$$\int_\Omega g(X) \, \mathrm{d}P = \int_{\mathbb{R}^n} f_X(x) g(x) \, \mathrm{d}x.$$

Proof

This is an immediate consequence of the above theorem and Theorem 4.41. $\square$

We conclude this section with a formula for a density of a function of a random variable with given density. Suppose that f_X is known and we want to find the density of $Y = g(X)$.

Theorem 4.48

If $g : \mathbb{R} \to \mathbb{R}$ is increasing and differentiable (thus invertible), then

$$f_{g(X)}(y) = f_X(g^{-1}(y)) \frac{\mathrm{d}}{\mathrm{d}y} g^{-1}(y).$$

Proof

Consider the distribution function:

$$F_{g(X)}(y) = P(g(X) \le y) = P(X \le g^{-1}(y)) = F_X(g^{-1}(y)).$$

Differentiate with respect to y to get the result. $\square$

Remark 4.49

A similar result holds if g is decreasing. The same argument as above gives

$$f_{g(X)}(y) = -f_X(g^{-1}(y))\frac{\mathrm{d}}{\mathrm{d}y}g^{-1}(y).$$

Example 4.50

If X has *standard* normal distribution

$$n(x) = \frac{1}{\sqrt{2\pi}}\mathrm{e}^{-\frac{1}{2}x^2}$$

(i.e. $\mu = 0$ and $\sigma = 1$ in (4.6)), then the density of $Y = \mu + \sigma X$ is given by
(4.6). This follows at once from Theorem 4.48: $g^{-1}(y) = \frac{\mu - x}{\sigma}$; its derivative is
equal to $\frac{1}{\sigma}$.

Exercise 4.16

Find the density of $Y = X^3$ where $f_X = \mathbf{1}_{[0,1]}$.

4.7.3 Expectation of a random variable

If X is a random variable defined on a probability space $(\Omega, \mathcal{F}, P)$ then we
introduce the following notation:

$$\mathbb{E}(X) = \int_\Omega X\,\mathrm{d}P$$

and we call this abstract integral the (*mathematical*) *expectation* of X.

Using the results from the previous section we immediately have the following formulae: the expectation can be computed using the probability distribution:

$$\mathbb{E}(X) = \int_{-\infty}^{\infty} x\,\mathrm{d}P_X(x),$$

and for absolutely continuous X we have

$$\mathbb{E}(X) = \int_{-\infty}^{\infty} x f_X(x)\,\mathrm{d}x.$$

Example 4.51

Suppose that $P_X = \frac{1}{2}P_1 + \frac{1}{2}P_2$, where $P_1 = \delta_a$, P_2 has a density f_2. Then

$$\mathbb{E}(X) = \frac{1}{2}a + \frac{1}{2}\int xf(x)\,\mathrm{d}x.$$

So, going back to Example 3.17 we can compute the expectation of the random variable considered there:

$$\mathbb{E}(X) = \frac{1}{2}\cdot 0 + \frac{1}{2}\frac{1}{25}\int_0^{25} x\,\mathrm{d}x = 6.25.$$

Exercise 4.17

Find the expectation of

(a) a constant random variable X, $X(\omega) = a$ for all ω,

(b) $X : [0,1] \to \mathbb{R}$ given by $X(\omega) = \min\{\omega, 1-\omega\}$ (the distance to the nearest endpoint of the interval $[0,1]$),

(c) $X : [0,1]^2 \to \mathbb{R}$, the distance to the nearest edge of the square $[0,1]^2$.

Exercise 4.18

Find the expectation of a random variable with

(a) uniform distribution over the interval $[a,b]$,

(b) triangle distribution,

(c) exponential distribution.

4.7.4 Characteristic function

In what follows we will need the integrals of some complex functions. The theory is a straightforward extension of the real case.

Let $Z = X + iY$ where X, Y are real-valued random variables and define

$$\int Z\,\mathrm{d}P = \int X\,\mathrm{d}P + i\int Y\,\mathrm{d}P.$$

Clearly, linearity of the integral and the dominated convergence theorem hold for the complex case. Another important relation which remains true is:

$$\left|\int Z\,\mathrm{d}P\right| \le \int |Z|\,\mathrm{d}P.$$

To see this consider the polar decomposition of $\int Z \, dP = |\int Z \, dP| e^{-i\theta}$. Then, with $\Re(z)$ as the real part of the complex number z, $|\int Z \, dP| = e^{i\theta} \int Z \, dP = \int e^{i\theta} Z \, dP$ is real, hence equal to $\int \Re(e^{i\theta} Z) \, dP$, but $\Re(e^{i\theta} Z) \le |e^{i\theta} Z| = |Z|$ and we are done.

The function we wish to integrate is $\exp\{itX\}$ where X is a real random variable, $t \in \mathbb{R}$. Then

$$\int \exp\{itX\} \, dP = \int \cos(tX) \, dP + i \int \sin(tX) \, dP$$

which always exists, by the boundedness of $x \mapsto \exp\{itx\}$.

Definition 4.52

For a random variable X we write

$$\varphi_X(t) = \mathbb{E}(e^{itX})$$

for $t \in \mathbb{R}$. We call φ_X the *characteristic function* of X.

To compute φ_X it is sufficient to know the distribution of X:

$$\varphi_X(t) = \int e^{itx} \, dP_X(x)$$

and in the absolutely continuous case

$$\varphi_X(t) = \int e^{itx} f_X(x) \, dx.$$

Some basic properties of the characteristic function are given below. Other properties are explored in Chapters 6 and 8.

Theorem 4.53

The function φ_X satisfies

(i) $\varphi_X(0) = 1$, $|\varphi_X(t)| \le 1$,

(ii) $\varphi_{aX+b}(t) = e^{itb} \varphi_X(at)$.

Proof

(i) The value at 0 is 1 since the expectation of the constant function is its value. The estimate follows from Proposition 4.23 (iii): $|\int e^{itx} \, dP_X(x)| \le \int |e^{itx}| \, dP_X(x) = 1$.

(ii) Here we use the linearity of the expectation:

$$\varphi_{aX+b}(t) = \mathbb{E}(e^{it(aX+b)}) = \mathbb{E}(e^{itaX}e^{itb}) = e^{itb}\mathbb{E}(e^{i(ta)X}) = e^{itb}\varphi_X(ta),$$

as required.   □

Exercise 4.19

Find the characteristic function of a random variable with

(a) uniform distribution over the interval $[a, b]$,

(b) exponential distribution,

(c) Gaussian distribution.

4.7.5 Applications to mathematical finance

Consider a derivative security of European type, that is, a random variable of the form $f(S(N))$, where $S(n)$, $n = 1, \ldots, N$, is the price of the underlying security, which we call a stock for simplicity. (Or we write $f(S(T))$, where the underlying security is described in continuous time $t \in [0, T]$ with prices $S(t)$.) One of the crucial problems in finance is to find the price $Y(0)$ of such a security. Here we assume that the reader is familiar with the following fact, which is true for certain specific models consisting of a probability space and random variables representing the stock prices:

$$Y(0) = \exp\{-rT\}\mathbb{E}(f(S(T))), \tag{4.7}$$

where r is the risk-free interest rate for continuous compounding. This will be explained in some detail in Section 7.4.4 but here we just want to draw some conclusions from this formula using the experience gathered in the present chapter.

In particular, taking account of the form of the payoff functions for the European call ($f(x) = (x - K)^+$) and put ($f(x) = (K - x)^+$) we have the following general formulae for the value of call and put, respectively:

$$C = \exp\{-rT\}\mathbb{E}(S(T) - K)^+,$$
$$P = \exp\{-rT\}\mathbb{E}(K - S(T))^+.$$

Without relying on any particular model one can prove the following relation, called call–put parity (see [4] for instance):

$$S(0) = C - P + K\exp\{-rT\}. \tag{4.8}$$

Proposition 4.54

The right-hand side of the call–put parity identity is independent of K.

Remark 4.55

This proposition allows us to make an interesting observation, which is a version of a famous result in finance, namely the Miller–Modigliani theorem which says that the value of a company does not depend on the way it is financed. Let us very briefly recall that the value of a company is the sum of equity (represented by stock) and debt, so the theorem says that the level of debt has no impact on company's value. Assume that the company borrowed $K \exp\{-rT\}$ at the rate equal to r and it has to pay back the amount of K at time T. Should it fail, the company goes bankrupt. So the stockholders, who control the company, can 'buy it back' by paying K. This will make sense only if the value $S(T)$ of the company exceeds K. The stock can be regarded as a call option so its value is C. The value of the debt is thus $K \exp\{-rT\} - P$, less than the present value of K, which captures the risk that the sum K may not be recovered in full.

We now evaluate the expectation to establish explicit forms of the general formula (4.7) in the two most widely used models.

Consider first the binomial model introduced in Section 2.6.3. Assume that the probability space is equipped with a measure determined by the probability $p = \frac{R-D}{U-D}$ for the up movement in single step, where $R = \exp\{rh\}$, h being the length of one step. (This probability is called risk-neutral; observe that $\mathbb{E}(\eta) = R$.) To ensure that $0 < p < 1$ we assume throughout that $D < R < U$.

Proposition 4.56

In the binomial model the price C of a call option with exercise time $T = hN$ is given by the Cox–Ross–Rubinstein formula

$$C = S(0)\Psi(A, N, pUe^{-rT}) - Ke^{-rT}\Psi(A, N, p)$$

where $\Psi(A, N, p) = \sum_{k=A}^{N} \binom{N}{k} p^k (1-p)^{N-k}$ and A is the first integer k such that $S(0)U^k D^{N-k} > K$.

In the famous continuous-time Black–Scholes model, the stock price at time T is of the form

$$S(T) = S(0) \exp\{(r - \frac{\sigma^2}{2})T + \sigma w(T)\},$$

where r is the risk-free rate, $\sigma > 0$ and $w(T)$ is a random variable with Gaussian distribution with mean 0 and variance T. (The reader familiar with finance will notice that we again assume that the probability space is equipped with a risk-neutral measure.)

Proposition 4.57

We have the following Black–Scholes formula for C :

$$C = S(0)N(d_1) - Ke^{-rT}N(d_2).$$

where

$$d_1 = \frac{\ln\frac{S(0)}{Ke^{-rT}} + \frac{1}{2}\sigma^2 T}{\sigma\sqrt{T}}, \quad d_2 = \frac{\ln\frac{S(0)}{Ke^{-rT}} - \frac{1}{2}\sigma^2 T}{\sigma\sqrt{T}}.$$

Exercise 4.20

Find the formula for the put option.

4.8 Proofs of propositions

Proof (of Proposition 4.5)

Let $f = \sum c_i \mathbf{1}_{A_i}$. We have to show that

$$\sum c_i m(A_i \cap E) = \sup Y(E, f).$$

First, we may take $\varphi = f$ in the definition of $Y(E, f)$, so the number on the left ($\sum c_i m(A_i \cap E)$) belongs to $Y(E, f)$ and so

$$\sum c_i m(A_i \cap E) \le \sup Y(E, f).$$

For the converse take any $a \in Y(E, f)$. So

$$a = \int_E \psi \, dm = \sum d_j m(E \cap B_j)$$

for some simple $\psi \le f$. Now

$$a = \sum_j \sum_i d_j m(E \cap B_j \cap A_i)$$

by the properties of measure (A_i form a partition of $\mathbb{R}$). For $x \in B_j \cap A_i$, $f(x) = c_i$ and $\psi(x) = d_j$ and so $d_j \leq c_i$ (if only $B_j \cap A_i \neq \emptyset$). Hence

$$a \leq \sum_i \sum_j c_i m(E \cap B_j \cap A_i) = \sum_i c_i m(E \cap A_i)$$

since B_j partition $\mathbb{R}$.   □

Proof (of Proposition 4.9)

Let $A = \{x : f(x) \leq g(x)\}$, then A^c is null and $f\mathbf{1}_A \leq g\mathbf{1}_A$. So $\int f\mathbf{1}_A \, dm \leq \int g\mathbf{1}_A \, dm$ by Theorem 4.7. But since A^c is null, $\int f\mathbf{1}_{A^c} \, dm = 0 = \int g\mathbf{1}_{A^c} \, dm$. So by (v) of the same theorem,

$$\int_{\mathbb{R}} f \, dm = \int_A f \, dm + \int_{A^c} f \, dm = \int_A f \, dm$$

$$\leq \int_A g \, dm = \int_A g \, dm + \int_{A^c} g \, dm = \int_{\mathbb{R}} g \, dm.$$

□

Proof (of Proposition 4.10)

If both f^+ and f^- are measurable then the same is true for f since $f = f^+ - f^-$. Conversely, $(f^+)^{-1}([a, \infty)) = \mathbb{R}$ if $a \leq 0$ and $(f^+)^{-1}([a, \infty)) = f^{-1}([a, \infty))$ otherwise; in each case a measurable set. Similarly for f^-.   □

Proof (of Proposition 4.15)

Put

$$s_n = \sum_{k=0}^{2^{2n}} \frac{k}{2^n} \mathbf{1}_{f^{-1}([\frac{k}{2^n}, \frac{k+1}{2^n}))},$$

which are measurable since the sets $A_k = f^{-1}([\frac{k}{2^n}, \frac{k+1}{2^n}))$ are measurable. The sequence increases since if we take $n + 1$, then each A_k is split in half, and to each component of the sum there correspond two new components. The two values of the fraction are equal to or greater than the old one, respectively. The convergence holds since for each x the values $s_n(x)$ will be a fraction of the form $\frac{k}{2^n}$ approximating $f(x)$. Figure 4.2 illustrates the above argument.   □

Proof (of Proposition 4.17)

If $f \leq g$, then $f^+ \leq g^+$ but $f^- \geq g^-$. These inequalities imply $\int f^+ \, dm \leq \int g^+ \, dm$ and $\int g^- \, dm \leq \int f^- \, dm$. Adding and rearranging gives the result.   □

Proof (of Proposition 4.20)

The claim for simple functions $f = \sum a_i \mathbf{1}_{A_i}$ is just elementary algebra. For non-negative measurable f, and positive c, take $s_n \nearrow f$, and note that $cs_n \nearrow cf$ and so

$$\int cf \, dm = \lim \int cs_n \, dm = \lim c \int s_n \, dm = c \lim \int s_n \, dm = c \int f \, dm.$$

Finally for any f and c we employ the usual trick introducing the positive and negative parts. $\square$

Proof (of Proposition 4.23)

(i) Suppose that $f(x) = \infty$ for $x \in A$ with $m(A) > 0$. Then the simple functions $s_n = n\mathbf{1}_A$ satisfy $s_n \le f$, but $\int s_n \, dm = nm(A)$ and the supremum here is ∞. Thus $\int f \, dm = \infty$ – a contradiction.

(ii) The simple function $s(x) = c\mathbf{1}_A$ with $c = \inf_A f$ has integral $\inf_A fm(A)$ and satisfies $s \le f$, which proves the first inequality. Put $t(x) = d\mathbf{1}_A$ with $d = \sup_A f$ and $f \le t$, so $\int f \, dm \le \int t \, dm$, which is the second inequality.

(iii) Note that $-|f| \le f \le |f|$ hence $-\int |f| \, dm \le \int f \, dm \le \int |f| \, dm$ and we are done.

(iv) Let $E_n = f^{-1}([\frac{1}{n}, \infty))$, and $E = \bigcup_{n=1}^{\infty} E_n$. The sets E_i are measurable and so is E. The function $s = \frac{1}{n}\mathbf{1}_{E_n}$ is a simple function with $s \le f$. Hence $\int s \, dm \le \int f \, dm = 0$, so $\int s_n \, dm = 0$, hence $\frac{1}{n}m(E_n) = 0$. Finally, $m(E_n) = 0$ for all n. Since $E_n \subset E_{n+1}$, $m(E) = \lim m(E_n) = 0$. But $E = \{x : f(x) > 0\}$, so f is zero outside the null set E. $\square$

Proof (of Proposition 4.28)

If $n \to \infty$ then $\mathbf{1}_{[-n,n]} \to 1$ hence $g_n = f\mathbf{1}_{[-n,n]} \to f$. The convergence is dominated: $g_n \le |f|$ and by the dominated convergence theorem we have $\int |f - g_n| \, dm \to 0$. Similarly, $h_n = \min(f, n) \to f$ as $n \to \infty$ and $h_n \le |f|$ so $\int |f - h_n| \, dm \to 0$. $\square$

Proof (of Proposition 4.29)

Using $\int (f + g) \, dm = \int f \, dm + \int g \, dm$ we can easily obtain (by induction)

$$\int \sum_{k=1}^{n} f_k \, dm = \sum_{k=1}^{n} \int f_k \, dm$$

for any n. The sequence $\sum_{k=1}^{n} f_k$ is increasing ($f_k \geq 0$) and converges to $\sum_{k=1}^{\infty} f_k$. So the monotone convergence theorem gives

$$\int \sum_{k=1}^{\infty} f_k \, dm = \lim_{n \to \infty} \int \sum_{k=1}^{n} f_k \, dm = \lim_{n \to \infty} \sum_{k=1}^{n} \int f_k \, dm = \sum_{k=1}^{\infty} \int f_k \, dm$$

as required.   □

Proof (of Proposition 4.32)

Continuous functions are measurable, and f is bounded on $[a, b]$, hence $f \in \mathcal{L}^1[a, b]$. Fix $a < x < x + h < b$, then $F(x + h) - F(x) = \int_x^{x+h} f \, dm$, since the intervals $[a, x]$ and $(x, x + h]$ are disjoint, so that the integral is additive with respect to the upper endpoint. By the mean value property the values of right-hand integrals are contained in the interval $[Ah, Bh]$, where $A = \inf\{f(t) : t \in [x, x + h]\}$ and $B = \sup\{f(t) : t \in [x, x + h]\}$. Both extrema are attained, as f is continuous, so we can find t_1, t_2 in $[x, x + h]$ with $A = f(t_1)$, $B = f(t_2)$. Thus

$$f(t_1) \leq \frac{1}{h} \int_x^{x+h} f \, dm \leq f(t_2).$$

The intermediate value theorem provides $\theta \in [0, 1]$ such that $f(x + \theta h) = \frac{1}{h} \int_x^{x+h} f \, dm = \frac{F(x+h) - F(x)}{h}$. Letting $h \to 0$, the continuity of f ensures that $F'(x) = f(x)$.   □

Proof (of Proposition 4.44)

(i) If $y_1 \leq y_2$, then $\{\omega : X(\omega) \leq y_1\} \subset \{\omega : X(\omega) \leq y_2\}$ and by the monotonicity of measure

$$F_X(y_1) = P(\{\omega : X(\omega) \leq y_1\}) \leq P(\{\omega : X(\omega) \leq y_2\}) = F_X(y_2).$$

(ii) Let $n \to \infty$; then $\bigcup_n \{\omega : X(\omega) \leq n\} = \Omega$ (the sets increase). Hence $P(\{\omega : X(\omega) \leq n\}) \to P(\Omega) = 1$ by Theorem 2.19 (i) and so $\lim_{y \to \infty} F_X(y) = 1$. For the second claim consider $F_X(-n) = P(\{\omega : X(\omega) \leq -n\})$ and note that $\lim_{y \to -\infty} F_X(y) = P(\bigcap_n \{\omega : X(\omega) \leq -n\}) = P(\emptyset) = 0$.

(iii) This follows directly from Theorem 2.19 (ii) with $A_n = \{\omega : X(\omega) \leq y_n\}$, $y_n \nearrow y$, because $F_X(y) = P(\bigcap_n \{\omega : X(\omega) \leq y_n\})$.   □

Proof (of Proposition 4.54)

Inserting the formulae for the option prices in call–put parity, we have

$$S(0) = \exp\{-rT\}\left(\int_\Omega (S(T) - K)^+ \mathrm{d}P - \int_\Omega (K - S(T))^+ \mathrm{d}P + K\right)$$

$$= \exp\{-rT\}\left(\int_{\{S(T)\geq K\}} (S(T) - K)\mathrm{d}P\right.$$

$$\left. - \int_{\{S(T)<K\}} (K - S(T))\mathrm{d}P + K\right)$$

$$= \exp\{-rT\}\int_\Omega S(T)\mathrm{d}P,$$

which is independent of K, as claimed. $\square$

Proof (of Proposition 4.56)

The general formula $C = \mathrm{e}^{-rT}\mathbb{E}(S(N) - K)^+$, where $S(N)$ has binomial distribution, gives

$$C = \mathrm{e}^{-rT}\sum_{k=1}^{N}\binom{N}{k}p^k(1 - q)^{N-k}(S(0)U^k D^{N-k} - K)^+$$

$$= \mathrm{e}^{-rT}\sum_{k=A}^{N}\binom{N}{k}p^k(1 - q)^{N-k}(S(0)U^k D^{N-k} - K)$$

We can rewrite this as follows: note that if we set $q = PU\mathrm{e}^{-rT}$ then $1 - q = (1 - p)U\mathrm{e}^{-rT}$, so that $0 \leq q \leq 1$ and

$$C = S(0)\sum_{k=A}^{N}\binom{N}{k}q^k(1 - q)^{N-k} - K\mathrm{e}^{-rT}\sum_{k=A}^{N}\binom{N}{k}p^k(1 - p)^{N-k},$$

which we write concisely as

$$C = S(0)\Psi(A, N, pU\mathrm{e}^{-rT}) - K\mathrm{e}^{-rT}\Psi(A, N, p) \tag{4.9}$$

where $\Psi(A, N, p) = \sum_{k=A}^{N}\binom{N}{k}p^k(1 - p)^{N-k}$ is the complementary binomial distribution function. $\square$

Proof (of Proposition 4.57)

To compute the expectation $\mathbb{E}(\mathrm{e}^{-rT}(S(T) - K)^+)$ we employ the density of the Gaussian distribution, hence

$$C = \frac{1}{\sqrt{2\pi}}\int_{\mathbb{R}} \mathrm{e}^{-\frac{1}{2}y^2}\left(S(0)\mathrm{e}^{-\frac{1}{2}\sigma^2 T}\mathrm{e}^{y\sigma\sqrt{T}} - K\mathrm{e}^{-rT}\right)^+\mathrm{d}y.$$

The integration reduces to the values of y satisfying

$$S(0)\mathrm{e}^{-\frac{1}{2}\sigma^2 T}\mathrm{e}^{y\sigma\sqrt{T}} - K\mathrm{e}^{-rT} \geq 0$$

since otherwise the integrand is zero. Solving for y gives

$$y \geq d = \frac{\ln\frac{K\mathrm{e}^{-rT}}{S(0)} + \frac{1}{2}\sigma^2 T}{\sigma\sqrt{T}}.$$

For those y we can drop the positive part and employ linearity. The first term on the right is

$$S(0)\int_d^{+\infty} \mathrm{e}^{y\sigma\sqrt{T}-\frac{1}{2}\sigma^2 T}\mathrm{e}^{-\frac{1}{2}y^2}\,\mathrm{d}y = S(0)\int_d^{+\infty} \mathrm{e}^{-\frac{1}{2}(y-\sigma\sqrt{T})^2}\,\mathrm{d}y.$$

Hence the substitution $z = y - \sigma\sqrt{T}$ yields

$$S(0)\int_d^{+\infty} \mathrm{e}^{-\frac{1}{2}z^2}\,\mathrm{d}z = S(0)N(d - \sigma\sqrt{T}),$$

where N is the cumulative distribution function of the Gaussian (normal) distribution.

The second term is

$$-K\mathrm{e}^{-rt}\int_d^{+\infty} \mathrm{e}^{-\frac{1}{2}y^2}\,\mathrm{d}y = -K\mathrm{e}^{-rT}N(-d),$$

so writing $d_1 = -d + \sigma\sqrt{T}$, $d_2 = -d$, we are done. $\square$

5

Spaces of integrable functions

Until now we have treated the points of the measure space $(\mathbb{R}, \mathcal{M}, m)$ and, more generally, of any abstract probability space $(\Omega, \mathcal{F}, P)$, as the basic objects, and regarded measurable or integrable functions as mappings associating real numbers with them. We now alter our point of view a little, by treating an integrable function as a 'point' in a function space, or, more precisely, as an element of a *normed vector space*. For this we need some extra structure on the space of functions we deal with, and we need to come to terms with the fact that the measure and integral cannot distinguish between functions which are almost everywhere equal.

The additional structure we require is to define a concept of *distance* (i.e. a *metric*) between given integrable functions – by analogy with the familiar Euclidean distance for vectors in $\mathbb{R}^n$ we shall obtain the distance between two functions as the length, or *norm*, of their difference – thus utilising the vector space structure of the space of functions. We shall be able to do this in a variety of ways, each with its own advantages – unlike the situation in $\mathbb{R}^n$, where all norms turn out to be equivalent, we now obtain genuinely different distance functions.

It is worth noting that the spaces of functions we shall discuss are all *infinite-dimensional* vector spaces: this can be seen already by considering the vector space $\mathcal{C}([a, b], \mathbb{R})$ of real-valued continuous functions defined on $[a, b]$ and noting that a polynomial function of degree n cannot be represented as a linear combination of polynomials of lower degree.

Finally, recall that in introducing characteristic functions at the end of the

125

previous chapter, we needed to extend the concept of integrability to complex-valued functions. We observed that for $f = u + iv$ the integral defined by $\int_E f \, dm = \int_E u \, dm + i \int_E v \, dm$ is linear, and that the inequality $|\int_E f \, dm| \leq \int_E |f| \, dm$ remains valid. When considering measurable functions $f : E \to \mathbb{C}$ in defining the appropriate spaces of integrable functions in this chapter, this inequality will show that $\int_E f \, dm \in \mathbb{C}$ is well defined.

The results proved below extend to the case of complex-valued functions, unless otherwise specified. When wishing to emphasise that we are dealing with complex-valued functions in particular applications or examples, we shall use notation such as $f \in \mathcal{L}^1(E, \mathbb{C})$ to indicate this. Complex-valued functions will have particular interest when we consider the important space $\mathcal{L}^2(E)$ of 'square-integrable' functions.

5.1 The space L^1

First we recall the definition of a general concept of 'distance' between points of a set:

Definition 5.1

Let X be any set. The function $d : X \times X \to \mathbb{R}$ is a *metric* on X (and (X, d) is called a *metric space*) if it satisfies:

(i) $d(x, y) \geq 0$ for all $x, y \in X$,

(ii) $d(x, y) = 0$ if and only if $x = y$,

(iii) $d(y, x) = d(x, y)$ for all $x, y \in X$,

(iv) $d(x, z) \leq d(x, y) + d(y, z)$ for all $x, y, z \in X$.

The final property is known as the *triangle inequality* and generalizes the well-known inequality of that name for vectors in $\mathbb{R}^n$. When X is a vector space (as will be the case in almost all our examples) then there is a very simple way to generate a metric by defining the distance between two vectors as the 'length' of their difference. For this we require a further definition:

Definition 5.2

Let X be a vector space over $\mathbb{R}$ (or $\mathbb{C}$). The function $x \mapsto \|x\|$ from X into $\mathbb{R}$ is a *norm* on X if it satisfies:

(i) $\|x\| \geq 0$ for all $x \in X$,

(ii) $\|x\| = 0$ if and only if $x = 0$,

(iii) $\|\alpha x\| = |\alpha| \|x\|$ for all $\alpha \in \mathbb{R}$ (or $\mathbb{C}$), $x \in X$,

(iv) $\|x + y\| \leq \|x\| + \|y\|$ for all $x, y \in X$.

Clearly a norm $x \mapsto \|x\|$ on X induces a metric by setting $d(x, y) = \|x - y\|$. The triangle inequality follows once we observe that $\|x - z\| = \|(x - y) + (y - z)\|$ and apply (iv).

We naturally wish to use the integral to define the concept of distance between functions in $\mathcal{L}^1(E)$, for measurable $E \subset \mathbb{R}$. The presence of null sets in $(\mathbb{R}, \mathcal{M}, m)$ means that the integral cannot distinguish between the function that is identically 0 and one which is 0 a.e. The natural idea of defining the 'length' of the vector f as $\int_E |f| \, dm$ thus runs into trouble, since it would be possible for non-zero elements of $\mathcal{L}^1(E)$ to have 'zero length'.

The solution adopted is to identify functions which are a.e. equal, by defining an equivalence relation on $\mathcal{L}^1(E)$ and defining the length function for the resulting equivalence classes of functions, rather than for the functions themselves.

Thus we define

$$L^1(E) = \mathcal{L}^1(E)/_{\equiv}$$

where the equivalence relation is given by:

$$f \equiv g \text{ if and only if } f(x) = g(x) \text{ for almost all } x \in E$$

(that is, $\{x \in E : f(x) \neq g(x)\}$ is null).

Write $[f]$ for the equivalence class containing the function $f \in \mathcal{L}^1(E)$. Thus $h \in [f]$ iff $h(x) = f(x)$ a.e.

Exercise 5.1

Check that $\equiv$ is an equivalence relation on $\mathcal{L}^1(E)$.

We now show that $L^1(E)$ is a vector space, since $\mathcal{L}^1$ is a vector space by Theorem 4.21. However, this requires that we explain what we mean by a linear combination of equivalence classes. This can be done quite generally for any equivalence relation. However, we shall focus on what is needed in our particular case: define the $[f] + [g]$ as the class $[f + g]$ of $f + g$, i.e. $h \in [f] + [g]$ iff $h(x) = f(x) + g(x)$ except possibly on some null set. This is consistent, since the union of two null sets is null. The definition clearly does not depend on the choice of representative taken from each equivalence class. Similarly for multiplication by constants: $a[f] = [af]$ for $a \in \mathbb{R}$. Hence $L^1(E)$ is a vector space with these operations.

Convention Strictly speaking we should continue to distinguish between the equivalence class $[f] \in L_1(E)$ and the function $f \in \mathcal{L}_1(E)$ which is a representative of this class. To do so consistently in all that follows would, however, obscure the underlying ideas, and there is no serious loss of clarity by treating f interchangeably as a member of L^1 and of $\mathcal{L}^1$, depending on the context. In other words, by treating the equivalence class $[f]$ as if it were the function f, we implicitly identify two functions as soon as they are a.e. equal. With this convention it will be clear that the 'length function' defined below is a genuine norm on $L^1(E)$.

For $f \in L^1(E)$ we write

$$\|f\|_1 = \int_E |f| \, dm.$$

We verify that this satisfies the conditions of Definition 5.2:

(i) $\|f\|_1 = 0$ if and only if $f = 0$ a.e., so that $f \equiv 0$ as an element of $L^1(E)$,

(ii) $\|cf\|_1 = \int_E |cf| \, dm = |c| \int_E |f| \, dm = |c| \cdot \|f\|_1, \ (c \in \mathbb{R})$,

(iii) $\|f + g\|_1 = \int_E |f + g| \, dm \leq \int_E |f| \, dm + \int_E |g| \, dm = \|f\|_1 + \|g\|_1$.

The most important feature of $L^1(E)$, from our present perspective, is the fact that it is a *complete* normed vector space. The precise definition is given below. Completeness of the real line $\mathbb{R}$ and Euclidean spaces $\mathbb{R}^n$ is what guides the analysis of real functions, and here we seek an analogue which has a similar impact in the infinite-dimensional context provided by function spaces. The definition will be stated for general normed vector spaces:

Definition 5.3

Let X be a vector space with norm $\|\cdot\|_X$. We say that a sequence $f_n \in X$ is *Cauchy* if

$$\forall \varepsilon > 0 \ \exists N : \forall n, m \geq N, \quad \|f_n - f_m\|_X < \varepsilon.$$

If each Cauchy sequence is convergent to some element of X, then we say that X is *complete*.

Example 5.4

Let $f_n(x) = \frac{1}{x} \mathbf{1}_{[n,n+1]}(x)$, and suppose that $n \neq m$.

$$\|f_n - f_m\|_1 = \int_0^\infty \frac{1}{x} |\mathbf{1}_{[n,n+1]} - \mathbf{1}_{[m,m+1]}| \, dx$$

$$= \int_n^{n+1} \frac{1}{x} \, dx + \int_m^{m+1} \frac{1}{x} \, dx$$

$$= \log \frac{n+1}{n} + \log \frac{m+1}{m}.$$

If $a_n \to 1$, then $\log a_n \to 0$ and the right-hand side can be as small as we wish: for $\varepsilon > 0$ take N such that $\log \frac{N+1}{N} < \frac{\varepsilon}{2}$. So f_n is a Cauchy sequence in $L^1(0,\infty)$. (When $E = (a,b)$, we write $L^1(a,b)$ for $L^1(E)$, etc.)

Exercise 5.2

Decide whether each of the following is Cauchy as a sequence in $L^1(0,\infty)$:

(a) $f_n = \mathbf{1}_{[n,n+1]}$,

(b) $f_n = \frac{1}{x} \mathbf{1}_{(0,n)}$,

(c) $f_n = \frac{1}{x^2} \mathbf{1}_{(0,n)}$.

The proof of the main result below makes essential use of the Beppo–Levi theorem in order to transfer the main convergence question to that of series of real numbers; its role is essentially to provide the analogue of the fact that in $\mathbb{R}$ (and hence in $\mathbb{C}$) absolutely convergent series will always converge. (The Beppo–Levi theorem clearly extends to complex-valued functions, just as we showed for the dominated convergence theorem, but we shall concentrate on the real case in the proof below, since the extension to $\mathbb{C}$ is immediate.)

We digress briefly to recall how this property of series ensures completeness in $\mathbb{R}$: let (x_n) be a Cauchy sequence in $\mathbb{R}$, and extract a subsequence (x_{n_k}) such that $|x_n - x_{n_k}| < 2^{-k}$ for all $n \geq n_k$ as follows:

- find n_1 such that $|x_n - x_{n_1}| < 2^{-1}$ for all $n \geq n_1$,
- find $n_2 > n_1$ such that $|x_n - x_{n_2}| < 2^{-2}$ for all $n \geq n_2$,
- ...
- find $n_k > n_{k-1}$ with $|x_n - x_{n_k}| < 2^{-k}$ for all $n \geq n_k$.

The Cauchy property ensures each time that such n_k can be found. Now consider the telescoping series with partial sums

$$y_k = x_{n_1} + (x_{n_2} - x_{n_1}) + \cdots + (x_{n_k} - x_{n_{k-1}}) = x_{n_k}$$

which has

$$|y_k| \leq |x_{n_1}| + \sum_{i=1}^{k} |x_{n_i} - x_{n_{i-1}}| < |x_{n_1}| + \sum_{i=1}^{k} \frac{1}{2^i}.$$

Thus this series converges, in other words (x_{n_k}) converges in $\mathbb{R}$, and its limit is also that of the whole Cauchy sequence (x_n).

To apply the Beppo–Levi theorem below we therefore need to extract a 'rapidly convergent sequence' from the given Cauchy sequence in $L^1(E)$. This provides an a.e. limit for the original sequence, and the Fatou lemma does the rest.

Theorem 5.5

The space $L^1(E)$ is complete.

Proof

Suppose that f_n is a Cauchy sequence. Let $\varepsilon = \frac{1}{2}$. There is N_1 such that for $n \geq N_1$

$$\|f_n - f_{N_1}\|_1 \leq \frac{1}{2}.$$

Next, let $\varepsilon = \frac{1}{2^2}$, and for some $N_2 > N_1$ we have

$$\|f_n - f_{N_2}\|_1 \leq \frac{1}{2^2}$$

for $n \geq N_2$. In this way we construct a subsequence f_{N_k} satisfying

$$\|f_{N_{n+1}} - f_{N_n}\|_1 \leq \frac{1}{2^n}$$

for all n. Hence the series $\sum_{n \geq 1} \|f_{N_{n+1}} - f_{N_n}\|_1$ converges and by the Beppo–Levi theorem, the series

$$f_{N_1}(x) + \sum_{n=1}^{\infty} [f_{N_{n+1}}(x) - f_{N_n}(x)]$$

converges a.e.; denote the sum by $f(x)$. Since

$$f_{N_1}(x) + \sum_{n=1}^{k} [f_{N_{n+1}}(x) - f_{N_n}(x)] = f_{N_{k+1}}$$

the left-hand side converges to $f(x)$, so $f_{N_{k+1}}(x)$ converges to $f(x)$. Since the sequence of real numbers $f_n(x)$ is Cauchy and the above subsequence converges, the whole sequence converges to the same limit $f(x)$.

We have to show that $f \in L^1$ and $\|f_k - f\|_1 \to 0$.

Let $\varepsilon > 0$. The Cauchy condition gives an N such that

$$\forall n, m \geq N, \ \|f_n - f_m\|_1 < \varepsilon.$$

By Fatou's lemma,

$$\|f - f_m\|_1 = \int |f - f_m| \, dm \leq \liminf_{k \to \infty} \int |f_{N_k} - f_m| \, dm = \liminf_{k \to \infty} \|f_{N_k} - f_m\|_1 < \varepsilon. \tag{5.1}$$

So $f - f_m \in L^1$ which implies $f = (f - f_m) + f_m \in L^1$, but (5.1) also gives $\|f - f_m\|_1 \to 0$. $\qquad\qquad\qquad\qquad\qquad\qquad\qquad\qquad\qquad\qquad\qquad\qquad\qquad\qquad\square$

5.2 The Hilbert space L^2

The space we now introduce plays a special role in the theory. It provides the closest analogue of the Euclidean space $\mathbb{R}^n$ among the spaces of functions, and its geometry is closely modelled on that of $\mathbb{R}^n$. It is possible, using the integral, to induce the norm via an inner product, which in turn provides a concept of orthogonality (and hence 'angles') between functions. This gives L^2 many pleasant properties, such as a 'Pythagoras theorem' and the concept of orthogonal projections, which plays vital role in many applications.

To define the norm, and hence the space $L^2(E)$ for a given measurable set $E \subset \mathbb{R}$, let

$$\|f\|_2 = \left(\int_E |f|^2 \, dm \right)^{\frac{1}{2}}$$

and *define* $\mathcal{L}^2(E)$ as the set of measurable functions for which this quantity is finite. (Note that, as for L^1, we require non-negative integrands; it is essential that the integral is non-negative in order for the square root to make sense. Although we always have $f^2(x) = (f(x))^2 \geq 0$ when $f(x)$ is real, the modulus is needed to include the case of complex-valued functions $f : E \to \mathbb{C}$. This also makes the notation consistent with that of the other L^p-spaces we shall consider below where $|f|^2$ is replaced by $|f|^p$ for arbitrary $p \geq 1$.)

We introduce $L^2(E)$ as the set of equivalence classes of elements of $\mathcal{L}^2(E)$, under the equivalence relation $f \equiv g$ iff $f = g$ a.e., exactly as for $L^1(E)$, and continue the convention of treating the equivalence classes as functions. If $f : E \to \mathbb{C}$ satisfies $\int_E |f|^2 \, dm < \infty$ we write $f \in L^2(E, \mathbb{C})$ – again using f interchangeably as a representative of its equivalence class and to denote the class itself.

It is straightforward to prove that $L^2(E)$ is a vector space: clearly, for $a \in \mathbb{R}$, $|af|^2$ is integrable if $|f|^2$ is, while

$$|f + g|^2 \leq 2^2 \max\{|f|^2, |g|^2\} \leq 4(|f|^2 + |g|^2)$$

shows that $L^2(E)$ is closed under addition.

5.2.1 Properties of the L^2-norm

We provide a simple proof that the map $f \mapsto \|f\|_2$ is a norm: to see that it satisfies the triangle inequality requires a little work, but the ideas will be very familiar from elementary analysis in $\mathbb{R}^n$, as is the terminology, though the context is rather different. We state and prove the result for the general case of $L^2(E, \mathbb{C})$.

Theorem 5.6 (Schwarz inequality)

If $f, g \in L^2(E, \mathbb{C})$ then $fg \in L^1(E, \mathbb{C})$ and

$$| \int_E f\bar{g} \, dm| \leq \|fg\|_1 \leq \|f\|_2 \|g\|_2 \qquad (5.2)$$

where $\bar{g}$ denotes the complex conjugate of g.

Proof

Replacing f, g by $|f|, |g|$ we may assume that f and g are non-negative (the first inequality has already been verified, since $\|fg\|_1 = \int_E |fg| \, dm$, and the second only involves the modulus in each case). Since we do not know in advance that $\int_E fg \, dm$ is finite, we shall first restrict attention to bounded measurable functions by setting $f_n = \min\{f, n\}$ and $g_n = \min\{g, n\}$, and confine our domain of integration to the bounded set $E \cap [-k, k] = E_k$.

For any $t \in \mathbb{R}$ we have

$$0 \leq \int_{E_k} (f_n + tg_n)^2 \, dm = \int_{E_k} f_n^2 \, dm + 2t \int_{E_k} f_n g_n \, dm + t^2 \int_{E_k} g_n^2 \, dm.$$

As a quadratic in t this does not have two distinct solutions, so the discriminant is non-positive. Thus for all $n \geq 1$

$$(2 \int_{E_k} f_n g_n \, dm)^2 \leq 4 \int_{E_k} f_n^2 \, dm \int_{E_k} g_n^2 \, dm$$

$$\leq 4 \int_E |f|^2 \, dm \int_E |g|^2 \, dm$$

$$= 4\|f\|_2^2 \|g\|_2^2.$$

Monotone convergence now yields

$$\left(\int_{E_k} fg\,\mathrm{d}m\right)^2 \le \|f\|_2^2\|g\|_2^2$$

for each k, and since $E = \bigcup_k E_k$ we obtain finally that

$$\left(\int_E fg\,\mathrm{d}m\right)^2 \le \|f\|_2^2\|g\|_2^2,$$

which is implies the Schwarz inequality.   □

The triangle inequality for the norm on $L^2(E, \mathbb{C})$ now follows at once – we need to show that $\|f + g\|_2 \le \|f\|_2 + \|g\|_2$ for $f, g \in L^2(E, \mathbb{C})$:

$$\|f+g\|_2^2 = \int_E |f+g|^2\,\mathrm{d}m = \int_E (f+g)\overline{(f+g)}\mathrm{d}m = \int_E (f+g)(\bar{f}+\bar{g})\,\mathrm{d}m.$$

The latter integral is

$$\int_E |f|^2\,\mathrm{d}m + \int_E (f\bar{g}+\bar{f}g)\,\mathrm{d}m + \int_E |g|^2\,\mathrm{d}m,$$

which is dominated by $(\|f\|_2 + \|g\|_2)^2$ since the Schwarz inequality gives

$$\int_E (f\bar{g}+g\bar{f})\,\mathrm{d}m \le 2\int_E |fg|\,\mathrm{d}m \le 2\|f\|_2\|g\|_2.$$

The result follows.

The other properties are immediate:

1. clearly $\|f\|_2 = 0$ means that $|f|^2 = 0$ a.e., hence $f = 0$ a.e.,

2. for $a \in \mathbb{C}$, $\|af\|_2 = (\int_E^2 |af|^2\,\mathrm{d}m)^{\frac{1}{2}} = |a|\|f\|_2$.

Thus the map $f \mapsto \|f\|_2$ is a norm on $L^2(E, \mathbb{C})$.

The proof that $L^2(E)$ is complete under this norm is similar to that for $L^1(E)$, and will be given in Theorem 5.24 below for arbitrary L^p-spaces ($1 < p < \infty$).

In general, without restriction of the domain set E, neither $L^1 \subseteq L^2$ nor $L^2 \subseteq L^1$. To see this consider $E = [1, \infty)$, $f(x) = \frac{1}{x}$. Then $f \in L^2(E)$ but $f \notin L^1(E)$. Next put $F = (0, 1)$, $g(x) = \frac{1}{\sqrt{x}}$. Now $g \in L^1(F)$ but $g \notin L^2(F)$.

For finite measure spaces – and hence for probability spaces! – we do have a useful inclusion:

Proposition 5.7

If the set D has finite measure (that is, $m(D) < \infty$), then $L^2(D) \subset L^1(D)$.

Hint Estimate $|f|$ by means of $|f|^2$ and then use the fact that the integral of $|f|^2$ is finite.

Before exploring the geometry induced on L^2 by its norm, we consider examples of sequences in L^2 to provide a little practice in determining which are Cauchy sequences for the L^2-norm, and compare this with their behaviour as elements of L^1.

Example 5.8

We show that the sequence $f_n = \frac{1}{x}\mathbf{1}_{[n,n+1]}$ is Cauchy in $L^2(0,\infty)$.

$$
\begin{aligned}
\|f_n - f_m\|_2 &= \int_0^\infty \frac{1}{x^2}|\mathbf{1}_{[n,n+1]} - \mathbf{1}_{[m,m+1]}|^2 \, dx \\
&= \int_n^{n+1} \frac{1}{x^2} \, dx + \int_m^{m+1} \frac{1}{x^2} \, dx \\
&= \left(\frac{1}{n} - \frac{1}{n+1}\right) + \left(\frac{1}{m} - \frac{1}{m+1}\right) \\
&\leq \frac{2}{n} + \frac{2}{m}
\end{aligned}
$$

and for $\varepsilon > 0$ let N be such that $\frac{2}{N} < \frac{\varepsilon}{2}$. Then $\|f_m - f_n\| < \varepsilon$ whenever $m, n \geq N$.

Exercise 5.3

Is the sequence

$$g_n(x) = \mathbf{1}_{(n,\infty)}(x)\frac{1}{x^2}$$

a Cauchy sequence in $L^2(R)$?

Exercise 5.4

Decide whether each of the following is Cauchy as a sequence in $L^2(0,\infty)$:

(a) $f_n = \mathbf{1}_{(0,n)}$,

(b) $f_n = \frac{1}{x}\mathbf{1}_{(0,n)}$,

(c) $f_n = \frac{1}{x^2}\mathbf{1}_{(0,n)}$.

5.2.2 Inner product spaces

We are ready for the additional structure specific (among the Lebesgue function spaces) to L^2:

$$\forall f, g \in L^2(E, \mathbb{C}) \qquad (f, g) = \int f\bar{g}\, dm \qquad (5.3)$$

defines an *inner product*, which induces the L^2-norm:

$$\sqrt{(f, f)} = \left(\int_E f\bar{f}\, dm\right)^{\frac{1}{2}} = \left(\int_E |f|^2\, dm\right)^{\frac{1}{2}} = \|f\|_2.$$

To explain what this means we verify the following properties, all of which follow easily from the integration theory we have developed:

Proposition 5.9

Linearity (in the first argument)

$$(f + g, h) = (f, h) + (g, h),$$

$$(cf, h) = c(f, h).$$

Conjugate symmetry

$$(f, g) = \overline{(g, f)}.$$

Positive definiteness

$$(f, f) \geq 0, \quad (f, f) = 0 \Leftrightarrow f = 0.$$

Hint Use the additivity of the integral in the first part, and recall for the last part that if $f = 0$ a.e. then f is the zero element of $L^2(E, \mathbb{C})$.

As an immediate consequence we get conjugate linearity with respect to the second argument

$$(f, cg + h) = \bar{c}(f, g) + (f, h).$$

Of course, if $f, g \in L^2$ are real-valued, the inner product is real and linear in the second argument also.

Examination of the proof of the Schwarz inequality reveals that the particular form of the inner product defined here on $L^2(E, \mathbb{C})$ is entirely irrelevant for this result: all we need for the proof is that the map defined in (5.3) has the properties proved for it in the last Proposition.

We shall therefore make the following important definition, which will be familiar from the finite-dimensional context of $\mathbb{R}^n$, and which we now wish to apply more generally.

Definition 5.10

An *inner product* on a vector space H over $\mathbb{C}$ is a map $(\cdot, \cdot) : H \times H \to \mathbb{C}$ which satisfies the three conditions listed in Proposition 5.9. The pair $(H, (\cdot, \cdot))$ is called an *inner product space*.

Example 5.11

The usual scalar product in $\mathbb{R}^n$ makes this space into a real inner product space, and $\mathbb{C}^n$ equipped with $(z, w) = \sum_{i=1}^{n} z_i \overline{w}_i$ is a complex one.

Proposition 5.9 shows that $L^2(E, \mathbb{C})$ is a (complex) inner product space. With the obvious simplifications in the definitions the vector space $L^2(E, \mathbb{R}) = L^2(E)$ is a *real* inner product space, i.e. with $\mathbb{R}$ as the set of scalars.

The following identities are immediate consequences of the above definitions.

Proposition 5.12

Let $(H, (\cdot, \cdot))$ be a complex inner product space, with induced norm $\| \cdot \|$. The following identities hold for all $h_1, h_2 \in H$:

(i) Parallelogram law:

$$\|h_1 + h_2\|^2 + \|h_1 - h_2\|^2 = 2(\|h_1\|^2 + \|h_2\|^2).$$

(ii) Polarisation identity:

$$4(h_1, h_2) = \|h_1 + h_2\|^2 - \|h_1 - h_2\|^2 + \mathrm{i}\{\|h_1 + \mathrm{i}h_2\|^2 - \|h_1 - \mathrm{i}h_2\|^2\}.$$

Remark 5.13

These identities, while trivial consequences of the definitions, are useful in checking that certain norms *cannot* be induced by inner products. An example is given in Exercise 5.5 below. With the addition of completeness, the identities serve to characterise inner product norms: it can be proved that in the class of complete normed spaces (known as *Banach* spaces), those whose norms are induced by inner products (i.e. are *Hilbert* spaces) are precisely those for which the parallelogram law holds, and the inner product is then recovered from the norm via the polarisation identity. We shall not prove this here.

Exercise 5.5

Show that it is impossible to define an inner product on the space $C[0,1]$ of continuous functions $f : [0,1] \to \mathbb{R}$ which will induce the sup norm $\|f\|_\infty = \sup\{|f(x)| : x \in [0,1]\}$.

Hint Try to verify the parallelogram law with the functions $f, g \in C[0,1]$ given by $f(x) = 1$, $g(x) = x$ for all x.

Exercise 5.6

Show that it is impossible to define an inner product on the space $L^1([0,1])$ with the norm $\|\cdot\|_1$.

Hint Try to verify the parallelogram law with the functions given by $f(x) = \frac{1}{2} - x$, $g(x) = x - \frac{1}{2}$.

5.2.3 Orthogonality and projections

We have introduced the concept of inner product space in a somewhat roundabout way, in order to emphasise that this structure is the natural additional tool available in the space L^2, which remains our principal source of interest. The additional structure does, however, allow us to simplify many arguments and prove results which are not available for other function spaces. In a sense, mathematical life in L^2 is 'as good as it gets' in an infinite-dimensional vector space, since the structure is so similar to that of the more familiar spaces $\mathbb{R}^n$ and $\mathbb{C}^n$.

As an example of the power of this new tool, recall that for vectors in $\mathbb{R}^n$ we have the important notion of orthogonality, which means that the scalar product of two vectors is zero. This extends to any inner product space, though we shall first state it and produce explicit examples for L^2: the functions f, g are *orthogonal* if
$$(f, g) = 0.$$

Example 5.14

If $f = \mathbf{1}_{[0,1]}$, then $(f, g) = 0$ if and only if $\int_0^1 g(x)\, dx = 0$, for example if $g(x) = x - \frac{1}{2}$.

Exercise 5.7

Show that $f(x) = \sin nx$, $g(x) = \cos mx$ for $x \in [-\pi, \pi]$, and 0 outside, are orthogonal.

Show that $f(x) = \sin nx$, $g(x) = \sin mx$ for $x \in [-\pi, \pi]$, and 0 outside, are orthogonal for $n \neq m$.

In fact, in any inner product space we can define the *angle* between the elements g, h by setting

$$\cos \theta = \frac{(g, h)}{\|g\| \|h\|}.$$

Note that by the Schwarz inequality this quantity – which, as we shall see below, also has a natural interpretation as the *correlation* between two (centred) random variables – lies in $[-1, 1]$ and that $(g, h) = 0$ means that $\cos \theta = 0$, i.e. θ is an odd multiple of $\frac{\pi}{2}$. It is therefore natural to say that g is *orthogonal to* h if $(g, h) = 0$.

Orthogonality of vectors in a complex inner product space $(H, (\cdot, \cdot))$ provides a way of formulating Pythagoras' theorem in H: since $\|g+h\|^2 = (g+h, g+h) = (g, g) + (h, h) + (g, h) + (h, g) = \|g\|^2 + \|h\|^2 + (g, h) + (h, g)$ we see at once that if g and h are orthogonal in H, then $\|g + h\|^2 = \|g\|^2 + \|h\|^2$.

Now restrict attention to the case where $(H, (\cdot, \cdot))$ is complete in the inner product norm $\|\cdot\|$ – recall that this means (see Definition 5.3) that if (h_n) is a Cauchy sequence in H then there exists $h \in H$ such that $\lim_{n \to \infty} \|h_n = h\| = 0$. As noted in Remark 5.13, we call H a *Hilbert space* if this holds. We content ourselves with one basic fact about such spaces:

Let K be a complete subspace of H, so that the above condition also holds for (h_n) in K and then yields $h \in K$. Just as the horizontal side of a right-angled triangle in standard position in the (x, y)-plane is the projection of the hypotenuse onto the horizontal axis, and the vertical side is orthogonal to that axis, we now prove the existence of orthogonal projections of a vector in H onto the subspace K.

Theorem 5.15

Let K be a complete subspace of the Hilbert space H. For each $h \in H$ we can find a unique $h' \in K$ such that $h'' = h - h'$ is orthogonal to every element of K. Equivalently, $\|h - h'\| = \inf\{\|h - k\| : k \in K\}$.

Proof

The two conditions defining h' are equivalent: assume that $h' \in K$ has been found so that $h'' = h - h'$ is orthogonal to every $k \in K$. Given $k \in K$, note that, as $(h' - k) \in K$, $(h'', h' - k) - 0$, so Pythagoras' theorem implies

$$\|h - k\|^2 = \|(h - h') + (h' - k)\|^2 = \|h''\|^2 + \|h' - k\|^2 > \|h''\|^2$$

unless $k = h'$. Hence $\|h''\| = \|h - h'\| = \inf\{\|h - k\| : k \in K\} = \delta_K$, say.

Conversely, having found $h' \in K$ such that $\|h - h'\| = \delta_K$, then for any real t and $k \in K$, $h' + tk \in K$, so that

$$\|h - (h' + tk)\|^2 \geq \|h - h'\|^2.$$

Multiplying out the inner products and writing $h'' = h - h'$, this means that $-t[(h'', k) + (k, h'')] + t^2 \|k\|^2 \geq 0$. This can only hold for all t near 0 if $(h'', k) = 0$, so that $h'' \perp k$ for every $k \in K$.

To find $h' \in K$ with $\|h - h'\| = \delta_K$, first choose a sequence (k_n) in K such that $\|h - k_n\| \to \delta_K$ as $n \to \infty$. Then apply the parallelogram law (Proposition 5.12 (i)) to the vectors $h_1 = h - \frac{1}{2}(k_m + k_n)$ and $h_2 = \frac{1}{2}(k_m - k_n)$. Note that $h_1 + h_2 = h - k_n$ and $h_1 - h_2 = h - k_m$. Hence the parallelogram law reads

$$\|h - k_n\|^2 + \|h - k_m\|^2 = 2(\|h - \frac{1}{2}(k_m + k_n)\|^2 + \|\frac{1}{2}(k_m - k_n)\|^2)$$

and since $\frac{1}{2}(k_m + k_n) \in K$, $\|h - \frac{1}{2}(k_m + k_n)\|^2 \geq \delta_K^2$. As $m, n \to \infty$ the left-hand side converges to $2\delta_K^2$, hence that final term on the right must converge to 0. Thus the sequence (k_n) is Cauchy in K, and so converges to an element h' of K. But since $\|h - k_n\| \to \delta_K$ while $\|k_n - h'\| \to 0$ as $n \to \infty$, $\|h - h'\| \leq \|h - k_n\| + \|k_n - h'\|$ shows that $\|h - h'\| = \delta_K$. This completes the proof. $\square$

In writing $h = h' + h''$ we have decomposed the vector $h \in H$ as the sum of two vectors, the first being its orthogonal projection onto K, while the second is orthogonal to all vectors in K. We say that h'' is *orthogonal to K*, and denote the set of all vectors orthogonal to K by $K^\perp$. This exhibits H as a *direct sum* $H = K \oplus K^\perp$ with each vector of the first factor being orthogonal to each vector in the second factor.

We shall use the existence of orthogonal projections onto subspaces of $L^2(\Omega, \mathcal{F}, P)$ to construct the conditional expectation of a random variable with respect to a σ-field in Section 5.4.3.

Remark 5.16

The foregoing discussion barely scratches the surface of the structure of inner product spaces, such as $L^2(E)$, which is elegantly explained, for example in [11]. On the one hand, the concept of orthogonality in an inner product space leads to consideration of *orthonormal sets*, i.e. families (e_α) in H that are mutually orthogonal and have norm 1. A natural question arises whether every element of H can be represented (or at least approximated) by linear combinations of the (e_α). In $L^2([-\pi, \pi])$ this leads, for example, to *Fourier series representations* of functions in the form $f(x) = \sum_{n=0}^{\infty} (f, \psi_n)\psi_n$, where the orthonormal functions are $\psi_0 = \frac{1}{\sqrt{2\pi}}$, $\psi_{2n}(x) = \frac{1}{\sqrt{\pi}}\cos nx$, $\psi_{2n-1}(x) = \frac{1}{\sqrt{\pi}}\sin nx$, and the series converges in L^2-norm. The completeness of L^2 is crucial in ensuring the existence of such an *orthonormal basis*.

5.3 The L^p spaces: completeness

More generally, the space $L^p(E)$ is obtained when we integrate the pth powers of $|f|$. For $p \geq 1$, we say that $f \in L^p$ (and similarly for $L^p(E)$ and $L^p(E, \mathbb{C})$) if $|f|^p$ is integrable (with the same convention of identifying f and g when they are a.e. equal). Some work will be required to check that L^p is a vector space and that the 'natural' generalisation of the norm introduced for L^2 is in fact a norm. We shall need $p \geq 1$ to achieve this.

Definition 5.17

For each $p \geq 1$, $p < \infty$, we define (identifying classes and functions)

$$L^p(E) = \{f : \int_E |f|^p \, dm \text{ is finite}\}$$

and the norm on L^p is defined by

$$\|f\|_p = \left(\int_E |f|^p \, dm\right)^{\frac{1}{p}}.$$

(With this in mind, we denoted the norm in $L^1(E)$ by $\|f\|_1$ and that in $L^2(E)$ by $\|f\|_2$.)

Recall Definition 3.14, where we considered non-negative measurable functions $f : E \to [0, \infty]$. Now we define the *essential supremum* of a measurable real-valued f on E by

$$\text{ess sup} f := \inf\{c : |f| \leq c \text{ a.e.}\}.$$

More precisely, if $F = \{c \geq 0 : m\{|f|^{-1}((c, \infty])\} = 0\}$, we set $\text{ess sup} f = \inf F$ (with the convention $\inf \emptyset = +\infty$). It is easy to see that the infimum belongs to F.

Definition 5.18

A measurable function f satisfying ess sup$|f| < \infty$ is said to be *essentially bounded* and the set of all essentially bounded functions on E is denoted by $L^\infty(E)$ (again with the usual identification of functions with a.e. equivalence classes), with the norm $\|f\|_\infty = $ ess sup$f$.

We shall need to justify the notation by showing that for each p ($1 \leq p \leq \infty$), $(L^p(E), \|\cdot\|)$ is a normed vector space.

First we observe that $L^p(E)$ is a vector space for $1 \leq p < \infty$. If f and g belong to L^p, then they are measurable and hence so are cf and $f+g$. We have $|cf(x)|^p = |c|^p |f(x)|^p$, hence

$$\|cf\|_p = \left(\int |cf(x)|^p \, dx \right)^{\frac{1}{p}} = |c| \left(\int |f(x)|^p \, dx \right)^{\frac{1}{p}} = |c| \|f\|_p.$$

Next $|f(x) + g(x)|^p \leq 2^p \max\{|f(x)|^p, |g(x)|^p\}$ and so $\|f+g\|_p$ is finite if $\|f\|_p$ and $\|g\|_p$ are. Moreover, if $\|f\|_p = 0$ then $|f(x)|^p = 0$ almost everywhere and so $f(x) = 0$ almost everywhere. The converse is obvious.

The triangle inequality

$$\|f + g\|_p \leq \|f\|_p + \|g\|_p$$

is by no means obvious for general $p \geq 1$: we need to derive a famous inequality due to Hölder, which is also extremely useful in many contexts, and generalises the Schwarz inequality.

Remark 5.19

Before tackling this, we observe that the case of $L^\infty(E)$ is rather easier: $|f+g| \leq |f| + |g|$ at once implies that $\|f + g\|_\infty \leq \|f\|_\infty + \|g\|_\infty$ (see Proposition 3.15) and similarly $|af| = |a||f|$ gives $\|af\|_\infty = |a|\|f\|_\infty$. Thus $L^\infty(E)$ is a vector space and since $\|f\|_\infty = 0$ obviously holds if and only if $f = 0$ a.e., it follows that $\|\cdot\|_\infty$ is a norm on $L^\infty(E)$. Exercises 3.6 and 5.6 show that this norm cannot be induced by any inner product on L^∞.

Lemma 5.20

For any non-negative real numbers x, y and all $\alpha, \beta \in (0,1)$ with $\alpha + \beta = 1$ we have

$$x^\alpha y^\beta \leq \alpha x + \beta y.$$

Proof

If $x = 0$ the claim is obvious. So take $x > 0$. Consider $f(t) = (1 - \beta) + \beta t - t^\beta$
for $t \geq 0$ and β as given. We have $f'(t) = \beta - \beta t^{\beta-1} = \beta(1 - t^{\beta-1})$ and since
$0 < \beta < 1$, $f'(t) < 0$ on $(0,1)$. So f decreases on $[0, 1]$ while $f'(t) > 0$ on $(1, \infty)$,
hence f increases on $[1, \infty)$. So $f(1) = 0$ is the only minimum point of f on
$[0, \infty)$, that is $f(t) \geq 0$ for $t \geq 0$. Now set $t = \frac{y}{x}$, then $(1 - \beta) + \beta \frac{y}{x} - \left(\frac{y}{x}\right)^\beta \geq 0$,
that is, $\left(\frac{y}{x}\right)^\beta \leq \alpha + \beta \frac{y}{x}$. Writing $x = x^{\alpha+\beta}$ we have $x^{\alpha+\beta} \left(\frac{y}{x}\right)^\beta \leq \alpha x + \beta x \frac{y}{x}$, so
that $x^\alpha y^\beta \leq \alpha x + \beta y$ as required. $\square$

Theorem 5.21 (Hölder's inequality)

If $\frac{1}{p} + \frac{1}{q} = 1$, $p > 1$, then for $f \in L^p(E)$, $g \in L^q(E)$, we have $fg \in L^1$ and

$$\|fg\|_1 \leq \|f\|_p \|g\|_q$$

that is

$$\int |f\bar{g}| \, dm \leq \left(\int |f|^p \, dm\right)^{\frac{1}{p}} \left(\int |g|^q \, dm\right)^{\frac{1}{q}}.$$

Proof

Step 1. Assume that $\|f\|_p = \|g\|_q = 1$, so we only need to show that $\|fg\|_1 \leq$
1. We apply Lemma 5.20 with $\alpha = \frac{1}{p}$, $\beta = \frac{1}{q}$, $x = |f|^p$, $y = |g|^q$, then we have

$$|f\bar{g}| = x^{\frac{1}{p}} y^{\frac{1}{q}} \leq \frac{1}{p}|f|^p + \frac{1}{q}|g|^q.$$

Integrating, we obtain

$$\int |f\bar{g}| \, dm \leq \frac{1}{p} \int |f|^p \, dm + \frac{1}{q} \int |g|^q \, dm = \frac{1}{p} + \frac{1}{q} = 1$$

since $\int |f|^p \, dm = 1$, $\int |g|^q \, dm = 1$. So we have $\|fg\|_1 \leq 1$ as required.

Step 2. For general $f \in L^p$ and $g \in L^q$ we write $\|f\|_p = a$, $\|g\|_q = b$ for
some $a, b > 0$. (If either a or b is zero, then one of the functions is zero almost
everywhere and the inequality is trivial.) Hence the functions $\tilde{f} = \frac{1}{a}f$, $\tilde{g} = \frac{1}{b}g$
satisfy the assumption of Step 1, and so $\|\tilde{f}\tilde{g}\|_1 \leq \|\tilde{f}\|_p \|\tilde{g}\|_q$. This yields

$$\frac{1}{ab}\|fg\|_1 \leq \frac{1}{a}\|f\|_p \frac{1}{b}\|g\|_q$$

and after multiplying by ab the result is proved. $\square$

Letting $p = q = 2$ and recalling the definition of the scalar product in L^2 we obtain the following now familiar special case of Hölder's inequality.

Corollary 5.22 (Schwarz inequality)

If $f, g \in L^2$, then
$$|(f, g)| \leq \|f\|_2 \|g\|_2.$$

We may now complete the verification that $\| \cdot \|_p$ is a norm on $L^p(E)$.

Theorem 5.23 (Minkowski's inequality)

For each $p \geq 1$, $f, g \in L^p(E)$
$$\|f + g\|_p \leq \|f\|_p + \|g\|_p.$$

Proof

Assume $1 < p < \infty$ (the case $p = 1$ was done earlier). We have
$$|f + g|^p = |(f + g)(f + g)^{p-1}| \leq |f||f + g|^{p-1} + |g||f + g|^{p-1},$$
and also taking q such that $\frac{1}{p} + \frac{1}{q} = 1$, in other words, $p + q = pq$, we obtain
$$|f + g|^{(p-1)q} = |f + g|^p < \infty.$$
Hence $(f + g)^{p-1} \in L^q$ and
$$\|(f + g)^{p-1}\|_q = \left(\int |f + g|^p \, dm \right)^{\frac{1}{q}}.$$

We may apply Hölder's inequality:
$$\int |f + g|^p \, dm \leq \int |f||f + g|^{p-1} \, dm + \int |g||f + g|^{p-1} \, dm$$
$$\leq \left(\int |f|^p \, dm \right)^{\frac{1}{p}} \left(\int |f + g|^p \, dm \right)^{\frac{1}{q}}$$
$$+ \left(\int |g|^p \, dm \right)^{\frac{1}{p}} \left(\int |f + g|^p \, dm \right)^{\frac{1}{q}}$$
$$= A \left(\left(\int |f|^p \, dm \right)^{\frac{1}{p}} + \left(\int |g|^p \, dm \right)^{\frac{1}{p}} \right)$$

with $A = \left(\int |f + g|^p \, dm \right)^{\frac{1}{q}}$. If $A = 0$ then $\|f + g\|_p = 0$ and there is nothing to prove. So suppose $A > 0$ and divide by A:

$$
\begin{aligned}
\|f + g\|_p &= \left(\int |f + g|^p \, dm \right)^{1 - \frac{1}{q}} \\
&= \frac{1}{A} \left(\int |f + g|^p \, dm \right) \\
&\leq \left(\int |f|^p \, dm \right)^{\frac{1}{p}} + \left(\int |g|^p \, dm \right)^{\frac{1}{p}} \\
&= \|f\|_p + \|g\|_p,
\end{aligned}
$$

which was to be proved. $\square$

Next we prove that $L^p(E)$ is an example of a complete normed space (i.e. a Banach space) for $1 < p < \infty$, i.e. that every Cauchy sequence in $L^p(E)$ converges in norm to an element of $L^p(E)$. We sometimes refer to convergence of sequences in the L^p-norm as *convergence in pth mean*.

The proof is quite similar to the case $p = 1$.

Theorem 5.24

The space $L^p(E)$ is complete for $1 < p < \infty$.

Proof

Given a Cauchy sequence f_n (that is, $\|f_n - f_m\|_p \to 0$ as $n, m \to \infty$), we find a subsequence f_{n_k} with

$$
\|f_n - f_{n_k}\|_p < \frac{1}{2^k}
$$

for all $k \geq 1$ and we set

$$
g_k = \sum_{i=1}^{k} |f_{n_{i+1}} - f_{n_i}|, \quad g = \lim_{k \to \infty} g_k = \sum_{i=1}^{\infty} |f_{n_{i+1}} - f_{n_i}|.
$$

The triangle inequality yields $\|g_k\|_p \leq \sum_{i=1}^{k} \frac{1}{2^i} < 1$ and we can apply Fatou's lemma to the non-negative measurable functions g_k^p, $k \geq 1$, so that

$$
\|g\|_p^p = \int \lim_{n \to \infty} g_k^p \, dm \leq \liminf_{k \to \infty} \int g_k^p \, dm \leq 1.
$$

Hence g is almost everywhere finite and $f_{n_1} + \sum_{i \geq 1} (f_{n_{i+1}} - f_{n_i})$ converges absolutely almost everywhere, defining a measurable function f as its sum.

We need to show that $f \in L^p$. Note first that $f = \lim_{k \to \infty} f_{n_k}$ a.e., and given $\varepsilon > 0$ we can find N such that $\|f_n - f_m\|_p < \varepsilon$ for $m, n \geq N$. Applying Fatou's lemma to the sequence $(|f_{n_i} - f_m|^p)_{i \geq 1}$, letting $i \to \infty$, we have

$$\int |f - f_m|^p \, dm \leq \liminf_{i \to \infty} \int |f_{n_i} - f_m|^p \, dm \leq \varepsilon^p.$$

Hence $f - f_m \in L^p$ and so $f = f_m + (f - f_m) \in L^p$ and we have $\|f - f_m\|_p < \varepsilon$ for all $m \geq N$. Thus $f_m \to f$ in L^p-norm as required.   □

The space $L^\infty(E)$ is also complete, since for any Cauchy sequence (f_n) in $L^\infty(E)$ the union of the null sets where $|f_k(x)| > \|f\|_\infty$ or $|f_n(x) - f_m(x)| > \|f_n - f_m\|_\infty$ for $k, m, n \in \mathbb{N}$, is still a null set, F say. Outside F the sequence (f_n) converges uniformly to a bounded function, f say. It is clear that $\|f_n - f\|_\infty \to 0$ and $f \in L^\infty(E)$, so we are done.

Exercise 5.8

Is the sequence

$$g_n(x) = \mathbf{1}_{(0, \frac{1}{n}]}(x) \frac{1}{\sqrt{x}}$$

Cauchy in L^4?

We have the following relations between the L^p spaces for different p which generalise Proposition 5.7.

Theorem 5.25

If E has finite Lebesgue measure, then $L^q(E) \subseteq L^p(E)$ when $1 \leq p \leq q \leq \infty$.

Proof

Note that $|f(x)|^p \leq 1$ if $|f(x)| \leq 1$. If $|f(x)| \geq 1$, then $|f(x)|^p \leq |f(x)|^q$. Hence

$$|f(x)|^p \leq 1 + |f(x)|^q,$$

$$\int_E |f|^p \, dm \leq \int_E 1 \, dm + \int_E |f|^q \, dm = m(E) + \int_E |f|^q \, dm < \infty,$$

so if $m(E)$ and $\int_E |f|^q \, dm$ are finite, the same is true for $\int_E |f|^p \, dm$.   □

5.4 Probability

5.4.1 Moments

Random variables belonging to spaces $L^p(\Omega)$, where the exponent $p \in \mathbb{N}$, play an important role in probability.

Definition 5.26

The moment of order n of a random variable $X \in L^n(\Omega)$ is the number

$$\mathbb{E}(X^n), \quad n = 1, 2, \dots .$$

Write $\mathbb{E}(X) = \mu$; then *central moments* are given by

$$\mathbb{E}(X - \mu)^n, \quad n = 1, 2, \dots .$$

By Theorem 4.41 moments are determined by the probability distribution:

$$\mathbb{E}(X^n) = \int x^n \, dP_X(x),$$

$$\mathbb{E}((X - \mu)^n) = \int (x - \mu)^n \, dP_X(x),$$

and if X has a density f_X then by Theorem 4.46 we have

$$\mathbb{E}(X^n) = \int x^n f_X(x) \, dx,$$

$$\mathbb{E}((X - \mu)^n) = \int (x - \mu)^n f_X(x) \, dx.$$

Proposition 5.27

If $\mathbb{E}(X^n)$ is finite for some n, then for $k \le n$, $\mathbb{E}(X^k)$ are finite. If $\mathbb{E}(X^n)$ is infinite, then the same is true for $\mathbb{E}(X^k)$ for $k \ge n$.

Hint Use Theorem 5.25.

Exercise 5.9

Find X, so that $\mathbb{E}(X^2) = \infty$, $\mathbb{E}(X) < \infty$. Can such an X have $\mathbb{E}(X) = 0$?

Hint You may use some previous examples in this chapter.

Definition 5.28

The *variance* of a random variable is the central moment of second order:

$$\text{Var}(X) = \mathbb{E}(X - \mathbb{E}(X))^2.$$

Clearly, writing $\mu = \mathbb{E}(X)$,

$$\text{Var}(X) = \mathbb{E}(X^2 - 2\mu X + \mu^2) = \mathbb{E}(X^2) - 2\mu\mathbb{E}(X) + \mu^2 = \mathbb{E}(X^2) - \mu^2.$$

This shows that the first two moments determine the second central moment. This may be generalised to arbitrary order and, what is more, this relationship also goes the other way round.

Proposition 5.29

Central moments of order n are determined by moments of order k for $k \le n$.

Hint Use the binomial theorem and linearity of the integral.

Proposition 5.30

Moments of order n are determined by central moments of order k for $k \le n$.

Hint Write $\mathbb{E}(X^n)$ as $\mathbb{E}((X - \mu + \mu)^n)$ and then use the binomial theorem.

Exercise 5.10

Find $\text{Var}(aX)$ in terms of $\text{Var}(X)$.

Example 5.31

If X has the uniform distribution on $[a, b]$, that is, $f_X(x) = \frac{1}{b-a}\mathbf{1}_{[a,b]}(x)$, then

$$\int x f_X(x)\,\mathrm{d}x = \frac{1}{b-a}\int_a^b x\,\mathrm{d}x = \frac{1}{b-a}\frac{1}{2}x^2\big|_a^b = \frac{1}{2}(a + b).$$

Exercise 5.11

Show that for uniformly distributed X, $\text{Var}X = \frac{1}{12}(b - a)^2$.

Exercise 5.12

Find the variance of

(a) a constant random variable X, $X(\omega) = a$ for all ω,

(b) $X : [0,1] \to \mathbb{R}$ given by $X(\omega) = \min\{\omega, 1-\omega\}$ (the distance to the nearest endpoint of the interval $[0,1]$),

(c) $X : [0,1]^2 \to \mathbb{R}$, the distance to the nearest edge of the square $[0,1]^2$.

We shall see that for the Gaussian distribution the first two moments determine the remaining ones. First we compute the expectation:

Theorem 5.32

$$\frac{1}{\sqrt{2\pi}\sigma} \int_{\mathbb{R}} x e^{-\frac{(x-\mu)^2}{2\sigma^2}}\, \mathrm{d}x = \mu.$$

Proof

Make the substitution $z = \frac{x-\mu}{\sigma}$, then, writing $\int$ for $\int_{\mathbb{R}}$,

$$\frac{1}{\sqrt{2\pi}\sigma} \int x e^{-\frac{(x-\mu)^2}{2\sigma^2}}\, \mathrm{d}x = \frac{\sigma}{\sqrt{2\pi}} \int z e^{-\frac{z^2}{2}}\, \mathrm{d}z + \frac{\mu}{\sqrt{2\pi}} \int e^{-\frac{z^2}{2}}\, \mathrm{d}z.$$

Notice that the first integral is zero since the integrand is an odd function. The second integral is $\sqrt{2\pi}$, hence the result. □

So the parameter μ in the density is the mathematical expectation. We show now that σ^2 is the variance.

Theorem 5.33

$$\frac{1}{\sqrt{2\pi}\sigma} \int_{\mathbb{R}} (x-\mu)^2 e^{-\frac{(x-\mu)^2}{2\sigma^2}}\, \mathrm{d}x = \sigma^2.$$

Proof

Make the same substitution as before: $z = \frac{x-\mu}{\sigma}$; then

$$\frac{1}{\sqrt{2\pi}\sigma} \int (x-\mu)^2 e^{-\frac{(x-\mu)^2}{2\sigma^2}}\, \mathrm{d}x = \frac{\sigma^2}{\sqrt{2\pi}} \int z^2 e^{-\frac{z^2}{2}}\, \mathrm{d}z.$$

Integrate by parts $u = z$, $v = ze^{-z^2/2}$, to get

$$\frac{\sigma^2}{\sqrt{2\pi}} \int z^2 e^{-\frac{z^2}{2}} \, dz = -\frac{\sigma^2}{\sqrt{2\pi}} z e^{-\frac{z^2}{2}} \Big|_{-\infty}^{+\infty} + \frac{\sigma^2}{\sqrt{2\pi}} \int e^{-\frac{z^2}{2}} \, dz = \sigma^2$$

since the first term vanishes. $\qquad\qquad\qquad\qquad\qquad\qquad\qquad\qquad\square$

Note that the odd central moments for a Gaussian random variable are zero: the integrals

$$\frac{1}{\sqrt{2\pi}\sigma} \int (x - \mu)^{2k+1} e^{-\frac{(x-\mu)^2}{2\sigma^2}} \, dx$$

vanish since after the above substitution we integrate an odd function. By repeating the integration by parts argument one can prove that

$$\mathbb{E}(X - \mu)^{2k} = 1 \cdot 3 \cdot 5 \cdots (2k - 1)\sigma^{2k}.$$

Example 5.34

Let us consider the Cauchy density $\frac{1}{\pi}\frac{1}{1+x^2}$ and try to compute the expectation (we shall see it is impossible):

$$\frac{1}{\pi} \int_{-\infty}^{+\infty} \frac{x}{1 + x^2} \, dx = \frac{1}{2\pi} \left(\lim_{x_n \to +\infty} \ln(1 + x_n^2) - \lim_{y_n \to -\infty} \ln(1 + y_n^2) \right)$$

for some sequences x_n, y_n. The result, if finite, should not depend on their choice, however if we set for example $x_n = ay_n$, then we have

$$\left(\lim_{x_n \to +\infty} \ln(1 + x_n^2) - \lim_{y_n \to -\infty} \ln(1 + y_n^2) \right) = \lim_{y_n \to \infty} \ln \frac{1 + ay_n^2}{1 + y_n^2} = \ln a,$$

which is a contradiction. As a consequence, we see that for the Cauchy density the moments do not exist.

Remark 5.35

We give without proof a simple relation between the characteristic function and the moments (recall that $\varphi_X(t) = \mathbb{E}(e^{itX})$ – see Definition 4.52):

If φ_X is k-times continuously differentiable then X has finite kth moment and

$$\mathbb{E}(X^k) = \frac{1}{i^k} \frac{d^k}{dt^k} \varphi_X(0).$$

Conversely, if X has kth moment finite then $\varphi_X(t)$ is k-times differentiable and the above formula holds.

5.4.2 Independence

The expectation provides a useful criterion for the independence of two random variables.

Theorem 5.36

The random variables X, Y are independent if and only if

$$\mathbb{E}(f(X)g(Y)) = \mathbb{E}(f(X))\mathbb{E}(g(Y)) \tag{5.4}$$

holds for all Borel measurable bounded functions f, g.

Proof

Suppose that (5.4) holds and take any Borel sets B_1, B_2. Let $f = \mathbf{1}_{B_1}$, $g = \mathbf{1}_{B_2}$ and application of (5.4) gives

$$\int_\Omega \mathbf{1}_{B_1}(X(\omega))\mathbf{1}_{B_2}(Y(\omega))\,\mathrm{d}P(\omega) = \int_\Omega \mathbf{1}_{B_1}(X(\omega))\,\mathrm{d}P(\omega) \int_\Omega \mathbf{1}_{B_2}(Y(\omega))\,\mathrm{d}P(\omega).$$

The left-hand side equals

$$\int_\Omega \mathbf{1}_{B_1 \times B_2}(X(\omega), Y(\omega))\,\mathrm{d}P(\omega) = P((X \in B_1) \cap (Y \in B_2)),$$

whereas the right-hand side is $P(X \in B_1)P(Y \in B_2)$, thus proving the independence of X and Y.

Suppose now that X, Y are independent. Then (5.4) holds for $f = \mathbf{1}_{B_1}$, $g = \mathbf{1}_{B_2}$, B_1, B_2 Borel sets, by the above argument. By linearity we extend the formula to simple functions: $\varphi = \sum b_i \mathbf{1}_{B_i}$, $\psi = \sum_j c_j \mathbf{1}_{C_j}$,

$$\begin{aligned}
\mathbb{E}(\varphi(X)\psi(Y)) &= \mathbb{E}(\sum b_i \mathbf{1}_{B_i}(X) \sum c_j \mathbf{1}_{C_j}(Y)) \\
&= \sum_{i,j} b_i c_j \mathbb{E}(\mathbf{1}_{B_i}(X)\mathbf{1}_{C_j}(Y)) \\
&= \sum_{i,j} b_i c_j \mathbb{E}(\mathbf{1}_{B_i}(X))\mathbb{E}(\mathbf{1}_{C_j}(Y)) \\
&= \sum_i b_i \mathbb{E}(\mathbf{1}_{B_i}(X)) \sum_j c_j \mathbb{E}(\mathbf{1}_{C_j}(Y)) \\
&= \mathbb{E}(\varphi(X))\mathbb{E}(\psi(Y)).
\end{aligned}$$

We approximate general f, g by simple functions and the dominated convergence theorem (f, g are bounded) extends the formula to f, g. $\square$

Proposition 5.37

Let X, Y be independent random variables. If $\mathbb{E}(X) = 0$, $\mathbb{E}(Y) = 0$, then $\mathbb{E}(XY) = 0$.

Hint The above theorem cannot be applied with $f(x) = x$, $g(x) = x$ (these functions are not bounded). So some approximation is required.

The expectation is nothing but an integral, so the number $(X, Y) = \mathbb{E}(XY)$ is the inner product in the space $L^2(\Omega)$ of random variables square integrable with respect to P. Hence independence implies orthogonality in this space. If the expectation of a random variable is non-zero, we modify the notion of orthogonality. The idea is that adding (or subtracting) a number does not destroy or improve independence.

Definition 5.38

For a random variable with finite $\mu = \mathbb{E}(X)$ we write $X_c = X - \mathbb{E}(X)$ and we call X_c a *centred* random variable (clearly $\mathbb{E}(X_c) = 0$). The *covariance* of X and Y is defined as

$$\text{Cov}(X, Y) = (X_c, Y_c) = \mathbb{E}\big((X - \mathbb{E}(X))(Y - \mathbb{E}(Y))\big).$$

The *correlation* is the cosine of the angle between X_c and Y_c: if $\|X\|_2$ and $\|Y\|_2$ are non-zero, then we write

$$\rho_{X,Y} = \frac{(X_c, Y_c)}{\|X\|_2 \|Y\|_2} = \frac{\text{Cov}(X, Y)}{\|X\|_2 \|Y\|_2}.$$

We say that X, Y are *uncorrelated* if $\rho = 0$.

Note that some elementary algebra gives a more convenient expression for the covariance:

$$\text{Cov}(X, Y) = \mathbb{E}(XY) - \mathbb{E}(X)\mathbb{E}(Y).$$

Thus uncorrelated X, Y satisfy $\mathbb{E}(XY) = \mathbb{E}(X)\mathbb{E}(Y)$. Clearly independent random variables are uncorrelated; it is sufficient to take $f(x) = x - \mathbb{E}(X)$, $g(x) = x - \mathbb{E}(Y)$ in Theorem 5.36. The converse is not true in general, although – as we shall see in Chapter 6 – it holds for Gaussian random variables.

Example 5.39

Let $\Omega = [-1, 1]$ with Lebesgue measure: $P = \frac{1}{2}m|_{[-1,1]}$, $X = x$, $Y = x^2$. Then $\mathbb{E}(X) = 0$, $\mathbb{E}(XY) = \int_{-1}^{1} x^3 \, dx = 0$, hence $\text{Cov}(X, Y) = 0$ and thus

$\rho_{X,Y} = 0$. However X, Y are not independent. Intuitively this is clear since $Y = X^2$, so that each of X, Y is a function of the other. Specifically, take $A = B = [-\frac{1}{2}, \frac{1}{2}]$ and compare (as required by Definition 3.18) the probabilities $P(X^{-1}(A) \cap Y^{-1}(A))$ and $P(X^{-1}(A))P(Y^{-1}(A))$. We obtain $X^{-1}(A) = A$, $Y^{-1}(A) = [-\frac{1}{\sqrt{2}}, \frac{1}{\sqrt{2}}]$, hence $P(X^{-1}(A) \cap Y^{-1}(A)) = \frac{1}{\sqrt{2}}$, $P(X^{-1}(A)) = \frac{1}{2}$, $P(Y^{-1}(A)) = \frac{1}{\sqrt{2}}$ and so X and Y are not independent.

Exercise 5.13

Find the correlation $\rho_{X,Y}$ if $X = 2Y + 1$.

Exercise 5.14

Take $\Omega = [0, 1]$ with Lebesgue measure and let $X(\omega) = \sin 2\pi\omega$, $Y(\omega) = \cos 2\pi\omega$. Show that X, Y are uncorrelated but not independent.

We close the section with two further applications.

Proposition 5.40

The variance of the sum of uncorrelated random variables is the sum of their variances:

$$\mathrm{Var}(\sum_{i=1}^{n} X_i) = \sum_{i=1}^{n} \mathrm{Var}(X_i).$$

Hint To avoid cumbersome notation first prove the formula for two random variables

$$\mathrm{Var}(X + Y) = \mathrm{Var}(X) + \mathrm{Var}(Y)$$

using the formula $\mathrm{Var}(X) = \mathbb{E}(X^2) - (\mathbb{E}X)^2$.

Proposition 5.41

Suppose that X, Y are independent random variables. Then we have the following formula for the characteristic function:

$$\varphi_{X+Y}(t) = \varphi_X(t)\varphi_Y(t).$$

More generally, if $X_1, \ldots, X_n$ are independent, then

$$\varphi_{X_1 + \cdots + X_n}(t) = \varphi_{X_1}(t) \cdots \varphi_{X_n}(t).$$

Hint Use the definition of characteristic functions and Theorem 5.36.

5.4.3 Conditional expectation (first construction)

The construction of orthogonal projections in complete inner product spaces, undertaken in Section 5.2.3, allows us to provide a preview of perhaps the most important concept in modern probability theory: the conditional expectation of an $\mathcal{F}$-measurable integrable random variable X, given a σ-field $\mathcal{G}$ contained in $\mathcal{F}$ (we call $\mathcal{G}$ a sub-σ-field of $\mathcal{F}$). We study this idea in detail in Chapter 7 where we will also justify the definition below by reference to more familiar concepts, but the construction of the conditional expectation as a $\mathcal{G}$-measurable random variable can be achieved for any integrable X with the tools we have readily to hand. Our argument owes much to the elegant construction given in [13].

We begin with $X \in L^2 = L^2(\Omega, \mathcal{F}, P)$. Thinking of σ-fields as containing 'information' about random events consider the subspace $L^2(\mathcal{G})$ of L^2 consisting of $\mathcal{G}$-measurable functions. (In accordance with our convention in Section 5.1, we work with functions rather than with equivalence classes.) By Theorem 5.24, the inner product space L^2 is complete, and the vector subspace $L^2(\mathcal{G})$ is a complete subspace. Thus the construction of the orthogonal projection in Section 5.2.3 applies and we denote by Y the orthogonal projection of X on $L^2(\mathcal{G})$. We can interpret Y as the 'best predictor' of X among the class of $\mathcal{G}$-measurable functions. We know that $X - Y$ is orthogonal to $L^2(\mathcal{G})$ and by definition of the inner product in L^2 this means that

$$(X - Y, Z)_{L^2} = \int_\Omega (X - Y)Z \, dP = 0$$

for every $Z \in L^2(\mathcal{G})$. In particular, since $\mathbf{1}_G \in L^2(\mathcal{G})$ for every $G \in \mathcal{G}$, we have

$$\int_G Y \, dP = \int_G X \, dP. \tag{5.5}$$

This motivates the following definition:

Definition 5.42

Let $(\Omega, \mathcal{F}, P)$ be a probability space and suppose that $\mathcal{G}$ is a sub-σ-field of $\mathcal{F}$. Suppose that for an $\mathcal{F}$-measurable X there exists a unique (up to a null set) $\mathcal{G}$-measurable Y such that (5.5) holds for every $G \in \mathcal{G}$. We write $Y = \mathbb{E}(X|\mathcal{G})$ and call Y the conditional expectation of X given $\mathcal{G}$.

The above considerations show that the conditional expectation is well defined for $X \in L^2$. However, condition (5.5) makes sense for X, Y from just L^1. This gives hope of extending the scope of the definition to such X.

We shall show how to construct $\mathbb{E}(X|\mathcal{G}) \in L^1(\mathcal{G})$ for an arbitrary $X \in L^1(\mathcal{F})$.

Step 1. The case of non-negative bounded X.

If X is bounded, it is in $L^2(\mathcal{F})$ by Theorem 5.25, since $P(\Omega)$ is finite. Hence it has a conditional expectation Y.

We shall see that $Y \geq 0$ P-a.s. To this end suppose that Y takes negative values with positive probability. Then there exists $n \geq 1$ such that the set $G = \{Y < -\frac{1}{n}\} \in \mathcal{G}$ has $P(G) > 0$. Thus $\int_G Y \, dP < -\frac{1}{n} P(G) < 0$. But $\int_G Y \, dP = \int_G X \, dP \geq 0$ by Proposition 4.17 – a contradiction.

Step 2. Approximating sequence for non-negative $X \in L^1$.

Take an arbitrary $X \geq 0$ in $L^1(\mathcal{F})$, and for $n \geq 1$ set $X_n = \min(X, n)$. Then X_n is bounded and non-negative, so Step 1 applies to X_n, yielding a non-negative $Y_n \in L^2(\mathcal{G})$ with $\int_G Y_n \, dP = \int_G X_n \, dP$.

Since (X_n) is increasing with n, so is $(\int_G X_n \, dP)$ for each $G \in \mathcal{G}$. Hence for each G, $\int_G Y_n \, dP \leq \int_G Y_{n+1} \, dP$, therefore $Y_{n+1} - Y_n \geq 0$-a.s., that is, the sequence (Y_n) increases a.s.

Step 3. Taking the limit.

Set $Y(\omega) = \limsup_{n \to \infty} Y_n(\omega)$ for each $\omega \in \Omega$. By Theorem 3.9 Y is $\mathcal{G}$-measurable.

Moreover, for $G \in \mathcal{G}$, $\int_G Y_n \, dP = \int_G X_n \, dP \leq \int_G X \, dP < \infty$ for all n. The monotone convergence theorem shows that the integrals $(\int_G Y_n \, dP)_{n \geq 1}$ increase to $\int_G Y \, dP$, and that the final integral is finite, so that $Y \in L^1(\mathcal{G})$.

On the other hand, $\int_G Y_n \, dP = \int_G X_n \, dP$ and the latter integrals also increase to $\int_G X \, dP$, so that this implies $\int_G Y \, dP = \int_G X \, dP$ for all $G \in \mathcal{G}$.

Step 4. General X (not necessarily non-negative).

We can consider $X = X^+ - X^-$. By Step 3 there are $Y^+, Y^- \in L^1(\mathcal{G})$) such that for $G \in \mathcal{G}$ both $\int_G Y^+ \, dP = \int_G X^+ \, dP$ and $\int_G Y^- \, dP = \int_G X^- \, dP$. Subtracting on both sides we obtain $\int_G Y \, dP = \int_G X \, dP$, where $Y = Y^+ - Y^- \in L^1(\mathcal{G})$.

Step 5. Uniqueness.

Theorem 4.22, applied to P instead of m and to $\mathcal{G}$ instead of $\mathcal{M}$, implies that Y is P-a.s. unique: if $Z \in L^1(\mathcal{G})$ also satisfies $\int_G Z \, dP = \int_G X \, dP$ for every $G \in \mathcal{G}$, then $Z = Y$, P-a.s.

This is often expressed by saying that Y is a *version* of $\mathbb{E}(X|\mathcal{G})$: by definition of $L^1(\Omega, \mathcal{G}, P)$ the uniqueness claim is that all versions belong to the same equivalence class in $L^1(\Omega, \mathcal{G}, P)$ under the equivalence relation: $f \equiv g$ iff $P(\{\omega \in \Omega : f(\omega) \neq g(\omega)\}) = 0$.

5.5 Proofs of propositions

Proof (of Proposition 5.7)

Suppose that $\int_D f^2(x)\,\mathrm{d}x$ is finite. Then using $a \leq 1 + a^2$ (which follows from $(a-1)^2 \geq 0$) we have

$$\int_D f(x)\,\mathrm{d}x \leq \int_D 1\,\mathrm{d}x + \int_D f^2(x)\,\mathrm{d}x = m(D) + \int_D f^2(x)\,\mathrm{d}(x) < \infty.$$

$\square$

Proof (of Proposition 5.9)

We verify the first two properties using the linearity of the integral:

$$(f+g, h) = \int (f(x) + g(x))h(x)\,\mathrm{d}x$$

$$= \int f(x)h(x)\,\mathrm{d}x + \int g(x)h(x)\,\mathrm{d}x$$

$$= (f, h) + (g, h),$$

$$(cf, h) = \int cf(x)h(x)\,\mathrm{d}x = c\int f(x)h(x)\,\mathrm{d}x = c(f, h).$$

The symmetry is obvious since $f(x)g(x) = g(x)f(x)$ under the integral. $\square$

Proof (of Proposition 5.12)

Parallelogram law: $\|h_1 + h_2\|^2 = (h_1 + h_2, h_1 + h_2) = (h_1, h_1) + (h_1, h_2) + (h_2, h_1) + (h_2, h_2)$, $\|h_1 - h_2\|^2 = (h_1 - h_2, h_1 - h_2) = (h_1, h_1) - (h_1, h_2) - (h_2, h_1) + (h_2, h_2)$, and adding we get the result.

Polarization identity: subtract the above $\|h_1 + h_2\|^2 - \|h_1 - h_2\|^2 = 2(h_1, h_2) + 2(h_2, h_1)$, replace h_2 by ih_2 to get $\|h_1 + ih_2\|^2 - \|h_1 - ih_2\|^2 = 2(h_1, ih_2) + 2(ih_2, h_1)$. Insert the obtained expressions into the right-hand side of the identity in question. On the left we have $2[(h_1, h_2) + (h_2, h_1) + i(h_1, ih_2) + i(ih_2, h_1)] = 2[(h_1, h_2) + (h_2, h_1) + i(-i)(h_1, h_2) + i^2(h_2, h_1)] = 4(h_1, h_2)$. $\square$

Proof (of Proposition 5.27)

Suppose $\mathbb{E}(X^n) < \infty$, which means that $X \in L^n(\Omega)$, then since the measure of Ω is finite we may apply Theorem 5.25 and so $X \in L^k(\Omega)$ for all $k \leq n$. If $\mathbb{E}(X^n) = \infty$ the same must be true for $\mathbb{E}(X^k)$ for $k \geq n$ since otherwise $\mathbb{E}(X^k) < \infty$ would imply $\mathbb{E}(X^n) < \infty$ – a contradiction. $\square$

Proof (of Proposition 5.29)

Using the binomial expansion we have

$$(X - \mu)^n = \sum_{i=0}^{n} \binom{n}{i} X^i (-\mu)^{n-i},$$

and so by linearity of the expectation

$$\mathbb{E}(X - \mu)^n = \sum_{i=0}^{n} \binom{n}{i} (-\mu)^{n-i} \mathbb{E}(X^i).$$

$\square$

Proof (of Proposition 5.30)

We have

$$\mathbb{E}(X^n) = \mathbb{E}((X - \mu + \mu)^n) = \sum_{i=1}^{n} \binom{n}{i} \mathbb{E}((X - \mu)^i) \mu^{n-i}.$$

$\square$

Proof (of Proposition 5.37)

Let $f_n(x) = \max\{-n, \min\{x, n\}\}$. By Theorem 5.36, since X, Y are independent, we have $\mathbb{E}(f_n(X)f_n(Y)) = \mathbb{E}(f_n(X))\mathbb{E}(f_n(Y))$. Integrability of X and Y enables us to pass to the limit, which is 0 on the right. $\square$

Proof (of Proposition 5.40)

Let X, Y be uncorrelated random variables. Then

$$\begin{aligned}
\text{Var}(X + Y) &= \mathbb{E}\big(((X + Y) - \mathbb{E}(X + Y))^2\big) \\
&= \mathbb{E}(X + Y)^2 - (\mathbb{E}(X) + \mathbb{E}(Y))^2 \\
&= \mathbb{E}X^2 + 2\mathbb{E}(XY) + \mathbb{E}Y^2 - (\mathbb{E}X)^2 - 2\mathbb{E}(X)\mathbb{E}(Y) - (\mathbb{E}Y)^2 \\
&= \mathbb{E}X^2 - (\mathbb{E}X)^2 + \mathbb{E}Y^2 - (\mathbb{E}Y)^2 + 2[\mathbb{E}(XY) - \mathbb{E}(X)\mathbb{E}(Y)] \\
&= \text{Var}(X) + \text{Var}(Y)
\end{aligned}$$

since $\mathbb{E}(XY) = \mathbb{E}(X)\mathbb{E}(Y)$. The general case for n random variables follows by induction or by repetitive use of the formula for two:

$$\begin{aligned}
\mathrm{Var}(X_1 + \cdots + X_n) &= \mathrm{Var}(X_1 + [X_2 + \cdots + X_n]) \\
&= \mathrm{Var}(X_1) + \mathrm{Var}(X_2 + [X_3 + \cdots + X_n]) \\
&\quad \cdots \\
&= \mathrm{Var}(X_1) + \mathrm{Var}(X_2) + \cdots + \mathrm{Var}(X_n).
\end{aligned}$$

$\square$

Proof (of Proposition 5.41)

By definition,

$$\begin{aligned}
\varphi_{X+Y}(t) &= \mathbb{E}(e^{it(X+Y)}) \\
&= \mathbb{E}(e^{itX} e^{itY}) \\
&= \mathbb{E}(e^{itX})\mathbb{E}(e^{itY}) \quad \text{(by Theorem 5.36)} \\
&= \varphi_X(t)\varphi_Y(t).
\end{aligned}$$

The generalization to n components is straightforward – induction or step-by-step application of the result for two. $\square$

6
Product measures

6.1 Multi-dimensional Lebesgue measure

In Chapter 2 we constructed Lebesgue measure on the real line. The basis for that was the notion of the length of an interval. Consider now the plane $\mathbb{R}^2$ in place of $\mathbb{R}$. Here by interval we understand a rectangle of any sort:

$$R = I_1 \times I_2$$

where I_1, I_2 are any intervals. The 'length' of a rectangle is its area

$$a(R) = l(I_1)l(I_2).$$

The concept of null set is introduced as in the one-dimensional case. As before, countable sets are null. It is worth noting that on the plane we have more sophisticated null sets such as, for example, a line segment or the graph of a function.

The whole construction goes through without change and the resulting measure is the Lebesgue measure m_2 on the plane defined on the σ-field generated by the rectangles.

A subtle point which clearly illustrates the difference from linear measure is the following: any set of the form $A \times \{a\}$, $a \in \mathbb{R}$, is null and hence Lebesgue-measurable on the plane. An interesting case of this is when A is a non-measurable set on the real line!

Next, we consider $\mathbb{R}^3$. By 'interval' here we mean a cube, and the 'length' is its volume:

$$C = I_1 \times I_2 \times I_3,$$

159

$$v(C) = l(I_1)l(I_2)l(I_3).$$

Now surfaces are examples of null sets. Following the same construction we obtain Lebesgue measure m_3 in $\mathbb{R}^3$.

Finally, we consider $\mathbb{R}^n$ (this includes the particular cases $n = 1, 2, 3$ so for a true mathematician this is the only case worth attention as it covers all the others). 'Intervals' are now n-dimensional cubes:

$$\mathbf{I} = I_1 \times \cdots \times I_n$$

and generalised 'length' is given by

$$l(\mathbf{I}) = l(I_1) \cdots l(I_n).$$

An interesting example of a null set is a hyperplane.

The above provides motivation for what follows. The multi-dimensional Lebesgue measures will emerge again from the considerations below where we will work with general measure spaces. Bearing in mind that we are principally interested in probabilistic applications, we stick to the notation of probability theory.

6.2 Product σ-fields

Let $(\Omega_1, \mathcal{F}_1, P_1)$, $(\Omega_2, \mathcal{F}_2, P_2)$ be two measure spaces. Put

$$\Omega = \Omega_1 \times \Omega_2.$$

We want to define a measure P on Ω to 'agree' with the measures given on Ω_1, Ω_2.

Before we construct P we need to specify its domain, that is, a σ-field on Ω.

Definition 6.1

Let $\mathcal{F}$ be the smallest σ-field of subsets of Ω containing the 'rectangles' $A_1 \times A_2$ for all $A_1 \in \mathcal{F}_1$, $A_2 \in \mathcal{F}_2$. We call $\mathcal{F}$ the *product σ-field* of $\mathcal{F}_1$ and $\mathcal{F}_2$. In other words, the product σ-field is generated by the family of sets ('rectangles')

$$\mathcal{R} = \{A_1 \times A_2 : A_1 \in \mathcal{F}_1, A_2 \in \mathcal{F}_2\}.$$

The notation used for the product σ-field is simply: $\mathcal{F} = \mathcal{F}_1 \times \mathcal{F}_2$.

There are many ways in which the same product σ-field may be generated, each of which will prove useful in the sequel.

Theorem 6.2

(i) The product σ-field $\mathcal{F}_1 \times \mathcal{F}_2$ is generated by the family of sets ('cylinders' or 'strips')

$$\mathcal{C} = \{A_1 \times \Omega_2 : A_1 \in \mathcal{F}_1\} \cup \{\Omega_1 \times A_2 : A_2 \in \mathcal{F}_2\}.$$

(ii) The product σ-field $\mathcal{F}$ is the smallest σ-field such that the projections

$$\mathrm{Pr}_1 : \Omega \to \Omega_1, \quad \mathrm{Pr}_1(\omega_1, \omega_2) = \omega_1$$

$$\mathrm{Pr}_2 : \Omega \to \Omega_2, \quad \mathrm{Pr}_2(\omega_1, \omega_2) = \omega_2$$

are measurable.

Proof

Recall that we write $\sigma(\mathcal{E})$ for the σ-field generated by a family $\mathcal{E}$.

Clearly $\mathcal{C}$ is contained in $\mathcal{R}$, hence the σ-field generated by $\mathcal{C}$ is smaller than the σ-field generated by $\mathcal{R}$. On the other hand,

$$A_1 \times A_2 = (A_1 \times \Omega_2) \cap (\Omega_1 \times A_2),$$

hence the rectangles belong to the σ-field generated by cylinders: $\mathcal{R} \subset \sigma(\mathcal{C})$. This implies $\sigma(\mathcal{R}) \subset \sigma(\sigma(\mathcal{C})) = \sigma(\mathcal{C})$, which completes the proof of (i).

For (ii) note that

$$\mathrm{Pr}_1^{-1}(A_1) = A_1 \times \Omega_2, \quad \mathrm{Pr}_2^{-1}(A_2) = \Omega_1 \times A_2,$$

hence the projections are measurable by (i). But the smallest σ-field such that they are both measurable is the smallest σ-field containing the cylinders, which is $\mathcal{F}$ by (i) again. $\qquad\square$

Consider a particular case of $\Omega_1, \Omega_2 = \mathbb{R}$, $\mathcal{F}_1 = \mathcal{F}_2 = \mathcal{B}$ Borel sets. Then we have two ways of producing $\mathcal{F} = \mathcal{B}_2$ — the Borel sets on the plane: we can use the family of products of Borel sets or the family of products of intervals. The following result shows that they give the same collection of subsets of $\mathbb{R}^2$.

Proposition 6.3

The σ-fields generated by

$$\mathcal{R} = \{B_1 \times B_2 : B_1, B_2 \in \mathcal{B}\},$$

$$\mathcal{I} = \{I_1 \times I_2 : I_1, I_2 \text{ are intervals}\}$$

are the same.

Hint Use the idea of the proof of the preceding theorem.

We may easily generalise to n factors: suppose we are given n measure spaces $(\Omega_i, \mathcal{F}_i, P_i)$, $i = 1, \ldots, n$, then the product σ-fields in $\Omega = \Omega_1 \times \cdots \times \Omega_n$ is the σ-field generated by the sets

$$\{A_1 \times \cdots \times A_n : A_i \in \mathcal{F}_i\}.$$

6.3 Construction of the product measure

Recall that $(\Omega_1, \mathcal{F}_1, P_1)$, $(\Omega_2, \mathcal{F}_2, P_2)$ are arbitrary measure spaces. We shall construct a measure P on $\Omega = \Omega_1 \times \Omega_2$ which is determined by P_1, P_2 in a natural way. A technical assumption is needed here: P_1 and P_2 are taken to be σ-finite, that is, there is a sequence of measurable sets A_n with $\bigcup_{n=1}^{\infty} A_n = \Omega_1$, $P_1(A_n)$ finite (and the same for P_2, Ω_2). This is of course true for probability measures and also for Lebesgue measure (in the latter case $A_n = [-n, n]$ will do for example). For simplicity we assume that P_1, P_2 are finite. (The extension to the case of σ-finite is obtained by a routine limit passage $n \to \infty$ in the results obtained for the restrictions of the measures to A_n.)

The motivation provided by construction of multi-dimensional Lebesgue measures gives the following natural condition on P:

$$P(A_1 \times A_2) = P_1(A_1)P_2(A_2) \tag{6.1}$$

for $A_1 \in \mathcal{F}_1$, $A_2 \in \mathcal{F}_2$.

We want to generalise (6.1) to all sets from the product σ-field. To do this we introduce the notion of a *section* of a subset A of $\Omega_1 \times \Omega_2$: for $\omega_2 \in \Omega_2$,

$$A_{\omega_2} = \{\omega_1 \in \Omega_1 : (\omega_1, \omega_2) \in A\} \subset \Omega_1.$$

A similar construction is carried out for $\omega_1 \in \Omega_1$:

$$A_{\omega_1} = \{\omega_2 \in \Omega_2 : (\omega_1, \omega_2) \in A\} \subset \Omega_2.$$

Theorem 6.4

If A is in the product σ-field $\mathcal{F}$, then for each ω_2, $A_{\omega_2} \in \mathcal{F}_1$, and for each ω_1, $A_{\omega_1} \in \mathcal{F}_2$.

Proof

Let

$$\mathcal{G} = \{A \in \mathcal{F} : \text{for all } \omega_2, A_{\omega_2} \in \mathcal{F}_1\}.$$

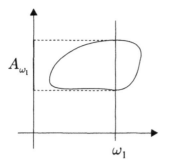

Figure 6.1 ω_1 section of a set

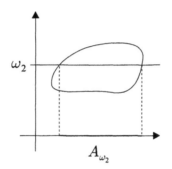

Figure 6.2 ω_2 section of a set

If we show that $\mathcal{G}$ is a σ-field containing rectangles, then $\mathcal{G} = \mathcal{F}$ since $\mathcal{F}$ is the smallest σ-field with this property.

If $A = A_1 \times A_2$, $A_1 \in \mathcal{F}_1$, then

$$A_{\omega_2} = \begin{cases} A_1 & \text{if } \omega_2 \in A_2 \\ \emptyset & \text{if } \omega_2 \notin A_2 \end{cases}$$

is in $\mathcal{F}_1$ so the rectangles are in $\mathcal{G}$.

If $A \in \mathcal{G}$, then $\Omega \setminus A \in \mathcal{G}$ since

$$\begin{aligned}
(\Omega \setminus A)_{\omega_2} &= \{\omega_1 : (\omega_1, \omega_2) \in (\Omega \setminus A)\} \\
&= \Omega_1 \setminus \{\omega_1 : (\omega_1, \omega_2) \in A\} \\
&= \Omega_1 \setminus A_{\omega_2}.
\end{aligned}$$

Finally, let $A_n \in \mathcal{G}$. Since

$$\left(\bigcup_{n=1}^{\infty} A_n \right)_{\omega_2} = \bigcup_{n=1}^{\infty} (A_n)_{\omega_2},$$

the union $\bigcup_{n=1}^{\infty} A_n$ is also in $\mathcal{G}$.

The proof for A_{ω_1} is exactly the same. $\square$

If $A = A_1 \times A_2$, then the function $\omega_2 \mapsto P(A_{\omega_2})$ is a step function:

$$P(A_{\omega_2}) = \begin{cases} P(A_1) & \text{if } \omega_2 \in A_2 \\ 0 & \text{if } \omega_2 \notin A_2 \end{cases}$$

and hence we have

$$P(A) = P_1(A_1)P_2(A_2) = \int_{\Omega_2} P(A_{\omega_2})\,dP_2(\omega_2).$$

This motivates the general formula; for any A we write

$$P(A) = \int_{\Omega_2} P_1(A_{\omega_2})\,dP_2(\omega_2). \tag{6.2}$$

We call P the *product measure* and we will sometimes denote it by $P = P_1 \times P_2$. We already know that $P_1(A_{\omega_2})$ makes sense since $A_{\omega_2} \in \mathcal{F}_1$ as shown before. For the integral to make sense we need more:

Theorem 6.5

Suppose that P_1, P_2 are finite. If $A \in \mathcal{F}$, then the functions

$$\omega_2 \mapsto P_1(A_{\omega_2}), \quad \omega_1 \mapsto P_2(A_{\omega_1})$$

are measurable with respect to $\mathcal{F}_2$, $\mathcal{F}_1$, respectively, and

$$\int_{\Omega_1} P_1(A_{\omega_2})\,dP_2(\omega_2) = \int_{\Omega_2} P_2(A_{\omega_1})\,dP_1(\omega_1). \tag{6.3}$$

Before we prove this theorem we allow ourselves a digression concerning the ways in which σ-fields can be produced. This will greatly simplify several proofs. Given a family of sets $\mathcal{A}$, let $\mathcal{F}$ be the smallest σ-field containing $\mathcal{A}$. Suppose that $\mathcal{A}$ is a field, i.e. it is closed with respect to finite unions and differences: $A, B \in \mathcal{A}$ implies $A \cup B$, $A \setminus B \in \mathcal{A}$. Then we have an alternative way of characterizing σ-fields by means of so-called monotone classes.

Definition 6.6

A *monotone class* is a family of sets closed under countable unions of increasing sets and countable intersections of decreasing sets. That is, $\mathcal{G}$ is a monotone class if

$$A_1 \subset A_2 \subset \cdots, \quad A_i \in \mathcal{G} \quad \Rightarrow \quad \bigcup_{i=1}^{\infty} A_i \in \mathcal{G},$$

$$A_1 \supset A_2 \supset \cdots, \quad A_i \in \mathcal{G} \quad \Rightarrow \quad \bigcap_{i=1}^{\infty} A_i \in \mathcal{G}.$$

Lemma 6.7 (monotone class theorem)

The smallest monotone class $\mathcal{G}_\mathcal{A}$ containing a field $\mathcal{A}$ coincides with the σ-field $\mathcal{F}_\mathcal{A}$ generated by $\mathcal{A}$.

Proof

A σ-field is a monotone class, so $\mathcal{G}_\mathcal{A} \subset \mathcal{F}_\mathcal{A}$ (since $\mathcal{G}_\mathcal{A}$ is the smallest monotone class containing $\mathcal{A}$).

To prove the converse, we first show that $\mathcal{G}_\mathcal{A}$ is a field. The family of sets

$$\{A : A^c \in \mathcal{G}_\mathcal{A}\}$$

contains $\mathcal{A}$ since $A \in \mathcal{A}$ implies $A^c \in \mathcal{A} \subset \mathcal{G}_\mathcal{A}$. We observe that it is a monotone class. Suppose that $A_1 \subset A_2 \subset \cdots$, are such that $A_i^c \in \mathcal{G}_\mathcal{A}$. We have to show that $(\bigcup A_i)^c \in \mathcal{G}_\mathcal{A}$. We have $A_1^c \supset A_2^c \supset \cdots$ and hence $\bigcap A_i^c \in \mathcal{G}_\mathcal{A}$ because $\mathcal{G}_\mathcal{A}$ is a monotone class. By de Morgan's law $\bigcap A_i^c = (\bigcup A_i)^c$ so the union satisfies the required condition. The proof for the intersection is similar: $A_1 \supset A_2 \supset \cdots$ implies $A_1^c \subset A_2^c \subset \cdots$, hence $\bigcup A_i^c \in \mathcal{G}_\mathcal{A}$ and so $(\bigcap A_i)^c = \bigcup A_i^c$ also belongs to $\mathcal{G}_\mathcal{A}$.

We conclude that

$$\mathcal{G}_\mathcal{A} \subset \{A : A^c \in \mathcal{G}_\mathcal{A}\}$$

so $\mathcal{G}_\mathcal{A}$ is closed with respect to taking complements.

Now consider unions. First fix $A \in \mathcal{A}$ and consider

$$\{B : A \cup B \in \mathcal{G}_\mathcal{A}\}.$$

This family contains $\mathcal{A}$ (if $B \in \mathcal{A}$, then $A \cup B \in \mathcal{A} \subset \mathcal{G}_\mathcal{A}$) and is a monotone class. For, let $B_1 \subset B_2 \subset \cdots$ be such that $A \cup B_i \in \mathcal{G}_\mathcal{A}$. Then $A \cup B_1 \subset A \cup B_2 \subset \cdots$, hence $\bigcup(A \cup B_i) \in \mathcal{G}_\mathcal{A}$, thus $A \cup \bigcup B_i \in \mathcal{G}_\mathcal{A}$. Similar arguments work for the intersection of a decreasing chain of sets, so for this fixed A,

$$\mathcal{G}_\mathcal{A} \subset \{B : A \cup B \in \mathcal{G}_\mathcal{A}\}.$$

This means that for $A \in \mathcal{A}$ and $B \in \mathcal{G}_\mathcal{A}$ we have $A \cup B \in \mathcal{G}_\mathcal{A}$.

Now take arbitrary $A \in \mathcal{G}_\mathcal{A}$. By what we have just observed,

$$\mathcal{A} \subset \{B : A \cup B \in \mathcal{G}_\mathcal{A}\}$$

and, by the same argument as before, the latter family is a monotone class. So

$$\mathcal{G}_A \subset \{B : A \cup B \in \mathcal{G}_A\}$$

this time for general A, which completes the proof that $\mathcal{G}_A$ is a field.

Now, having shown that $\mathcal{G}_A$ is a field, we observe that it is a σ-field. This is obvious since for a sequence $A_i \in \mathcal{G}_A$ we have $A_1 \subset A_1 \cup A_2 \subset \cdots$, they all are in $\mathcal{G}_A$ (by the field property) and so is their union (since $\mathcal{G}_A$ is a monotone class).

Therefore $\mathcal{G}_A$ is a σ-field containing $\mathcal{A}$, so it contains the σ-field generated by $\mathcal{A}$:

$$\mathcal{F}_A \subset \mathcal{G}_A,$$

which completes the proof. $\square$

The family $\mathcal{R}$ of rectangles introduced above is not a field, so it cannot be used in the above result. Therefore we take $\mathcal{A}$ to be the family of all unions of disjoint rectangles.

Proof (of Theorem 6.5)

Write

$$\mathcal{G} = \Big\{ A : \omega_2 \mapsto P_1(A_{\omega_2}), \quad \omega_1 \mapsto P_2(A_{\omega_1}) \text{ are measurable and}$$

$$\int_{\Omega_2} P_1(A_{\omega_2})\, \mathrm{d}P_2(\omega_2) = \int_{\Omega_1} P_2(A_{\omega_1})\, \mathrm{d}P_1(\omega_1) \Big\}.$$

The idea of the proof is this. First we show that $\mathcal{R} \subset \mathcal{G}$, then $\mathcal{A} \subset \mathcal{G}$, and finally we show that $\mathcal{G}$ is a monotone class. By Lemma 6.7, $\mathcal{G} = \mathcal{F}$, which means that the claim of the theorem holds for all sets from $\mathcal{F}$.

If A is a rectangle, $A = A_1 \times A_2$, then as we noticed before, $\omega_2 \mapsto P_1(A_{\omega_2})$, $\omega_1 \mapsto P_2(A_{\omega_1})$ are indicator functions multiplied by some constants and (6.3) holds, each side being equal to $P_1(A_1)P_2(A_2)$.

Next let $A = (A_1 \times A_2) \cup (B_1 \times B_2)$ be the union of disjoint rectangles. Disjoint means that either $A_1 \cap B_1 = \emptyset$ or $A_2 \cap B_2 = \emptyset$. Assume the former, for example. Then

$$A_{\omega_2} = \begin{cases} A_1 \cup B_1 & \text{if } \omega_2 \in A_2 \cap B_2 \\ A_1 & \text{if } \omega_2 \in A_2 \setminus B_2 \\ B_1 & \text{if } \omega_2 \in B_2 \setminus A_2 \\ \emptyset & \text{otherwise} \end{cases}$$

and

$$\int_{\Omega_2} P_1(A_{\omega_2})\, dP_2(\omega_2) = [P_1(A_1) + P_1(B_1)]P_2(A_2 \cap B_2)$$

$$+ P_1(A_1)P_2(A_2 \setminus B_2) + P_1(B_1)P_2(B_2 \setminus A_2)$$

$$= P_1(A_1)[P_2(A_2 \cap B_2) + P_2(A_2 \setminus B_2)]$$

$$+ P_1(B_1)[P_2(A_2 \cap B_2) + P_2(B_2 \setminus A_2)]$$

$$= P_1(A_1)P_2(A_2) + P_1(B_1)P_2(B_2).$$

On the other hand,

$$A_{\omega_1} = \begin{cases} A_2 & \text{if } \omega_1 \in A_1 \\ B_2 & \text{if } \omega_1 \in B_1 \\ \varnothing & \text{otherwise} \end{cases}$$

and

$$\int_{\Omega_1} P_2(A_{\omega_1})\, dP_1(\omega_1) = P_1(A_1)P_2(A_2) + P_1(B_1)P_2(B_2)$$

as before.

The general case of finitely many rectangles can be proved in the same way. This is easy but tedious and we skip this argument, hoping that presenting it in detail for two rectangles is sufficient to guide the reader in the general case. It remains true that the functions $\omega_2 \mapsto P_1(A_{\omega_2})$, $\omega_1 \mapsto P_2(A_{\omega_1})$ are simple functions, and so the verification of (6.3) is just simple algebra.

It remains to verify that $\mathcal{G}$ is a monotone class. Let $A_1 \subset A_2 \subset \cdots$ be sets from $\mathcal{G}$; hence the functions $\omega_2 \mapsto P_1((A_i)_{\omega_2})$, $\omega_1 \mapsto P_2((A_i)_{\omega_1})$ are measurable. They increase with i since the sections $(A_i)_{\omega_2}$, $(A_i)_{\omega_1}$ are increasing. If $i \to \infty$, then

$$P_1((A_i)_{\omega_2}) \to P_1(\bigcup_i (A_i)_{\omega_2}) = P_1((\bigcup_i A_i)_{\omega_2})$$

and so the function $\omega_2 \mapsto P_1((\bigcup_i A_i)_{\omega_2})$ is measurable. The same argument shows that the function $\omega_1 \mapsto P_2((\bigcup_i A_i)_{\omega_1})$ is measurable. The equality (6.3) holds for each i and by the monotone convergence theorem it is preserved in the limit. Thus (6.3) holds for unions $\bigcup A_i$.

For intersections the argument is similar. The sequences in question are decreasing; the functions $\omega_2 \mapsto P_1((\bigcap_i A_i)_{\omega_2})$, $\omega_1 \mapsto P_2((\bigcap_i A_i)_{\omega_1})$ are measurable as their limits and (6.3) holds by the monotone convergence theorem. □

Theorem 6.8

Suppose that P_1, P_2 are finite measures. The set function P given by (6.2) is countably additive. Any other measure coinciding with P on rectangles is equal to P on the product σ-field.

Proof

Let $A_i \in \mathcal{F}$ be pairwise disjoint. Then $(A_i)_{\omega_2}$ are also pairwise disjoint and

$$
\begin{aligned}
P(\bigcup A_i) &= \int_{\Omega_2} P_1((\bigcup_i A_i)_{\omega_2}) \, \mathrm{d}P_2(\omega_2) \\
&= \int_{\Omega_2} P_1(\bigcup_i (A_i)_{\omega_2}) \, \mathrm{d}P_2(\omega_2) \\
&= \int_{\Omega_2} \sum_i P_1((A_i)_{\omega_2}) \, \mathrm{d}P_2(\omega_2) \\
&= \sum_i \int_{\Omega_2} P_1((A_i)_{\omega_2}) \, \mathrm{d}P_2(\omega_2) \\
&= \sum_i P(A_i)
\end{aligned}
$$

where we have employed the fact that the section of the union is the union of the sections (see Figure 6.3) and the Beppo–Levi theorem.

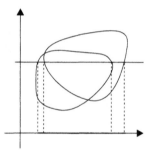

Figure 6.3 Section of a union

For uniqueness let Q be a measure defined on the product σ-field $\mathcal{F}$ such that $P(A_1 \times A_2) = Q(A_1 \times A_2)$, $A_1 \in \mathcal{F}_1$, $A_2 \in \mathcal{F}_2$. Let

$$
\mathcal{H} = \{A \subset \Omega_1 \times \Omega_2 : A \in \mathcal{F}, P(A) = Q(A)\}.
$$

This family contains all rectangles by the hypothesis. It contains unions of disjoint rectangles by the additivity of both P and Q; in other words $\mathcal{H}$ contains the field $\mathcal{A}$.

It remains to show that it is a monotone class since then it coincides with $\mathcal{F}$ by Lemma 6.7. This is quite straightforward, using again the fact that P and Q are measures. If $A_1 \subset A_2 \subset \cdots$, $A_i \in \mathcal{H}$, then $P(A_i) = Q(A_i)$, $P(\bigcup A_i) = \lim P(A_i)$, $Q(\bigcup A_i) = \lim Q(A_i)$, hence $P(\bigcup A_i) = Q(\bigcup A_i)$, which means that $\mathcal{H}$ is closed with respect to monotone unions. The argument for monotone intersections is exactly the same: if $A_1 \supset A_2 \supset \cdots$, then $P(\bigcap A_i) = \lim P(A_i)$, $Q(\bigcap A_i) = \lim Q(A_i)$, hence $P(A_i) = Q(A_i)$ implies $P(\bigcap A_i) = Q(\bigcap A_i)$. $\square$

Remark 6.9

The uniqueness part of the proof of Theorem 6.8 illustrates an important technique: in order to show that two measures on a σ-field coincide it suffices to prove that they coincide on the generating sets of that σ-field, by an application of the monotone class theorem.

As an immediate consequence of Theorem 6.5 we have

$$P(A) = \int_{\Omega_1} P_2(A_{\omega_1}) \, \mathrm{d}P_1(\omega_1).$$

The completion of the product σ-field $\mathcal{M} \times \mathcal{M}$ built from the σ-field of Lebesgue-measurable sets is the σ-field $\mathcal{M}_2$ on which m_2 is defined.

It easy to see that two-dimensional Lebesgue measure coincides with the completion of the product of one-dimensional ones. First, they agree on rectangles built from intervals. As a consequence, they agree on the σ-field generated by such rectangles, which is the Borel σ-field on the plane. The completion of Borel sets gives the σ-field of Lebesgue-measurable sets in the same way as in the one-dimensional case.

6.4 Fubini's theorem

We wish to integrate functions defined on the product of the spaces $(\Omega_1, \mathcal{F}_1, P_1)$, $(\Omega_2, \mathcal{F}_2, P_2)$ by exploiting the integration with respect to the measures P_1, P_2 individually.

We tackle the issue of measurability first.

Theorem 6.10

If a non-negative function $f : \Omega_1 \times \Omega_2 \to \mathbb{R}$ is measurable with respect to $\mathcal{F}_1 \times \mathcal{F}_2$, then for each $\omega_1 \in \Omega_1$ the function (which we shall call a *section* of f, see Figure 6.4) $\omega_2 \mapsto f(\omega_1, \omega_2)$ is $\mathcal{F}_2$-measurable, and for each $\omega_2 \in \Omega_2$ the section $\omega_1 \mapsto f(\omega_1, \omega_2)$ is $\mathcal{F}_1$-measurable.

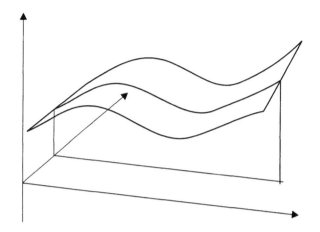

Figure 6.4 Section of f

Proof

First we approximate f by simple functions in similar fashion to Proposition 4.15; we write

$$f_n(\omega_1, \omega_2) = \begin{cases} \frac{k}{n} & \text{if } f(\omega_1, \omega_2) \in [\frac{k}{n}, \frac{k+1}{n}),\ k < n^2 \\ n & \text{if } f(\omega_1, \omega_2) > n \end{cases}$$

and as $n \to \infty$, $f_n \nearrow f$.

The sections of simple measurable functions are simple and measurable. This is clear for the indicator functions as observed above, and next we use the fact that the section of the sum is the sum of the sections.

Finally, it is clear that the sections of f_n converge to the sections of f and since measurability is preserved in the limit, the theorem is proved.   □

Corollary 6.11

The functions

$$\omega_1 \mapsto \int_{\Omega_2} f(\omega_1, \omega_2)\, dP_2(\omega_2), \quad \omega_2 \mapsto \int_{\Omega_1} f(\omega_1, \omega_2)\, dP_1(\omega_1)$$

are $\mathcal{F}_1$, $\mathcal{F}_2$-measurable, respectively.

Proof

The integrals may be taken (being possibly infinite) due to measurability of the functions in question. By the monotone convergence theorem, they are limits of the integrals of the sections of f_n. The integrals $\int_{\Omega_1} f_n(\omega_1, \omega_2) \, dP_1(\omega_1)$, $\int_{\Omega_2} f_n(\omega_1, \omega_2) \, dP_2(\omega_2)$ are simple functions, and hence the limits are measurable. $\qquad\square$

Theorem 6.12

Let f be a measurable non-negative function defined on $\Omega_1 \times \Omega_2$. Then

$$\int_{\Omega_1 \times \Omega_2} f(\omega_1, \omega_2) \, d(P_1 \times P_2)(\omega_1, \omega_2) = \int_{\Omega_1} \left(\int_{\Omega_2} f(\omega_1, \omega_2) \, dP_2(\omega_2) \right) dP_1(\omega_1)$$

$$= \int_{\Omega_2} \left(\int_{\Omega_1} f(\omega_1, \omega_2) \, dP_1(\omega_1) \right) dP_2(\omega_2). \tag{6.4}$$

Proof

For the indicator function of a rectangle $A_1 \times A_2$ each side of (6.4) just becomes $P_1(A_1)P_2(A_2)$. Then by additivity of the integral the formula is true for simple functions. Monotone approximation of any measurable f by simple functions allows us to extend this formula to the general case. $\qquad\square$

Theorem 6.13 (Fubini's theorem)

If $f \in L^1(\Omega_1 \times \Omega_2)$ then the sections are integrable in appropriate spaces, the functions

$$\omega_1 \mapsto \int_{\Omega_2} f(\omega_1, \omega_2) \, dP_2(\omega_2), \quad \omega_2 \mapsto \int_{\Omega_1} f(\omega_1, \omega_2) \, dP_1(\omega_1)$$

are in $L^1(\Omega_1)$, $L^1(\Omega_2)$, respectively, and (6.4) holds: in concise form it reads

$$\int_{\Omega_1 \times \Omega_2} f \, d(P_1 \times P_2) = \int_{\Omega_1} \left(\int_{\Omega_2} f \, dP_2 \right) dP_1 = \int_{\Omega_2} \left(\int_{\Omega_1} f \, dP_1 \right) dP_2.$$

Proof

This relation is immediate by the decomposition $f = f^+ - f^-$ and the result proved for non-negative functions. The integrals are finite since if $f \in L^1$ then $f^+, f^- \in L^1$ and all the integrals on the right are finite. $\qquad\square$

Remark 6.14

The whole procedure may be extended to the product of an arbitrary finite number of spaces. In particular, we have a method of constructing n-dimensional Lebesgue measure as the completion of the product of n copies of one-dimensional Lebesgue measure.

Example 6.15

Let $\Omega_1 = \Omega_2 = [0,1]$, $P_1 = P_2 = m_{[0,1]}$,

$$f(x,y) = \begin{cases} \frac{1}{x^2} & \text{if } 0 < y < x < 1 \\ 0 & \text{otherwise.} \end{cases}$$

We shall see that the integral of f over the square is infinite. For this we take a non-negative simple function dominated by f and compute its integral. Let $\varphi(x,y) = n$ if $f(x,y) \in [n, n+1)$. Then $\varphi(x,y) = n$ if $x > y$, $x \in (\frac{1}{\sqrt{n+1}}, \frac{1}{\sqrt{n}}]$. The area of this set is $\frac{1}{2}(\frac{1}{n} - \frac{1}{n+1})$ and

$$\int_{[0,1]^2} \varphi \, dm_2 = \sum_{n=1}^{\infty} n \frac{1}{2} (\frac{1}{n} - \frac{1}{n+1}) = \sum_{n=1}^{\infty} \frac{1}{2} \frac{1}{n+1} = \infty.$$

Hence the function

$$g(x,y) = \begin{cases} \frac{1}{x^2} & \text{if } 0 < y < x < 1 \\ -\frac{1}{y^2} & \text{if } 0 < x < y < 1 \\ 0 & \text{otherwise} \end{cases}$$

is not integrable since the integral of g^+ is infinite (the same is true for the integral of g^-).

Exercise 6.1

For g from the above example show that

$$\int_0^1 \int_0^1 g(x,y) \, dx \, dy = -1, \qquad \int_0^1 \int_0^1 g(x,y) \, dy \, dx = 1,$$

which shows that the iterated integrals may not be equal if Fubini's theorem condition is violated.

The following proposition opens the way for many applications of product measures and Fubini's theorem.

Proposition 6.16

Let $f : \mathbb{R} \to \mathbb{R}$ be measurable and positive. Consider the set of all points in the upper half-plane being below the graph of f:

$$A_f = \{(x, y) : 0 \le y < f(x)\}.$$

Show that A_f is m_2-measurable and $m_2(A_f) = \int f(x) \, \mathrm{d}x$.

Hint For measurability 'fill' A_f with rectangles using the approximation of f by simple functions. Then apply the definition of the product measure.

Exercise 6.2

Compute $\int_{[0,3] \times [-1,2]} x^2 y \, \mathrm{d}m_2$.

Exercise 6.3

Compute the area of the region inside the ellipse $\frac{x^2}{a^2} + \frac{y^2}{b^2} = 1$.

6.5 Probability

6.5.1 Joint distributions

Let X, Y be two random variables defined on the same probability space $(\Omega, \mathcal{F}, P)$. Consider the random vector

$$(X, Y) : \Omega \to \mathbb{R}^2.$$

Its distribution is the measure defined for the Borel sets on the plane given by

$$P_{(X,Y)}(B) = P((X, Y) \in B), \quad B \subset \mathbb{R}^2.$$

If this measure can be written as

$$P_{(X,Y)}(B) = \int_B f_{(X,Y)}(x, y) \, \mathrm{d}m_2(x, y)$$

for some integrable $f_{(X,Y)}$, then we say that X, Y have a *joint density*.

The joint distribution determines the distributions of one-dimensional random variables X, Y:

$$P_X(A) = P_{(X,Y)}(A \times \mathbb{R}),$$

$$P_Y(A) = P_{(X,Y)}(\mathbb{R} \times A),$$

for Borel $A \subset \mathbb{R}$; these are called *marginal distributions*. If X, Y have a joint density, then both X and Y are absolutely continuous with densities given by

$$f_X(x) = \int_{\mathbb{R}} f_{(X,Y)}(x,y) \, dy,$$

$$f_Y(y) = \int_{\mathbb{R}} f_{(X,Y)}(x,y) \, dx.$$

The following example shows that the converse is not true in general.

Example 6.17

Let $\Omega = [0,1]$ with $P = m_{[0,1]}$ and let $(X,Y)(\omega) = (\omega, \omega)$. This vector does not have density since $P_{(X,Y)}(\{(x,y) : x = y\}) = 1$ and for any integrable function $f : \mathbb{R}^2 \to \mathbb{R}$, $\int_{\{(x,y):x=y\}} f(x,y) \, dm_2(x,y) = 0$; a contradiction. However the marginal distributions P_X, P_Y are absolutely continuous with the densities $f_X = f_Y = \mathbf{1}_{[0,1]}$.

Example 6.18

A simple example of joint density is the uniform one: $f = \frac{1}{m(A)} \mathbf{1}_A$, with Borel $A \subset \mathbb{R}^2$. A particular case is $A = [0,1] \times [0,1]$, then clearly the marginal densities are $\mathbf{1}_{[0,1]}$.

Exercise 6.4

Take A to be the square with corners at $(0,1)$, $(1,0)$, $(2,1)$, $(1,2)$. Find the marginal densities of $f = \mathbf{1}_A$.

Exercise 6.5

Let $f_{X,Y}(x,y) = \frac{1}{50}(x^2 + y^2)$ if $0 < x < 2$, $1 < y < 4$ and zero otherwise. Find $P(X + Y > 4)$, $P(Y > X)$.

The two-dimensional standard Gaussian (normal) density is given by

$$n(x,y) = \frac{1}{2\pi\sqrt{1-\rho^2}} \exp\left\{ -\frac{1}{2(1-\rho^2)}(x^2 - 2\rho xy + y^2) \right\}. \qquad (6.5)$$

It can be shown that ρ is the correlation of X, Y, random variables whose densities are the marginal densities of $n(x,y)$ (see [10]).

Joint densities enable us to compute the distributions of various functions of random variables. Here is an important example.

Theorem 6.19

If X, Y have joint density $f_{X,Y}$, then the density of their sum is given by

$$f_{X+Y}(z) = \int_{\mathbb{R}} f_{X,Y}(x, z - x) \, dx. \tag{6.6}$$

Proof

We employ the distribution function:

$$
\begin{aligned}
F_{X+Y}(z) &= P(X + Y \le z) \\
&= P_{X,Y}(\{(x, y) : x + y \le z\}) \\
&= \int \int_{\{(x,y):x+y\le z\}} f_{X,Y}(x, y) \, dxdy \\
&= \int_{\mathbb{R}} \int_{-\infty}^{z-x} f_{X,Y}(x, y) \, dydx \\
&= \int_{-\infty}^{z} \int_{\mathbb{R}} f_{X,Y}(x, y' - x) \, dxdy'
\end{aligned}
$$

(we have used the substitution $y' = y + x$ and Fubini's theorem), which by differentiation gives the result. $\qquad \square$

Exercise 6.6

Find f_{X+Y} if $f_{X,Y} = \mathbf{1}_{[0,1]\times[0,1]}$.

6.5.2 Independence again

Suppose that the random variables X, Y are independent. Then for a Borel rectangle $B = B_1 \times B_2$ we have

$$
\begin{aligned}
P_{(X,Y)}(B_1 \times B_2) &= P((X, Y) \in B_1 \times B_2) \\
&= P((X \in B_1) \cap (Y \in B_2)) \\
&= P(X \in B_1)P(Y \in B_2) \\
&= P_X(B_1)P_Y(B_2)
\end{aligned}
$$

and so the distribution $P_{(X,Y)}$ coincides with the product measure $P_X \times P_Y$ on rectangles; therefore they are the same. The converse is also true:

Theorem 6.20

The random variables X, Y are independent if and only if

$$P_{(X,Y)} = P_X \times P_Y.$$

Proof

The 'only if' part was shown above. Suppose that $P_{(X,Y)} = P_X \times P_Y$ and take any Borel sets B_1, B_2. The same computation shows that $P((X \in B_1) \cap (Y \in B_2)) = P(X \in B_1)P(Y \in B_2)$, i.e. X and Y are independent. $\qquad\square$

We have a useful version of this theorem in the case of absolutely continuous random variables.

Theorem 6.21

If X, Y have a joint density, then they are independent if and only if

$$f_{(X,Y)}(x,y) = f_X(x)f_Y(y). \tag{6.7}$$

If X and Y are absolutely continuous and independent, then they have a joint density and it is given by (6.7).

Proof

Suppose $f_{(X,Y)}$ is the joint density of X, Y. If they are independent, then

$$\int_{B_1 \times B_2} f_{(X,Y)}(x,y)\,dm_2(x,y) = P((X,Y) \in B_1 \times B_2)$$
$$= P(X \in B_1)P(Y \in B_2)$$
$$= \int_{B_1} f_X(x)\,dm(x) \int_{B_2} f_Y(y)\,dm(y)$$
$$= \int_{B_1 \times B_2} f_X(x)f_Y(y)\,dm_2(x,y),$$

which implies (6.7). The same computation shows the converse:

$$P((X,Y) \in B_1 \times B_2) = \int_{B_1 \times B_2} f_{(X,Y)}(x,y)\,dm_2(x,y)$$
$$= \int_{B_1 \times B_2} f_X(x)f_Y(y)\,dm_2(x,y)$$
$$= P(X \in B_1)P(Y \in B_2).$$

For the final claim note that the function $f_X(x)f_Y(y)$ plays the role of the joint density if X and Y are independent. $\qquad\square$

Corollary 6.22

If Gaussian random variables are orthogonal, then they are independent.

Proof

Inserting $\rho = 0$ into (6.5), we immediately see that the two-dimensional Gaussian density is the product of the one-dimensional ones. $\qquad\square$

Proposition 6.23

The density of the sum of independent random variables with densities f_X, f_Y is given by

$$f_{X+Y}(z) = \int_{\mathbb{R}} f_X(x)f_Y(z-x)\,\mathrm{d}x.$$

Exercise 6.7

Suppose that the joint density of X, Y is $\mathbf{1}_A$ where A is the square with corners at $(0,1)$, $(1,0)$, $(2,1)$, $(1,2)$. Are X, Y independent?

Exercise 6.8

Find $P(Y > X)$ and $P(X + Y > 1)$, if X, Y are independent with $f_X = \mathbf{1}_{[0,1]}$, $f_Y = \frac{1}{2}\mathbf{1}_{[0,2]}$.

6.5.3 Conditional probability

We consider the case of two random variables X, Y with joint density $f_{X,Y}(x,y)$. Given Borel sets A, B, we compute

$$
\begin{aligned}
P(Y \in B | X \in A) &= \frac{P(X \in A, Y \in B)}{P(X \in A)} \\
&= \frac{\int_{A \times B} f_{(X,Y)}(x,y)\,\mathrm{d}m_2(x,y)}{\int_A f_X(x)\,\mathrm{d}m(x)} \\
&= \int_B \frac{\int_A f_{(X,Y)}(x,y)\,\mathrm{d}x}{\int_A f_X(x)\,\mathrm{d}x}\,\mathrm{d}y
\end{aligned}
$$

using Fubini's theorem. So the conditional distribution of Y given $X \in A$ has a density

$$h(y|X \in A) = \frac{\int_A f_{(X,Y)}(x,y)\,\mathrm{d}x}{\int_A f_X(x)\,\mathrm{d}x}.$$

The case where $A = \{a\}$ does not make sense here since then we would have zero in the denominator. However, formally we may put

$$h(y|X = a) = \frac{f_{(X,Y)}(a,y)}{f_X(a)},$$

which makes sense if only $f_X(a) \neq 0$. This restriction turns out to be not relevant since

$$
\begin{aligned}
P((X,Y) \in \{(x,y) : f_X(x) = 0\}) &= \int_{\{(x,y):f_X(x)=0\}} f_{(X,Y)}(x,y)\,\mathrm{d}x\,\mathrm{d}y \\
&= \int_{\{x:f_X(x)=0\}} \int_{\mathbb{R}} f_{(X,Y)}(x,y)\,\mathrm{d}y\,\mathrm{d}x \\
&= \int_{\{x:f_X(x)=0\}} f_X(x)\,\mathrm{d}x \\
&= 0.
\end{aligned}
$$

We may thus define the conditional probability of $Y \in B$ given $X = a$ by means of $h(y|X = a)$, which we briefly write as $h(y|a)$:

$$P(Y \in B|X = a) = \int_B h(y|a)\,\mathrm{d}y$$

and the conditional expectation

$$\mathbb{E}(Y|X = a) = \int_{\mathbb{R}} y h(y|a)\,\mathrm{d}y.$$

This can be viewed as a random variable with X as the source of randomness. Namely, for $\omega \in \Omega$ we write

$$\mathbb{E}(Y|X)(\omega) = \int_{\mathbb{R}} y h(y|X(\omega))\,\mathrm{d}y.$$

This function is of course measurable with respect to the σ-field generated by X.

The expectation of this random variable can be computed using Fubini's theorem:

$$\mathbb{E}(\mathbb{E}(Y|X)) = \mathbb{E}(\int_{\mathbb{R}} y h(y|X(\omega)) \, dy)$$
$$= \int_{\mathbb{R}} \int_{\mathbb{R}} y h(y|x) \, dy f_X(x) \, dx$$
$$= \int_{\mathbb{R}} \int_{\mathbb{R}} y f_{(X,Y)}(x, y) \, dx \, dy$$
$$= \int_{\mathbb{R}} y f_Y(y) \, dy$$
$$= \mathbb{E}(Y).$$

More generally, for $A \subset \Omega$, $A = X^{-1}(B)$, B Borel,

$$\int_A \mathbb{E}(Y|X) \, dP = \int_\Omega \mathbf{1}_B(X) \mathbb{E}(Y|X) \, dP$$
$$= \int_\Omega \mathbf{1}_B(X(\omega))(\int_{\mathbb{R}} y h(y|X(\omega)) \, dy) \, dP(\omega)$$
$$= \int_{\mathbb{R}} \int_{\mathbb{R}} \mathbf{1}_B(x) y h(y|x) \, dy f_X(x) \, dx$$
$$= \int_{\mathbb{R}} \int_{\mathbb{R}} \mathbf{1}_B(x) y f_{(X,Y)}(x, y) \, dx \, dy$$
$$= \int_\Omega \mathbf{1}_A(X) Y \, dP$$
$$= \int_A Y \, dP.$$

This provides a motivation for a general notion of conditional expectation of a random variable Y, given random variable X: $\mathbb{E}(Y|X)$ is a random variable measurable with respect to the σ-field $\mathcal{F}_X$ generated by X and such that for all $A \in \mathcal{F}_X$

$$\int_A \mathbb{E}(Y|X) \, dP = \int_A Y \, dP.$$

We will pursue these ideas further in the next chapter. Recall that we have already outlined a general construction in Section 5.4.3, using the existence of orthogonal projections in L^2.

Exercise 6.9

Let $f_{X,Y} = \mathbf{1}_A$, where A is the triangle with corners at $(0,0)$, $(2,0)$, $(0,1)$. Find the conditional density $h(y|x)$ and conditional expectation $\mathbb{E}(Y|X = 1)$.

Exercise 6.10

Let $f_{X,Y}(x,y) = (x+y)\mathbf{1}_A$, where $A = [0,1] \times [0,1]$. Find $\mathbb{E}(X|Y = y)$ for each $y \in \mathbb{R}$.

6.5.4 Characteristic functions determine distributions

We have now sufficient tools to prove a fundamental property of characteristic functions.

Theorem 6.24 (inversion formula)

If the cumulative distribution function of a random variable X is continuous at $a, b \in \mathbb{R}$, then

$$F_X(b) - F_X(a) = \lim_{c \to \infty} \frac{1}{2\pi} \int_{-c}^{c} \frac{e^{-iua} - e^{-iub}}{iu} \varphi_X(u)\, du.$$

Proof

First, by the definition of φ_X,

$$\frac{1}{2\pi} \int_{-c}^{c} \frac{e^{-iua} - e^{-iub}}{iu} \varphi_X(u)\, du = \frac{1}{2\pi} \int_{-c}^{c} \frac{e^{-iua} - e^{-iub}}{iu} \int_{\mathbb{R}} e^{iux}\, dP_X(x)\, du.$$

We may apply Fubini's theorem, since

$$\left| \frac{e^{-iua} - e^{-iub}}{iu} e^{iux} \right| = \left| \int_a^b e^{iux}\, d(x) \right| \le b - a,$$

which is integrable with respect to $P_X \times m|_{[-c,c]}$. We compute the integral in u:

$$\frac{1}{2\pi} \int_{-c}^{c} \frac{e^{-iua} - e^{-iub}}{iu} e^{iux}\, du = \frac{1}{2\pi} \int_{-c}^{c} \frac{e^{iu(x-a)} - e^{iu(x-b)}}{iu}\, du =$$

$$\frac{1}{2\pi} \int_{-c}^{c} \frac{\sin u(x - a) - \sin u(x - b)}{u}\, du + \frac{1}{2\pi} \int_{-c}^{c} \frac{\cos u(x - a) - \cos u(x - b)}{iu}\, du.$$

The second integral vanishes since the integrand is an odd function. We change variables in the first: $y = u(x - a)$, $z = u(x - b)$ and then it takes the form

$$I(x,c) = \frac{1}{2\pi} \int_{-c(x-a)}^{c(x-a)} \frac{\sin y}{y}\, dy - \frac{1}{2\pi} \int_{-c(x-b)}^{c(x-b)} \frac{\sin z}{z}\, dz = I_1(x,c) - I_2(x,c),$$

say. We employ the following elementary fact without proof:

$$\int_s^t \frac{\sin y}{y}\,\mathrm{d}y \to \pi \quad \text{as } t \to \infty, s \to -\infty.$$

Consider the following cases:

1. $x < a$, then also $x < b$ and $c(x-a) \to -\infty$, $c(x-b) \to -\infty$, $-c(x-a) \to \infty$, $-c(x-b) \to \infty$ as $c \to \infty$. Hence $I_1(x,c) \to -\frac{1}{2}$, $I_2(x,c) \to -\frac{1}{2}$ and so $I(x,c) \to 0$.

2. $x > b$, then also $x > a$, and $c(x-a) \to \infty$, $c(x-b) \to \infty$, $-c(x-a) \to -\infty$, $-c(x-b) \to -\infty$, as $c \to \infty$, so $I_1(x,c) \to \frac{1}{2}$, $I_2(x,c) \to \frac{1}{2}$ and the result is the same as in 1.

3. $a < x < b$, hence $I_1(x,c) \to \frac{1}{2}$, $I_2(x,c) \to -\frac{1}{2}$ and the limit of the whole expression is 1.

Write $f(x) = \lim_{c\to\infty} I(x,c)$ (we have not discussed the values $x = a$, $x = b$, but they are irrelevant, as will be seen).

$$\lim_{c\to\infty} \frac{1}{2\pi} \int_{-c}^{c} \frac{e^{-iua} - e^{-iub}}{iu} \varphi_X(u)\,\mathrm{d}u = \lim_{c\to\infty} \int_{\mathbb{R}} I(x,c)\,\mathrm{d}P_X(x)$$

$$= \int_{\mathbb{R}} f(x)\,\mathrm{d}P_X(x)$$

by Lebesgue's dominated convergence theorem. The integral of f can be easily computed since f is a simple function:

$$\int_{\mathbb{R}} f(x)\,\mathrm{d}P_X(x) = P_X((a,b]) = F_X(b) - F_X(a)$$

$(P_X(\{a\}) = P_X(\{b\}) = 0$ since F_X is continuous at a and b). $\qquad\square$

Corollary 6.25

The characteristic function determines the probability distribution.

Proof

Since F_X is monotone, it is continuous except (possibly) at countably many points where it is right-continuous. Its values at discontinuity points can be approximated from above by the values at continuity points. The latter are determined by the characteristic function via the inversion formula.

Finally, we see that F_X determines the measure P_X. This is certainly so for $B = (a,b]$: $P_X((a,b]) = F_X(b) - F_X(a)$. Next we show the same for any interval, then for finite unions of intervals, and the final extension to any Borel set is via the monotone class theorem.   □

Theorem 6.26

If φ_X is integrable, then X has a density which is given by

$$f_X(x) = \frac{1}{2\pi} \int_{-\infty}^{\infty} e^{-iux} \varphi_X(u)\, du.$$

Proof

The function f is well defined. To show that it is a density of X we first show that it gives the right values of the probability distribution of intervals $(a, b]$ where F_X is continuous:

$$
\begin{aligned}
\int_a^b f_X(x)\, dx &= \frac{1}{2\pi} \int_{-\infty}^{\infty} \varphi_X(u) \left(\int_a^b e^{-iux}\, dx \right) du \\
&= \lim_{c \to \infty} \frac{1}{2\pi} \int_{-c}^{c} \varphi_X(u) \left(\int_a^b e^{-iux}\, dx \right) du \\
&= \lim_{c \to \infty} \frac{1}{2\pi} \int_{-c}^{c} \varphi_X(u) \frac{e^{-iua} - e^{-iub}}{iu}\, du \\
&= F_X(b) - F_X(a)
\end{aligned}
$$

by the inversion formula. This extends to all a, b since F_X is right-continuous and the integral on the left is continuous with respect to a and b. Moreover, F_X is non-decreasing, so $\int_a^b f_X(x)\, dx \geq 0$ for all $a \leq b$, hence $f_X \geq 0$. Finally

$$\int_{-\infty}^{\infty} f_X(x)\, dx = \lim_{b \to \infty} F_X(b) - \lim_{a \to -\infty} F_X(a) = 1,$$

so f_X is a density.   □

6.5.5 Application to mathematical finance

Classical portfolio theory is concerned with an analysis of the balance between risk and return. This balance is of fundamental importance, particularly in corporate finance, where the key concept is the cost of capital, which is a rate of return based on the level of risk of an investment. In probabilistic terms, return is represented by the expectation and risk by the variance. A theory which deals only with two moments of a random variable is relevant if we assume the normal (Gaussian) distribution of random variables in question, since in that case these two moments determine the distribution uniquely. We give a brief account of basic facts of portfolio theory under this assumption.

Let k be a return on some investment in single period, that is, $k(\omega) = \frac{V(1,\omega)-V(0)}{V(0)}$ where $V(0)$ is the known amount invested at the beginning, and $V(1)$ is the random terminal value. A typical example which should be kept in mind is buying and selling one share of some stock. With a number of stocks available, we are facing a sequence (k_i) of returns on stock S_i, $k_i = \frac{S_i(1,\omega)-S_i(0)}{S_i(0)}$, but for simplicity we restrict our attention to just two, k_1, k_2. A portfolio is formed by deciding the percentage split, between holdings in S_1 and S_2, of the initial wealth $V(0)$ by choosing the weights $\mathbf{w} = (w_1, w_2)$, $w_1 + w_2 = 1$. Then, as is well known and elementary to verify, the portfolio of $n_1 = \frac{w_1 V(0)}{S_1(0)}$ shares of stock number one and $n_2 = \frac{w_2 V(0)}{S_2(0)}$ shares of stock number two has return

$$k_\mathbf{w} = w_1 k_1 + w_2 k_2.$$

We assume that the vector (k_1, k_2) is jointly normal with correlation coefficient ρ. We denote the expectations and variances of the ingredients by $\mu_i = \mathbb{E}(k_i)$, $\sigma_i^2 = \mathrm{Var}(k_i)$. It is convenient to introduce the following matrix:

$$C = \begin{bmatrix} c_{11} & c_{12} \\ c_{21} & c_{22} \end{bmatrix} = \begin{bmatrix} \sigma_1^2 & \rho\sigma_1\sigma_2 \\ \rho\sigma_1\sigma_2 & \sigma_2^2 \end{bmatrix}$$

where $c_{12} = c_{21}$ is the covariance between k_1 and k_2. Assume (which is not elegant to do but saves us an algebraic detour) that C is invertible, with $C^{-1} = [d_{ij}]$. By definition, the joint density has the form

$$f(x_1, x_2) = \frac{1}{2\pi\sqrt{\det C}} \exp\{-\frac{1}{2}\sum_{i,j=1}^{2} d_{ij}(x_i - \mu_i)(x_j - \mu_j)\}$$

It is easy to see that (6.5) is a particular case of this formula with $\mu_i = 0$, $\sigma_i = 1$, $-1 < \rho < 1$. It is well known that the characteristic function $\varphi(t_1, t_2) = \mathbb{E}(\exp\{i(t_1 k_1 + t_2 k_2)\})$ of the vector (k_1, k_2) is of the form

$$\varphi(t_1, t_2) = \exp\{i\sum_{i=1}^{2} t_i\mu_i - \frac{1}{2}\sum_{i,j=1}^{2} c_{ij}t_it_j\}. \tag{6.8}$$

We shall show that the return on the portfolio is also normally distributed and we shall find the expectation and standard deviation. This can all be done in one step.

Theorem 6.27

The characteristic $\varphi_{\mathbf{w}}$ function of $k_{\mathbf{w}}$ is of the form

$$\varphi_{\mathbf{w}}(t) = \exp\{it(w_1\mu_1 + w_2\mu_2)) - \frac{1}{2}t^2(w_1^2\sigma_1^2 + w_2^2\sigma_2^2 + 2w_1w_2\rho\sigma_1\sigma_2)\}.$$

Proof

By definition $\varphi_{\mathbf{w}}(t) = \mathbb{E}(\exp\{itk_{\mathbf{w}}\})$, and using the form of $k_{\mathbf{w}}$ we have

$$\begin{aligned}
\varphi_{\mathbf{w}}(t) &= \mathbb{E}(\exp\{it(w_1k_1 + w_2k_2)\} \\
&= \mathbb{E}(\exp\{itw_1k_1 + itw_2k_2\}) \\
&= \varphi(tw_1, tw_2)
\end{aligned}$$

by the definition of the characteristic function of a vector. Since the vector is normal, (6.8) immediately gives the result. $\qquad\square$

The multi-dimensional version of Corollary 6.25 (which is easy to believe after mastering the one-dimensional case, but slightly tedious to prove, so we take it for granted, referring the reader to any probability textbook) shows that $k_{\mathbf{w}}$ has normal distribution with

$$\begin{aligned}
\mu_{\mathbf{w}} &= w_1\mu_1 + w_2\mu_2 \\
\sigma_{\mathbf{w}}^2 &= w_1^2\sigma_1^2 + w_2^2\sigma_2^2 + 2w_1w_2\rho\sigma_1\sigma_2.
\end{aligned}$$

The fact that the variance of a portfolio can be lower than the variances of the components is crucial. These formulae are valid in the general case (i.e. without the assumption of a normal distribution) and can be easily proved using the formula for $k_{\mathbf{w}}$. The main goal of this section was to see that the portfolio return is normally distributed.

Example 6.28

Suppose that the second component is not random, i.e. $S_2(1)$ is a constant independent of ω. Then the return k_2 is risk-free and it is denoted by r (the notation is usually reserved for the case where the length of the period is one year). It can be thought of as a bank account and it is convenient to assume that $S_2(0) = 1$. Then the portfolio of n shares purchased at the price $S_1(0)$ and m units of the bank account has the value $V(1) = nS_1(1) + m(1+r)$ at the end of the period and the expected return is $k_{\mathbf{w}} = w_1\mu_1 + w_2r$, $w_1 = \frac{nS_1(0)}{V(0)}$, $w_2 = 1 - w_1$. The assumption of normal joint returns is violated but the standard deviation of this portfolio can be easily computed directly from the definition, giving $\sigma_{\mathbf{w}} = w_1\sigma_1$ ($\sigma_2 = 0$ of course and the formula is consistent with the above).

Remark 6.29

The above considerations can be immediately generalised to portfolios built of any finite number of ingredients with the following key formulae:

$$k_{\mathbf{w}} = \sum w_i k_i,$$

$$\mu_{\mathbf{w}} = \sum w_i \mu_i,$$

$$\sigma_{\mathbf{w}}^2 = \sum_{i,j} w_i w_j c_{ij}.$$

This is just the beginning of the story started in the 1950s by Nobel prize winner Harry Markowitz. A vast number of papers and books on this topic have been written since, proving the general observation that 'simple is beautiful'.

6.6 Proofs of propositions

Proof (of Proposition 6.3)

Denote by $\mathcal{F}_{\mathcal{R}}$ the σ-field generated by the Borel 'rectangles' $\mathcal{R} = \{B_1 \times B_2 : B_1, B_2 \in \mathcal{B}\}$, and by $\mathcal{F}_{\mathcal{I}}$ the σ-field generated by the true rectangles $\mathcal{I} = \{I_1 \times I_2 : I_1, I_2 \text{ are intervals}\}$.

Since $\mathcal{I} \subset \mathcal{R}$, obviously $\mathcal{F}_{\mathcal{I}} \subset \mathcal{F}_{\mathcal{R}}$.

To show the inverse inclusion we show that Borel cylinders $B_1 \times \Omega_2$ and $\Omega_1 \times B_2$ are in $\mathcal{F}_{\mathcal{I}}$. For that write $\mathcal{D} = \{A : A \times \Omega_2 \in \mathcal{F}_{\mathcal{I}}\}$; note that this is a σ-field containing all intervals, hence $\mathcal{B} \subset \mathcal{D}$ as required. $\square$

Proof (of Proposition 6.16)

Let $s_n = \sum c_k \mathbf{1}_{A_k}$ be an increasing sequence of simple functions convergent to f. Let $R_k = A_k \times [0, c_k]$ and the union of such rectangles is in fact $\int s_n \mathrm{d}m$. Then $\bigcup_{n=1}^{\infty} \bigcup_k R_k = A_f$, so A_f is measurable.

For the second claim take a y section of A_f which is the interval $[0, f(x))$. Its measure is $f(x)$ and by the definition of the product measure $m_2(A_f) = \int f(x)\, \mathrm{d}x$. $\square$

Proof (of Proposition 6.23)

The joint density is the product of the densities: $f_{X,Y}(x,y) = f_X(x)f_Y(y)$ and substituting this to (6.6) immediately gives the result. $\qquad\square$

The Radon–Nikodym theorem

In this chapter we shall consider the relationship between a real Borel measure ν and the Lebesgue measure m. Key to such relationships is Theorem 4.24, which shows that for each non-negative integrable real function f, the set function

$$A \mapsto \nu(A) = \int_A f \, dm \qquad (7.1)$$

defines a (Borel) measure ν on $(\mathbb{R}, \mathcal{M})$. The natural question to ask is the converse: exactly which real Borel measures can be found in this way? We shall find a complete answer to this question in this chapter, and in keeping with our approach in Chapters 5 and 6, we shall phrase our results in terms of general measures on an abstract set Ω.

7.1 Densities and conditioning

The results we shall develop in this chapter also allow us to study probability densities (introduced in Section 4.7.2), conditional expectations and conditional probabilities (see Sections 5.4.3 and 6.5.3) in much greater detail. For ν as defined above to be a probability measure, we clearly require $\int f \, dm = 1$. In particular, if $\nu = P_X$ is the distribution of a random variable X, the function $f = f_X$ corresponding to ν in (7.1) was called the *density* of X.

In similar fashion we defined the joint density $f_{(X,Y)}$ of two random variables in Section 6.5.1, by reference of their joint distribution to two-dimensional

Lebesgue measure m_2: if X and Y are real random variables defined on some probability space $(\Omega, \mathcal{F}, P)$ their joint distribution is the measure defined on Borel subsets B of $\mathbb{R}^2$ by $P_{(X,Y)}(B) = P((X,Y) \in B)$. In the special case where this measure, relative to m_2, is given as above by an integrable function $f_{(X,Y)}$, we say that X and Y have this function as their joint density.

This, in turn, leads naturally (see Section 6.5.3) to the concepts of conditional density

$$h(y|a) = h(y|X = a) = \frac{f_{(X,Y)}(a,y)}{f_X(a)}$$

and conditional expectation

$$\mathbb{E}(Y|X = a) = \int_{\mathbb{R}} y h(y|a) \, dy.$$

Recalling that $X : \Omega \to \mathbb{R}$, the last equation can be written as $\mathbb{E}(Y|X)(\omega) = \int_{\mathbb{R}} y h(y|X(\omega)) \, dy$, displaying the conditional expectation as a random variable $\mathbb{E}(Y|X) : \Omega \to \mathbb{R}$, measurable with respect to the σ-field $\mathcal{F}_X$ generated by X. An application of Fubini's theorem leads to a fundamental identity, valid for all $A \in \mathcal{F}_X$

$$\int_A \mathbb{E}(Y|X) \, dP = \int_A Y \, dP. \tag{7.2}$$

The existence of this random variable in the general case, irrespective of the existence of a joint density, is of great importance in both theory and applications – Williams [13] calls it 'the central definition of modern probability'. It is essential for the concept of martingale, which plays such a crucial role in many applications, and which we introduce at the end of this chapter.

As we described in Section 5.4.3, the existence of orthogonal projections in L^2 allows one to extend the scope of the definition further still: instead of restricting ourselves to random variables measurable with respect to σ-fields of the form $\mathcal{F}_X$ we specify any sub-σ-field $\mathcal{G}$ of $\mathcal{F}$ and ask for a $\mathcal{G}$-measurable random variable $\mathbb{E}(Y|\mathcal{G})$ to play the role of $\mathbb{E}(Y|X) = \mathbb{E}(Y|\mathcal{F}_X)$ in (7.2). As was the case for product measures, the most natural context for establishing the properties of the conditional expectation is that of general measures; note that the proof of Theorem 4.24 simply required monotone convergence to establish the countable additivity of P. We therefore develop the comparison of abstract measures further, as always guided by the specific examples of random variables and distributions.

7.2 The Radon–Nikodym theorem

In the special case where the measure ν has the form $\nu(A) = \int_A f \, dm$ for some non-negative integrable function f we said (Section 4.7.2) that ν is absolutely continuous with respect to m. It is immediate that $\int_A f \, dm = 0$ whenever $m(A) = 0$ (see Theorem 4.7 (iv)). Hence $m(A) = 0$ implies $\nu(A) = 0$ when the measure ν is given by a density. We use this as a definition for the general case of two given measures.

Definition 7.1

Let Ω be a set and let $\mathcal{F}$ be a σ-field of its subsets. (The pair $(\Omega, \mathcal{F})$ is a *measurable space*.) Suppose that ν and μ are measures on $(\Omega, \mathcal{F})$. We say that ν is *absolutely continuous with respect to* μ if $\mu(A) = 0$ implies $\nu(A) = 0$ for $A \in \mathcal{F}$. We write this as $\nu \ll \mu$.

Exercise 7.1

Let λ_1, λ_2 and μ be measures on $(\Omega, \mathcal{F})$. Show that if $\lambda_1 \ll \mu$ and $\lambda_2 \ll \mu$ then $(\lambda_1 + \lambda_2) \ll \mu$.

It will not be immediately obvious what this definition has to do with the usual notion of continuity of functions. We shall see later in this chapter how it fits with the concept of absolute continuity of real functions. For the present, we note the following reformulation of the definition, which is not needed for the main result we will prove, but serves to bring the relationship between ν and μ a little 'closer to home' and is useful in many applications:

Proposition 7.2

Let ν and μ be finite measures on the measurable space $(\Omega, \mathcal{F})$. Then $\nu \ll \mu$ if and only if for every $\varepsilon > 0$ there exists a $\delta > 0$ such that for $F \in \mathcal{F}$, $\mu(F) < \delta$ implies $\nu(F) < \varepsilon$.

Hint Suppose the (ε, δ)-condition fails. We can then find $\varepsilon > 0$ and sets (F_n) such that for all $n \geq 1$, $\mu(F_n) < \frac{1}{2^n}$ but $\nu(F_n) > \varepsilon$. Consider $\mu(A)$ and $\nu(A)$ for $A = \bigcap_{n \geq 1}(\bigcup_{i \geq n} F_i)$.

We generalise from the special case of Lebesgue measure: if μ is any measure on $(\Omega, \mathcal{F})$ and $f : \Omega \to \mathbb{R}$ is a measurable function for which $\int f \, d\mu$ exists, then $\nu(F) = \int_F f \, d\mu$ defines a measure $\nu \ll \mu$. (This follows exactly as for m, since $\mu(F) = 0$ implies $\int_F f \, d\mu = 0$. Note that we employ the convention $0 \times \infty = 0$.)

For σ-finite measures, the following key result asserts the converse:

Theorem 7.3 (Radon–Nikodym)

Given two σ-finite measures ν, μ on a measurable space $(\Omega, \mathcal{F})$, with $\nu \ll \mu$, then there is a non-negative measurable function $h : \Omega \to \mathbb{R}$ such that $\nu(F) = \int_F h \, d\mu$ for every $F \in \mathcal{F}$. The function h is unique up to μ-null sets: if g also satisfies $\nu(F) = \int_F g \, d\mu$ for all $F \in \mathcal{F}$, then $g = h$ a.e. (μ).

Since the most interesting case for applications arises for probability spaces and then $h \in L^1(\mu)$, we shall initially restrict attention to the case where μ and ν are finite measures. In fact, it is helpful initially to take μ to be a probability measure, i.e. $\mu(\Omega) = 1$. From among several different approaches to this very important theorem, we base our argument on one given by R.C. Bradley in the *American Mathematical Monthly* (Vol. 96, no 5, May 1989, pp. 437–440), since it offers the most 'constructive' and elementary treatment of which we are aware.

It is instructive to begin with a special case: Suppose (until further notice) that $\mu(\Omega) = 1$. We say that the measure μ *dominates* ν when $0 \leq \nu(F) \leq \mu(F)$ for every $F \in \mathcal{F}$. This obviously implies $\nu \ll \mu$. In this simplified situation we shall construct the required function h explicitly. First we generalise the idea of partitions and their refinements, which we used to good effect in constructing the Riemann integral, to measurable subsets in $(\Omega, \mathcal{F})$.

Definition 7.4

Let $(\Omega, \mathcal{F})$ be a measurable space. A finite (measurable) *partition* of Ω is a finite collection of disjoint subsets $\mathcal{P} = (A_i)_{i \leq n}$ in $\mathcal{F}$ whose union is Ω. The finite partition $\mathcal{P}'$ is a *refinement* of $\mathcal{P}$ if each set in $\mathcal{P}$ is a disjoint union of sets in $\mathcal{P}'$.

Exercise 7.2

Let $\mathcal{P}_1$ and $\mathcal{P}_2$ be finite partitions of Ω. Show that the coarsest partition (i.e. with least number of sets) which refines them both consists of all intersections $A \cap B$, where $A \in \mathcal{P}_1$, $B \in \mathcal{P}_2$.

The following is a simplified 'Radon–Nikodym theorem' for dominated measures:

Theorem 7.5

Suppose that $\mu(\Omega) = 1$ and $0 \leq \nu(F) \leq \mu(F)$ for every $F \in \mathcal{F}$. Then there exists a non-negative $\mathcal{F}$-measurable function h on Ω such that $\nu(F) = \int_F h \, d\mu$ for all $F \in \mathcal{F}$.

We shall prove this in three steps: in Step 1 we define the required function $h_{\mathcal{P}}$ for sets in a finite partition $\mathcal{P}$ and compare the functions $h_{\mathcal{P}_1}$ and $h_{\mathcal{P}_2}$ when the partition $\mathcal{P}_2$ refines $\mathcal{P}_1$. This enables us to show that the integrals $\int_\Omega h_{\mathcal{P}}^2 \, d\mu$ are non-decreasing if we take successive refinements. Since they are also bounded above (by 1), $c = \sup \int_\Omega h_{\mathcal{P}}^2 \, d\mu$ exists in $\mathbb{R}$. In Step 2 we then construct the desired function h by a careful limit argument, using the convergence theorems of Chapter 4. In Step 3 we show that h has the desired properties.

Step 1. The function $h_{\mathcal{P}}$ for a finite partition.

Suppose that $0 \leq \nu(F) \leq \mu(F)$ for every $F \in \mathcal{F}$. Let $\mathcal{P} = \{A_1, A_2, \dots, A_k\}$ be a finite partition of Ω such that each $A_i \in \mathcal{F}$. Define the simple function $h_{\mathcal{P}} : \Omega \to \mathbb{R}$ by setting

$$h_{\mathcal{P}}(\omega) = c_i = \frac{\nu(A_i)}{\mu(A_i)} \text{ for } \omega \in A_i \text{ when } \mu(A_i) > 0, \text{ and } h_{\mathcal{P}}(\omega) = 0 \text{ otherwise.}$$

Since $h_{\mathcal{P}}$ is constant on each 'atom' A_i, $\nu(A_i) = \int_{A_i} h_{\mathcal{P}} \, d\mu$. Then $h_{\mathcal{P}}$ has the following properties:

(i) For each finite partition $\mathcal{P}$ of Ω, $0 \leq h_{\mathcal{P}}(\omega) \leq 1$ for all $\omega \in \Omega$.

(ii) If $A = \bigcup_{j \in J} A_j$ for an index set $J \subset \{1, 2, \dots, k\}$ then $\nu(A) = \int_F h_{\mathcal{P}} \, d\mu$. Thus $\nu(\Omega) = \int_\Omega h_{\mathcal{P}} \, d\mu$.

(iii) If $\mathcal{P}_1$ and $\mathcal{P}_2$ are finite partitions of Ω and $\mathcal{P}_2$ refines $\mathcal{P}_1$ then, with $h_n = h_{\mathcal{P}_n}$, $(n = 1, 2)$ we have

    (a) for all $A \in \mathcal{P}_1$, $\int_A h_1 \, d\mu = \nu(A) = \int_A h_2 \, d\mu$,

    (b) for all $A \in \mathcal{P}_1$, $\int_A h_1 h_2 \, d\mu = \int_A h_1^2 \, d\mu$.

(iv) $\int_\Omega (h_2^2 - h_1^2) \, d\mu = \int_\Omega (h_2 - h_1)^2 \, d\mu$ and therefore

$$\int_\Omega h_2^2 \, d\mu = \int_\Omega h_1^2 \, d\mu + \int_\Omega (h_2 - h_1)^2 \, d\mu \geq \int_\Omega h_1^2 \, d\mu.$$

We now prove these assertions in turn.

(i) This is trivial by construction of $h_{\mathcal{P}}$, since μ dominates ν.

(ii) Let $A = \bigcup_{j \in J} A_j$ for some index set $J \subset \{1, 2, \ldots, k\}$. Since the $\{A_j\}$ are disjoint and $\nu(A_j) = 0$ whenever $\mu(A_j) = 0$, we have

$$\nu(A) = \sum_{j \in J} \nu(A_j) = \sum_{j \in J, \mu(A_i) > 0} \frac{\nu(A_j)}{\mu(A_j)} \mu(A_j)$$

$$= \sum_{j \in J, \mu(A_i) > 0} c_j \mu(A_j) = \sum_{j \in J} \int_{A_j} h_{\mathcal{P}} \, d\mu$$

$$= \int_A h_{\mathcal{P}} \, d\mu.$$

In particular, since $\mathcal{P}$ partitions Ω, this holds for $A = \Omega$.

(iii) (a) With the $\mathcal{P}_n, h_n$ as above ($n = 1, 2$) we can write $A = \bigcup_{j \in J} B_j$ for each $A \in \mathcal{P}_1$, where J is a finite index set and $B_j \in \mathcal{P}_2$. The sets B_j are pairwise disjoint, and again $\nu(B_j) = 0$ when $\mu(B_j) = 0$, so that

$$\int_A h_1 \, d\mu = \nu(A) = \sum_{j \in J} \nu(B_j) = \sum_{j \in J, \mu(B_j) > 0} \frac{\nu(B_j)}{\mu(B_j)} \mu(B_j)$$

$$= \sum_{j \in J} \int_{B_j} h_2 \, d\mu = \int_A h_2 \, d\mu.$$

(b) With A as in part (a) and $\mu(A) > 0$, note that $h_1 = \frac{\nu(A)}{\mu(A)}$ is constant on A, so that

$$\int_A h_1 h_2 \, d\mu = \frac{\nu(A)}{\mu(A)} \int_A h_2 \, d\mu = \frac{(\nu(A))^2}{\mu(A)} = \int_A \left(\frac{\nu(A)}{\mu(A)}\right)^2 d\mu = \int_A h_1^2 \, d\mu.$$

(iv) By (iii) (b), $\int_A h_1(h_2 - h_1) \, d\mu = 0$ for every $A \in \mathcal{P}_1$. Since the $A_i \in \mathcal{P}_1$ partition Ω, we also have

$$\int_\Omega h_1(h_2 - h_1) \, d\mu = \sum_{i=1}^k \int_{A_i} h_1(h_2 - h_1) \, d\mu = 0.$$

Hence

$$\int_\Omega (h_2 - h_1)^2 \, d\mu = \int_\Omega (h_2^2 - 2h_1 h_2 + h_1^2) \, d\mu$$

$$= \int_\Omega [h_2^2 - 2h_1(h_2 - h_1) - h_1^2] \, d\mu$$

$$= \int_\Omega (h_2^2 - h_1^2) \, d\mu,$$

and thus

$$\int_\Omega h_2^2 \, d\mu = \int_\Omega h_1^2 \, d\mu + \int_\Omega (h_2 - h_1)^2 \, d\mu \geq \int_\Omega h_1^2 \, d\mu.$$

Step 2. Passage to the limit – construction of h.

In Step 1 we showed that the integrals $\int_\Omega h_{\mathcal{P}}^2 \, \mathrm{d}\mu$ are non-decreasing over successive refinements of a finite partition of Ω. Moreover, by (i) above, each function $h_{\mathcal{P}}$ satisfies $0 \leq h_{\mathcal{P}}(\omega) \leq 1$ for all $\omega \in \Omega$. Thus, setting $c = \sup \int_\Omega h_{\mathcal{P}}^2 \, \mathrm{d}\mu$, where the supremum is taken over all finite partitions of Ω, we have $0 \leq c \leq 1$. (Here we use the assumption that $\mu(\Omega) = 1$.)

For each $n \geq 1$ let $\mathcal{P}_n$ be a finite measurable partition of Ω such that $\int_\Omega h_{\mathcal{P}_n}^2 \, \mathrm{d}\mu > c - \frac{1}{4^n}$. Let $\mathcal{Q}_n$ be the smallest common refinement of the partitions $\mathcal{P}_1, \mathcal{P}_2, \ldots, \mathcal{P}_n$. For each n, $\mathcal{Q}_n$ refines $\mathcal{P}_n$ by construction, and $\mathcal{Q}_{n+1}$ refines $\mathcal{Q}_n$ since each $\mathcal{Q}_k$ consists of all intersections $A_1 \cap A_2 \cap \cdots \cap A_k$, where $A_i \in \mathcal{P}_i$, $i \leq k$. Hence each set in $\mathcal{Q}_n$ is a disjoint union of sets in $\mathcal{Q}_{n+1}$. We therefore have the inequalities:

$$c - \frac{1}{4^n} < \int_\Omega h_{\mathcal{P}_n}^2 \, \mathrm{d}\mu \leq \int_\Omega h_{\mathcal{Q}_n}^2 \, \mathrm{d}\mu \leq \int_\Omega h_{\mathcal{Q}_{n+1}}^2 \, \mathrm{d}\mu \leq c.$$

Using the identity proved in Step 1 (iv), we now have

$$\int_\Omega (h_{\mathcal{Q}_{n+1}} - h_{\mathcal{Q}_n})^2 \, \mathrm{d}\mu = \int_\Omega (h_{\mathcal{Q}_{n+1}}^2 - h_{\mathcal{Q}_n}^2) \, \mathrm{d}\mu < \frac{1}{4^n}.$$

The Schwarz inequality applied with $f = |h_{\mathcal{Q}_{n+1}} - h_{\mathcal{Q}_n}|$ and $g \equiv 1$ then yields for each $n \geq 1$,

$$\int_\Omega |h_{\mathcal{Q}_{n+1}} - h_{\mathcal{Q}_n}| \, \mathrm{d}\mu < \frac{1}{2^n}.$$

By the Beppo–Levi theorem, since $\sum_{n \geq 1} \int_\Omega |h_{\mathcal{Q}_{n+1}} - h_{\mathcal{Q}_n}| \, \mathrm{d}\mu$ is finite, we conclude that the series $\sum_{n \geq 1} (h_{\mathcal{Q}_{n+1}} - h_{\mathcal{Q}_n})$ converges almost everywhere (μ), so that the limit function

$$h = h_{\mathcal{P}_1} + \sum_{n \geq 1} (h_{\mathcal{Q}_{n+1}} - h_{\mathcal{Q}_n}) = \lim_n h_{\mathcal{Q}_n}$$

(noting that $\mathcal{Q}_1 = \mathcal{P}_1$) is well-defined almost everywhere (μ). We complete the construction by setting $h = 0$ on the exceptional μ-null set.

Step 3. Verification of the properties of h.

By Step 1 (i) it follows that $0 \leq h(\omega) \leq 1$, and it is clear from its construction that h is $\mathcal{F}$-measurable.

We need to show that $\nu(F) = \int_F h \, \mathrm{d}\mu$ for every $F \in \mathcal{F}$. Fix any such measurable set F and let $n \geq 1$. Define $\mathcal{R}_n$ as the smallest common refinement of the two partitions $\mathcal{Q}_n$ (defined as in Step 2) and $\{F, F^c\}$. Since F is a finite disjoint union of sets in $\mathcal{R}_n$, we have $\nu(F) = \int_F h_{\mathcal{R}_n} \, \mathrm{d}\mu$ from Step 1 (ii).

By Step 2, $c - \frac{1}{4^n} < \int_\Omega h_{\mathcal{Q}_n}^2 \, d\mu \le \int_\Omega h_{\mathcal{R}_n}^2 \, d\mu \le c$, so, as before, we can conclude that $\int_\Omega (h_{\mathcal{R}_n} - h_{\mathcal{Q}_n})^2 \, d\mu < \frac{1}{4^n}$, and using the Schwarz inequality once more, this time with $g = \mathbf{1}_F$, we have

$$\left| \int_F (h_{\mathcal{R}_n} - h_{\mathcal{Q}_n}) \, d\mu \right| \le \int_F |h_{\mathcal{R}_n} - h_{\mathcal{Q}_n}| \, d\mu < \frac{1}{2^n}.$$

For all n, $\nu(F) = \int_F h_{\mathcal{R}_n} \, d\mu = \int_F (h_{\mathcal{R}_n} - h_{\mathcal{Q}_n}) \, d\mu + \int_F h_{\mathcal{Q}_n} \, d\mu$. The first integral on the right converges to 0 as $n \to \infty$, while the second converges to $\int_F h \, d\mu$ by dominated convergence theorem (since for all $n \ge 1$, $0 \le h_{\mathcal{Q}_n} \le 1$ and $\mu(\Omega)$ is finite). Thus we have verified that $\nu(F) = \int_F h \, d\mu$, as required.

It is straightforward to check that the assumption $\mu(\Omega) = 1$ is not essential since for any finite positive measure μ we can repeat the above arguments using $\frac{\mu}{\mu(\Omega)}$ instead of μ. We write the function h defined above as $\frac{d\nu}{d\mu}$ and call it the *Radon–Nikodym derivative* of ν with respect to μ. Its relationship to derivatives of functions will become clear when we consider real functions of bounded variation.

Exercise 7.3

Let $\Omega = [0,1]$ with Lebesgue measure and consider measures μ, ν given by densities $\mathbf{1}_A$, $\mathbf{1}_B$ respectively. Find a condition on the sets A, B, so that μ dominates ν and find the Radon–Nikodym derivative $\frac{d\nu}{d\mu}$ applying the above definition of the function h.

Exercise 7.4

Suppose Ω is a finite set equipped with the algebra of all subsets. Let μ and ν be two measures on Ω such that $\mu(\{\omega\}) \ne 0$, $\nu(\{\omega\}) \ne 0$, for all $\omega \in \Omega$. Decide under which conditions μ dominates ν and find $\frac{d\nu}{d\mu}$.

The next observation is an easy application of the general procedure highlighted in Remark 4.18:

Proposition 7.6

If μ and φ are finite measures with $0 \le \mu \le \varphi$, and if $h_\mu = \frac{d\mu}{d\varphi}$ is constructed as above, then for any non-negative $\mathcal{F}$-measurable function g on Ω we have

$$\int_\Omega g \, d\mu = \int_\Omega g h_\mu \, d\varphi.$$

The same identity holds for any $g \in L^1(\mu)$.

Hint Begin with indicator functions, use linearity of the integral to extend to simple functions, and monotone convergence for general non-negative g. The rest is obvious from the definitions.

For finite measures we can now prove the general result announced earlier:

Theorem 7.7 (Radon–Nikodym)

Let ν and μ be finite measures on the measurable space $(\Omega, \mathcal{F})$ and suppose that $\nu \ll \mu$. Then there is a non-negative $\mathcal{F}$-measurable function h on Ω such that $\nu(A) = \int_A h \, d\mu$ for all $A \in \mathcal{F}$.

Proof

Let $\varphi = \nu + \mu$. Then φ is a positive finite measure which dominates both ν and μ. Hence the Radon–Nikodym derivatives $h_\nu = \frac{d\nu}{d\varphi}$ and $h_\mu = \frac{d\mu}{d\varphi}$ are well defined by the earlier constructions. Consider the sets $F = \{h_\mu > 0\}$ and $G = \{h_\mu = 0\}$ in $\mathcal{F}$. Clearly $\mu(G) = \int_G h_\mu \, d\varphi = 0$, hence also $\nu(G) = 0$, since $\nu \ll \mu$. Define $h = \frac{h_\nu}{h_\mu} \mathbf{1}_F$, and let $A \in \mathcal{F}$, $A \subset F$. By the previous proposition, with $h\mathbf{1}_A$ instead of g, we have

$$\nu(A) = \int_A h_\nu \, d\varphi = \int_A h h_\mu \, d\varphi = \int_A h \, d\mu$$

as required. Since μ and ν are both null on G this proves the theorem. □

Exercise 7.5

Let $\Omega = [0,1]$ with Lebesgue measure and consider probability measures μ, ν given by densities f, g respectively. Find a condition characterising the absolute continuity $\nu \ll \mu$ and find the Radon–Nikodym derivative $\frac{d\nu}{d\mu}$.

Exercise 7.6

Suppose Ω is a finite set equipped with the algebra of all subsets and let μ and ν be two measures on Ω. Characterise the absolute continuity $\nu \ll \mu$ and find $\frac{d\nu}{d\mu}$.

You can now easily complete the picture for σ-finite measures and verify that the function h is 'essentially unique':

Proposition 7.8

The Radon–Nikodym theorem remains valid if the measures ν and μ are σ-finite: for any two such measures with $\nu \ll \mu$ we can find a finite-valued non-negative measurable function f on Ω such that $\nu(F) = \int_F h \, d\mu$ for all $F \in \mathcal{F}$. The function h so defined is unique up to μ-null sets, i.e. if $g : \Omega \to \mathbb{R}^+$ also satisfies $\nu(F) = \int_\Omega g \, d\mu$ for all $F \in \mathcal{F}$ then $g = h$ a.e. (with respect to μ).

Hint There are sequences $(A_n), (B_m)$ of sets in $\mathcal{F}$ with $\mu(A_n), \nu(B_m)$ finite for all $m, n \geq 1$ and $\bigcup_{n \geq 1} A_n = \Omega = \bigcup_{m \geq 1} B_m$. We can choose these to be sequences of disjoint sets (why?). Hence display Ω as the disjoint union of the sets $A_n \cap B_m$ $(m, n \geq 1)$.

Radon–Nikodym derivatives of measures obey simple combination rules which follow from the uniqueness property. We illustrate this with the sum and composition of two Radon–Nikodym derivatives, and leave the 'inverse rule' as an exercise.

Proposition 7.9

Assume we are given σ-finite measures λ, ν, μ satisfying $\lambda \ll \mu$ and $\nu \ll \mu$ with Radon–Nikodym derivatives $\frac{d\lambda}{d\mu}$ and $\frac{d\nu}{d\mu}$, respectively.

(i) With $\phi = \lambda + \nu$ we have $\frac{d\phi}{d\mu} = \frac{d\lambda}{d\mu} + \frac{d\nu}{d\mu}$ a.s. (μ).

(ii) If $\lambda \ll \nu$ then $\frac{d\lambda}{d\mu} = \frac{d\lambda}{d\nu} \frac{d\nu}{d\mu}$ a.s. (μ).

Exercise 7.7

Show that if μ, ν are equivalent measures, i.e. both $\nu \ll \mu$ and $\mu \ll \nu$ are true, then
$$\frac{d\mu}{d\nu} = \left(\frac{d\nu}{d\mu}\right)^{-1} \text{ a.s. } (\mu).$$

Given a pair of σ-finite measures λ, μ on $(\Omega, \mathcal{F})$ it is natural to ask whether we can identify the sets for which $\mu(E) = 0$ implies $\lambda(E) = 0$. This would mean that we can split the mass of λ into two pieces, one being represented by a μ-integral, and the other 'concentrated' on μ-null sets, i.e. away from the mass of μ. We turn this idea of 'separating' the masses of two measures into the following:

Definition 7.10

If there is a set $E \in \mathcal{F}$ such that $\lambda(F) = \lambda(E \cap F)$ for every $F \in \mathcal{F}$ then λ is *concentrated on* E. If two measures μ, ν are concentrated on disjoint subsets of Ω, we say that they are *mutually singular* and write $\mu \perp \nu$.

Clearly, if λ is concentrated on E and $E \cap F = \varnothing$, then $\lambda(F) = \lambda(E \cap F) = 0$. Conversely, if for all $F \in \mathcal{F}$, $F \cap E = \varnothing$ implies $\lambda(F) = 0$, consider $\lambda(F) = \lambda(F \cap E) + \lambda(F \backslash E)$. Since $(F \backslash E) \cap E = \varnothing$ we must have $\lambda(F \backslash E) = 0$, so $\lambda(F) = \lambda(F \cap E)$. We have proved that λ is concentrated on E if and only if for all $F \in \mathcal{F}$, $F \cap E = \varnothing$ implies $\lambda(F) = 0$. We gather some simple facts about mutually singular measures:

Proposition 7.11

If $\mu, \nu, \lambda_1, \lambda_2$ are measures on a σ-field $\mathcal{F}$, the following are true:

(i) If $\lambda_1 \perp \mu$ and $\lambda_2 \perp \mu$ then also $(\lambda_1 + \lambda_2) \perp \mu$.

(ii) If $\lambda_1 \ll \mu$ and $\lambda_2 \perp \mu$ then $\lambda_2 \perp \lambda_1$.

(iii) If $\nu \ll \mu$ and $\nu \perp \mu$ then $\nu = 0$.

Hint For (i), with $i = 1, 2$ let A_i, B_i be disjoint sets with λ_i concentrated on A_i, μ on B_i. Consider $A_1 \cup A_2$ and $B_1 \cap B_2$. For (ii) use the remark preceding the proposition.

The next result shows that a unique 'mass-splitting' of a σ-finite measure relative to another is always possible:

Theorem 7.12 (Lebesgue decomposition)

Let λ, μ be σ-finite measures on $(\Omega, \mathcal{F})$. Then λ can be expressed uniquely as a sum of two measures, $\lambda = \lambda_a + \lambda_s$ where $\lambda_a \ll \mu$ and $\lambda_s \perp \mu$.

Proof

Existence: we consider finite measures; the extension to the σ-finite case is routine. Since $0 \le \lambda \le \lambda + \mu = \varphi$, i.e. ϕ dominates λ, there is $0 \le h \le 1$ such that $\lambda(E) = \int_E h \, d\varphi$ for all measurable E. Let $A = \{\omega : h(\omega < 1\}$ and $B = \{\omega : h(\omega) = 1\}$. Set $\lambda_a(E) = \lambda(A \cap E)$ and $\lambda_s(E) = \lambda(B \cap E)$ for every $E \in \mathcal{F}$.

Now if $E \subset A$ and $\mu(E) = 0$ then $\lambda(E) = \int_E h \, d\varphi = \int_E h \, d\lambda$, so that $\int_E (1 - h) \, d\lambda = 0$. But $h < 1$ on A, hence also on E. Therefore we must have

$\lambda(E) = 0$. Hence if $E \in \mathcal{F}$ and $\mu(E) = 0$, $\lambda_a(E) = \lambda(A \cap E) = 0$ as $A \cap E \subset A$. So $\lambda_a \ll \lambda$. On the other hand, if $E \subset B$ we obtain $\lambda(E) = \int_E h \, d\varphi = \int_E \mathbf{1} \, d(\lambda + \mu) = \lambda(E) + \mu(E)$, so that $\mu(E) = 0$. As $A = B^c$ we have shown that $\mu(E) = 0$ whenever $E \cap A = \emptyset$, so that μ is concentrated on A. Since λ_s is concentrated on B this shows that λ_s and μ are mutually singular.

Uniqueness is left to the reader. (Hint: Employ Proposition 7.11.)   □

Combining this with the Radon–Nikodym theorem, we can describe the structure of λ with respect to μ as 'basis measure':

Corollary 7.13

With $\mu, \lambda, \lambda_a, \lambda_s$ as in the theorem, there is a μ-a.s. unique non-negative measurable function h such that

$$\lambda(E) = \int_E h \, d\mu + \lambda_s(E)$$

for every $E \in \mathcal{F}$.

Remark 7.14

This result is reminiscent of the structure theory of finite-dimensional vector spaces: if $x \in \mathbb{R}^n$ and $m < n$, we can write $x = y + z$, where $y = \sum_{i=1}^m y_i e_i$ is the orthogonal projection onto $\mathbb{R}^m$ and z is orthogonal to this subspace. We also exploited similar ideas for Hilbert space. In this sense the measure μ has the role of a 'basis' providing the 'linear combination' which describes the projection of the measure λ onto a subspace of the space of measures on Ω.

Exercise 7.8

Consider the following measures on the real line: $P_1 = \delta_0$, $P_2 = \frac{1}{25} m|_{[0,25]}$, $P_3 = \frac{1}{2} P_1 + \frac{1}{2} P_2$ (see Example 3.17). For which $i \neq j$ do we have $P_i \ll P_j$? Find the Radon–Nikodym derivative in each such case.

Exercise 7.9

Let $\lambda = \delta_0 + m|_{[1,3]}$, $\mu = \delta_1 + m|_{[2,4]}$ and find λ_a, λ_s, and h as in Corollary 7.13.

7.3 Lebesgue–Stieltjes measures

Recall (see Section 3.5.3) that given any random variable $X : \Omega \longrightarrow \mathbb{R}$, we define its probability distribution as the measure $P_X = P \circ X^{-1}$ on Borel sets on $\mathbb{R}$ (i.e. we set $P(X \leq x) = P \circ X^{-1}((-\infty, x]) = P_X((-\infty, x])$ and extend this to $\mathcal{B}$). Setting $F_X(x) = P_X((-\infty, x])$, we verified in Proposition 4.44 that the distribution function F_X so defined is monotone increasing, right-continuous, with limits at infinity $F_X(-\infty) = \lim_{x \to -\infty} F_X(x) = 0$ and $F_X(+\infty) = \lim_{x \to \infty} F_X(x) = 1$.

In Chapter 4 we studied the special case where $F_X(x) = P_X((-\infty, x]) = \int_{-\infty}^{x} f_X \, dm$ for some real function f_X, the density of P_X with respect to Lebesgue measure m. Proposition 4.32 showed that if f_X is continuous, then F_X is differentiable and has the density f_X as its derivative at every $x \in \mathbb{R}$. On the other hand, the Lebesgue function in Example 4.43 illustrated that continuity of F_X is not sufficient to guarantee the existence of a density.

Moreover, when F_X has a density f_X, the measure P_X was said to be 'absolutely continuous' with respect to m. In the context of the Radon–Nikodym theorem we should reconcile the terminology of this special case with the general one considered in the present chapter. Trivially, when

$$P_X(B) = \int_B f_X \, dm$$

we have $P_X \ll m$, so that P_X has a Radon–Nikodym derivative $\frac{dP_X}{dm}$ with respect to m. The a.s. uniqueness ensures that $\frac{dP_X}{dm} = f_X$ a.s.

Later in this chapter we shall establish the precise analytical requirements on the cumulative distribution function F_X which will guarantee the existence of a density.

7.3.1 Construction of Lebesgue–Stieltjes measures

To do this we first study, only slightly more generally, measures defined on $(\mathbb{R}, \mathcal{B})$ which correspond in similar fashion to increasing, right-continuous functions on $\mathbb{R}$. Their construction mirrors that of Lebesgue measure, with only a few changes, by generalising the concept of 'interval length'. The measures we obtain are known as *Lebesgue–Stieltjes measures*. In this context we call a function $F : \mathbb{R} \longrightarrow \mathbb{R}$ a *distribution function* if F is monotone increasing and right-continuous. It is clear that every finite measure μ defined on $(\mathbb{R}, \mathcal{B})$ defines such a function by $F(x) = \mu((-\infty, x])$, with $F(-\infty) = \lim_{x \to -\infty} F(x) = 0$, $F(+\infty) = \lim_{x \to \infty} F(x) = \mu(\Omega)$.

Our principal concern, however, is with the converse: given a monotone right-continuous $F : \mathbb{R} \longrightarrow \mathbb{R}$, can we always associate with F a measure on $(\Omega, \mathcal{B})$, and if so, what is its relation to Lebesgue measure?

The first question is answered by looking back carefully at the construction of Lebesgue measure m on $\mathbb{R}$ in Chapter 2: first we defined the natural concept of interval length, $l(I) = b - a$, for any interval I with endpoints a, b ($a < b$), and by analogy with our discussion of null sets, we defined Lebesgue outer measure m^* for an arbitrary subset of $\mathbb{R}$ as the infimum of the total lengths $\sum_{n=1}^{\infty} l(I_n)$ of all sequences $(I_n)_{n \geq 1}$ of intervals covering A. To generalise this idea, we should clearly replace $b - a$ by $F(b) - F(a)$ to obtain a 'generalised interval length' relative to F, but since F is only right-continuous we will need to take care of possible discontinuities. Thus we need to identify the possible discontinuities of monotone increasing functions – fortunately these functions are rather well behaved, as you can easily verify in the following:

Proposition 7.15

If $F : \mathbb{R} \longrightarrow \mathbb{R}$ is monotone increasing (i.e. $x_1 \leq x_2$ implies $F(x_1) \leq F(x_2)$) then the left-limit $F(x-)$ and the right-limit $F(x+)$ exist at every $x \in \mathbb{R}$ and $F(x-) \leq F(x) \leq F(x+)$. Hence F has at most countably many discontinuities, and these are jump discontinuities, i.e. $F(x-) < F(x+)$.

Hint For any x, consider $\sup\{F(y) : y < x\}$ and $\inf\{F(y) : x < y\}$ to verify the first claim. For the second, note that $F(x-) < F(x+)$ if F has a discontinuity at x. Use the fact that $\mathbb{Q}$ is dense in $\mathbb{R}$ to show that there can only be countably many such points.

Since F is monotone, it remains bounded on bounded sets. For simplicity we assume that $\lim_{x \to -\infty} F(x) = 0$. We define the 'length relative to F' of the bounded interval $(a, b]$ by

$$l_F(a, b] = F(b) - F(a).$$

Note that we have restricted ourselves to left-open, right-closed intervals. Since F is right-continuous, $F(x+) = F(x)$ for all x, including a, b. Thus $l_F(a, b] = F(b+) - F(a+)$, and all jumps of F have the form $F(x) - F(x-)$. By restricting to intervals of this type we also ensure that l_F is additive over adjoining intervals: if $a < c < b$ then $l_F(a, b] = l_F(a, c] + l_F(c, b]$.

We generalise Definition 2.2 as follows:

Definition 7.16

The *F-outer measure* of any set $A \subseteq \mathbb{R}$ is the element of $[0, \infty]$

$$m_F^*(A) = \inf Z_F(A)$$

where

$$Z_F(A) = \{\sum_{n=1}^{\infty} l_F(I_n) : I_n = (a_n, b_n], a_n \leq b_n, A \subseteq \bigcup_{n=1}^{\infty} I_n\}.$$

Our 'covering intervals' are now also restricted to be left-open and right-closed. This is essential to 'make things fit together', but does not affect measurability: recall (Theorem 2.25) that the Borel σ-field is generated whether we start from the family of all intervals or from various sub-families.

Now consider the proof of Theorem 2.6 in more detail: our purpose there was to prove that the outer measure of an interval equals its length. We show how to adapt the proof to make this claim valid for m_F^* and l_F applied to intervals of the form $(a, b]$. It will be therefore helpful to review the proof of Theorem 2.6 before reading on!

Step 1. The proof that $m_F^*((a, b]) \leq l_F(a, b]$ remains much the same:

To see that $l_F(a, b] \in Z_F((a, b])$, we cover $(a, b]$ by (I_n) with $I_1 = (a, b]$, $I_n = (a, a] = \emptyset$, $n > 1$. The total length of this sequence is $F(b) - F(a) = l_F(a, b]$, hence the result follows by definition of inf.

Step 2. It remains to show that $l_F(a, b] \leq m_F^*((a, b])$. Here we need to be careful always to 'approach points from the right' in order to make use of the right-continuity of F and thus to avoid its jumps.

Fix $\varepsilon > 0$ and $0 < \delta < b - a$. By definition of inf we can find a covering of $I = (a, b]$ by intervals $I_n = (a_n, b_n]$ such that $\sum_{n=1}^{\infty} l_F(I_n) < m_F^*(I) + \frac{\varepsilon}{2}$. Next, let $J_n = (a_n, b_n')$, where by right-continuity of F, for each $n \geq 1$ we can choose $b_n' > b_n$ and $F(b_n') - F(b_n) < \frac{\varepsilon}{2^{n+1}}$. Then $F(b_n') - F(a_n) < \{F(b_n) - F(a_n)\} + \frac{\varepsilon}{2^{n+1}}$.

The $(J_n)_{n \geq 1}$ then form an open cover of the compact interval $[a + \delta, b]$, so that by the Heine–Borel theorem there is a finite subfamily $(J_n)_{n \leq N}$, which also covers $[a + \delta, b]$. Re-ordering these N intervals J_n we can assume that their right-hand endpoints form an increasing sequence and then

$$F(b) - F(a + \delta) = l_F(a + \delta, b] \leq \sum_{n=1}^{N} \{F(b_n') - F(a_n)\}$$

$$< \sum_{n=1}^{N} \{F(b_n) - F(a_n) + \frac{\varepsilon}{2^{n+1}}\} < \sum_{n=1}^{\infty} l_F(I_n) + \frac{\varepsilon}{2}$$

$$< m_F^*(I) + \varepsilon.$$

This holds for all $\varepsilon > 0$, hence $F(b) - F(a + \delta) \leq m_F^*(I)$ for every $\delta > 0$. By right-continuity of F, letting $\delta \downarrow 0$ we obtain $l_F(a, b] = \lim_{\delta \downarrow 0} l_F(a + \delta, b] \leq m_F^*(a, b]$. This completes the proof that $m_F^*((a, b]) = l_F(a, b]$.

This is the only substantive change needed from the construction that led to Lebesgue measure. The proof that m_F^* is an outer measure, i.e.

$$m_F^*(A) \geq 0, \quad m_F^*(\emptyset) = 0, \quad m_F^*(A) \leq m_F^*(B) \text{ if } A \subseteq B,$$

$$m_F^*(\bigcup_{i=1}^{\infty} A_i) \leq \sum_{i=1}^{\infty} m_F^*(A_i),$$

is word-for-word identical with that given for Lebesgue outer measure (Proposition 2.5, Theorem 2.7). Hence, as in Definition 2.9 we say that a set E is *measurable* (for the outer measure m_F^*) if for each $A \subseteq \mathbb{R}$,

$$m_F^*(A) = m_F^*(A \cap E) + m_F^*(A \cap E^c).$$

Again, the proof of Theorem 2.11 goes through verbatim, and we denote the resulting Lebesgue–Stieltjes measure, i.e. m_F^* restricted to the σ-field $\mathcal{M}_F$ of *Lebesgue–Stieltjes measurable sets*, by m_F. By construction, just like Lebesgue measure, m_F is a complete measure: subsets of m_F-null sets are in $\mathcal{M}_F$. However, as we shall see later, $\mathcal{M}_F$ does not always coincide with the σ-field $\mathcal{M}$ of Lebesgue-measurable sets, although both contain all the Borel sets. It is also straightforward to verify that the properties of Lebesgue measure proved in Section 2.4 hold for general Lebesgue–Stieltjes measures, with one exception: the outer measure m_F^* will not, in general, be translation-invariant. We can see this at once for intervals, since $l_F((a + t, b + t]) = F(b + t) - F(a + t)$ will not usually equal $F(b) - F(a)$; simply take $F(x) = \arctan x$, for example. In fact, it can be shown that Lebesgue measure is the unique translation-invariant measure on $\mathbb{R}$.

Note, moreover, that a singleton $\{a\}$ is now not necessarily a null set for m_F: we have, by the analogue of Theorem 2.19, that

$$m_F(\{a\}) = \lim_{n \to \infty} m_F((a - \frac{1}{n}, a]) = F(a) - \lim_{n \to \infty} F(a - \frac{1}{n}) = F(a) - F(a-).$$

Thus, the measure of the set $\{a\}$ is precisely the size of the jump at a (if any). From this it is easy to see by similar arguments how the 'length' of an interval depends on the presence or absence of its endpoints: given that $m_F((a, b]) = F(b) - F(a)$, we see that: $m_F((a, b)) = F(b-) - F(a)$, $m_F([a, b]) = F(b) - F(a-)$, $m_F([a, b)) = F(b-) - F(a-)$.

Example 7.17

When $F = \mathbf{1}_{[a,\infty)}$ we obtain $m_F = \delta_a$, the Dirac measure concentrated at a. Similarly, we can describe a general discrete probability distribution, where the random variable X takes the values $\{a_i : i = 1, 2, \ldots, n\}$ with probabilities $\{p_i = 1, 2, \ldots, n\}$ as the Lebesgue–Stieltjes measure arising from the function $F = \sum_{i=1}^{n} p_i \mathbf{1}_{[a_i,\infty)}$.

Mixtures of discrete and continuous distributions, such as described in Example 3.17, clearly also fit into this picture. Of course, Lebesgue measure m is the special case where the distribution is uniform, i.e. if $F(x) = x$ for all $x \in \mathbb{R}$ then $m_F = m$.

Example 7.18

Only slightly more generally, every finite Borel measure μ on $\mathbb{R}$ corresponds to a Lebesgue–Stieltjes measure, since the distribution function $F(x) = \mu((-\infty, x])$ is obviously increasing and is right-continuous by Theorem 2.19 applied to μ and the intervals $I_n = (-\infty, x + \frac{1}{n}]$. The corresponding Lebesgue–Stieltjes measure m_F is equal to μ since they coincide on the generating family of intervals of the above form. Hence they coincide on the σ-field $\mathcal{B}$ of Borel sets. By our construction of m_F as a complete measure it follows that m_F is the completion of μ.

Example 7.19

Return to the Lebesgue function F discussed in Example 4.43. Since F is continuous and monotone increasing, it induces a Lebesgue–Stieltjes measure m_F on the interval $[0, 1]$, whose properties we now examine. On each 'middle thirds' set F is constant, hence these intervals are null sets for m_F, and as there are countably many of them, so is their union, the 'middle thirds' set, D. Hence the Cantor set $C = D^c$ satisfies

$$1 = F(1) - F(0) = m_F([0, 1]) = m_F(C),$$

(Note that since F is continuous, $m_F(\{0\}) = F(0) - F(0-) = 0$; in fact, each singleton is m_F-null.) We thus conclude that m_F is concentrated on a null set for Lebesgue measure m, i.e. $m_F \perp m$, and that in the Lebesgue decomposition of m_F relative to m there is no absolutely continuous component (by uniqueness of the decomposition).

Exercise 7.10

Suppose the monotone increasing function F is non-constant at most countably many points (as would be the case for a discrete distribution). Show that every subset of $\mathbb{R}$ is m_F-measurable.

Hint Consider m_F over the bounded interval $[-M, M]$ first.

Exercise 7.11

Find the Lebesgue–Stieltjes measure m_F generated by

$$F(x) = \begin{cases} 0 \text{ if } x < 0 \\ 2x \text{ if } x \in [0, 1] \\ 2 \text{ if } x \geq 1. \end{cases}$$

7.3.2 Absolute continuity of functions

We now address the requirements on a distribution F which ensure that it has a density. As we saw in Example 4.43, continuity of a probability distribution function does not guarantee the existence of a density. The following stronger restriction, however, does the trick:

Definition 7.20

A real function F is *absolutely continuous* on the interval $[a, b]$ if, given $\varepsilon > 0$, there is $\delta > 0$ such that for every finite set of disjoint intervals $J_k = (x_k, y_k)$, $k = 1, 2, \ldots, n$, contained in $[a, b]$ and with $\sum_{k=1}^{n}(y_k - x_k) < \delta$, we have $\sum_{k=1}^{n}|F(x_k) - F(y_k)| < \varepsilon$.

This condition will allow us to identify those distribution functions which generate Lebesgue–Stieltjes measures that are absolutely continuous (in the sense of measures) relative to Lebesgue measure. We will see shortly that absolutely continuous functions are also 'of bounded variation': this describes functions which do not 'vary too much' over small intervals. First we verify that the indefinite integral (see Proposition 4.32) relative to a density is absolutely continuous.

Proposition 7.21

If $f \in L^1([a, b])$, where the interval $[a, b]$ is finite, then the function $F(x) = \int_a^x f \, dm$ is absolutely continuous.

Hint Use the absolute continuity of $\mu(G) = \int_G |f| \, dm$ with respect to Lebesgue measure m.

Exercise 7.12

Decide which of the following functions are absolutely continuous:

(a) $f(x) = |x|$, $x \in [-1, 1]$,

(b) $g(x) = \sqrt{x}$, $x \in [0, 1]$,

(c) the Lebesgue function.

The next result is the important converse to the above example, and shows that all Stieltjes integrals arising from absolutely continuous functions lead to measures which are absolutely continuous relative to Lebesgue measure, and hence have a density. Together with Example 7.19 this characterises the distributions arising from densities (under the conditions we have imposed on distribution functions).

Theorem 7.22

If F is monotone increasing and absolutely continuous on $\mathbb{R}$, let m_F be the Lebesgue–Stieltjes measure it generates. Then every Lebesgue-measurable set is m_F-measurable, and on these sets $m_F \ll m$.

Proof

We first show that if the Borel set B has $m(B) = 0$, then also $m_F(B) = 0$. Recall that, given $\delta > 0$ we can find an open set O containing B with $m(O) < \delta$ (Theorem 2.17), and there is a sequence of disjoint open intervals $(I_k)_{k \geq 1}$, $I_k = (a_k, b_k)$ with union O. Since the intervals are disjoint, their total length is less than δ. By the absolute continuity of F, given any $\varepsilon > 0$, we can find $\delta > 0$ such that for every finite sequence of intervals $J_k = (x_k, y_k)$, $k \leq n$, with total length $\sum_{k=1}^{n} (y_k - x_k) < \delta$, we have

$$\sum_{k=1}^{n} \{F(y_k) - F(x_k)\} < \frac{\varepsilon}{2}.$$

Applying this to the sequence $(I_k)_{k \leq n}$ for a fixed n we obtain

$$\sum_{k=1}^{n} \{F(b_k) - F(a_k)\} < \frac{\varepsilon}{2}.$$

As this holds for every n, we also have $\sum_{k=1}^{\infty}\{F(b_k) - F(a_k)\} \leq \frac{\varepsilon}{2} < \varepsilon$. This is the total length of a sequence of disjoint intervals covering $O \supset B$, hence $m_F(B) < \varepsilon$ for every $\varepsilon > 0$, so $m_F(B) = 0$.

Now for every Lebesgue-measurable set E with $m(E) = 0$ we can find a Borel set $B \supseteq E$ with $m(B) = 0$. Thus also $m_F(B) = 0$. Now E is a subset of an m_F-null set, hence it is also m_F-null. Hence all m-measurable sets are m_F-measurable and m-null sets are m_F-null, i.e. $m_F \ll m$ when both are regarded as measures on $\mathcal{M}$.   □

Together with the Radon–Nikodym theorem, the above result helps to clarify the structural relationship between Lebesgue measure and Lebesgue–Stieltjes measures generated by monotone increasing right-continuous functions, and thus, in particular, for probability distributions: when the function F is absolutely continuous it has a density f, and can therefore be written as its 'indefinite integral'. Since the Lebesgue–Stieltjes measure $m_F \ll m$, the Radon–Nikodym derivative $\frac{dm_F}{dm}$ is well defined. Conversely, for the density f of F to exist, the function F must be absolutely continuous. It now remains to clarify the relationship between the Radon–Nikodym derivative $\frac{dm_F}{dm}$ and the density f. It is natural to expect from the example of a continuous f (Proposition 4.32) that f should be the derivative of F (at least m-a.e.). So we need to understand which conditions on F will ensure that $F'(x)$ exists for m-almost all $x \in \mathbb{R}$.

We shall address this question in the somewhat wider context where the 'integrator' function F is no longer necessarily monotone increasing, but has bounded variation, as introduced in the next section.

7.3.3 Functions of bounded variation

Since in general we need to handle set functions that can take negative values, for example, the map

$$E \longrightarrow \int_E g \, dm, \text{ where } g \in L^1(m),$$

(clearly, $L^1(m)$ denotes the space of L^1 functions with respect to the measure m) we therefore need a concept of 'generalised length functions' which are expressed as the difference of two monotone increasing functions. We need first to characterise such functions. This is done by introducing the following:

Definition 7.23

A real function F is of *bounded variation* on $[a, b]$ (briefly $F \in BV[a, b]$) if $T_F[a, b] < \infty$, where for any $x \in [a, b]$

$$T_F[a, x] = \sup\{\sum_{k=1}^{n} |F(x_k) - F(x_{k-1})|\}$$

with the supremum taken over all finite partitions of $[a, x]$ with $a = x_0 < x_1 < \cdots < x_n = x$.

We introduce two further non-negative functions by setting

$$P_F[a, x] = \sup\{\sum_{k=1}^{n} [F(x_k) - F(x_{k-1})]^+\}$$

and

$$N_F[a, x] = \sup\{\sum_{k=1}^{n} [F(x_k) - F(x_{k-1})]^-\}$$

where the supremum is again taken over all partitions of $[a, x]$. The functions $T_F(P_F, N_F)$ are known respectively as the *total (positive, negative) variation functions* of F. We shall keep a fixed in what follows, and consider these as functions of x for $x \geq a$.

We can easily verify the following basic relationships between these definitions:

Proposition 7.24

If F is of bounded variation on $[a, b]$, we have $F(x) - F(a) = P_F(x) - N_F(x)$, while $T_F(x) = P_F(x) + N_F(x)$ for $x \in [a, b]$.

Hint Consider $p(x) = \sum_{k=1}^{n} [F(x_k) - F(x_{k-1})]^+$ and $n(x) = \sum_{k=1}^{n} [F(x_k) - F(x_{k-1})]^-$ for a fixed partition of $[a, x]$ and note that $F(x) - F(a) = p(x) - n(x)$. Now use the definition of the supremum. For the second identity consider $T_F(x) \geq p(x) + n(x) = 2p(x) - F(x) + F(a)$ and use the first identity.

Proposition 7.25

If F is of bounded variation and $a \leq x \leq b$ then $T_F[a, b] = T_F[a, x] + T_F[x, b]$. Similar results hold for P_F and N_F. Hence all three variation functions are monotone increasing in x for fixed $a \in \mathbb{R}$. Moreover, if F has bounded variation on $[a, b]$, then it has bounded variation on any $[c, d] \subset [a, b]$.

Hint Adding a point to a partition will increase all three sums. On the other hand, putting together partitions of $[a, c]$ and $[c, b]$ we obtain a partition of $[a, b]$.

We show that bounded variation functions on finite intervals are exactly what we are looking for:

Theorem 7.26

Let $[a, b]$ be a finite interval. A real function is of bounded variation on $[a, b]$ if and only if it is the difference of two monotone increasing real functions on $[a, b]$.

Proof

If F is of bounded variation, use $F(x) = [F(a) + P_F(x)] - N_F(x)$ from Proposition 7.24 to represent F as the difference of two monotone increasing functions. Conversely, if $F = g - h$ is the difference of two monotone increasing functions, then for any partition $a = x_0 < x_1 < \cdots < x_n = b$ of $[a, b]$ we obtain, since g, h are increasing,

$$\sum_{i=1}^{n} |F(x_i) - F(x_{i-1})| = \sum_{i=1}^{n} |g(x_i) - h(x_i) - g(x_{i-1}) + h(x_{i-1})|$$

$$\leq \sum_{i=1}^{n} [g(x_i) - g(x_{i-1})] + \sum_{i=1}^{n} [h(x_i) - h(x_{i-1})]$$

$$\leq g(b) - g(a) + h(b) - h(a).$$

Thus $M = g(b) - g(a) + h(b) - h(a)$ is an upper bound independent of the choice of partition, and so $T_F[a, b] \leq M < \infty$, as required. $\qquad\square$

This decomposition is minimal: if $F = F_1 - F_2$ and F_1, F_2 are increasing, then for any partition $a = x_0 < x_1 \ldots < x_n = b$ we can write, for fixed $i \leq n$,

$$\{F(x_i) - F(x_{i-1})\}^+ - \{F(x_i) - F(x_{i-1})\}^- = F(x_i) - F(x_{i-1})$$

$$= \{F_1(x_i) - F_1(x_{i-1})\} - \{F_2(x_i) - F_2(x_{i-1})\}$$

which shows from the minimality property of $x = x^+ - x^-$ that each term in the difference on the right dominates its counterpart of the left. Adding and taking suprema we conclude that P_F is dominated by the total variation of F_1 and N_F by that of F_2. In other words, in the collection of increasing functions whose difference is F, the functions $(F(a) + P_F)$ and N_F have the smallest sum at every point of $[a, b]$.

Exercise 7.13

(a) Let F be monotone increasing on $[a, b]$. Find $T_F[a, b]$.

(b) Prove that if $F \in BV[a, b]$ then F is continuous a.e. (m) and Lebesgue-measurable.

(c) Find a differentiable function which is not in $BV[0, 1]$.

(d) Show that if there is a (Lipschitz) constant $M > 0$ such that $|F(x) - F(y)| \leq M|x - y|$ for all $x, y \in [a, b]$, then $F \in BV[a, b]$.

The following simple facts link bounded variation and absolute continuity for functions on a bounded interval $[a, b]$:

Proposition 7.27

Suppose the real function F is absolutely continuous on $[a, b]$; then we have:

(i) $F \in BV[a, b]$,

(ii) If $F = F_1 - F_2$ is the minimal decomposition of F as the difference of two monotone increasing functions described in Theorem 7.26, then both F_1 and F_2 are absolutely continuous on $[a, b]$.

Hint Given $\varepsilon > 0$ choose $\delta > 0$ as in Definition 7.20. In (i), starting with an arbitrary partition (x_i) of $[a, b]$ we cannot use the absolute continuity of F unless we know that the subintervals are of length δ. So add enough new partition points to guarantee this and consider the sums they generate. For (ii), compare the various variation functions when summing over a partition where the sum of intervals lengths is bounded by δ.

Definition 7.28

If $F \in BV[a, b]$, where $a, b \in \mathbb{R}$, let $F = F_1 - F_2$ be its minimal decomposition into monotone increasing functions. Define the Lebesgue–Stieltjes *signed measure* determined by F as the countably additive set function m_F given on the σ-field $\mathcal{B}$ of Borel sets by $m_F = m_{F_1} - m_{F_2}$, where m_{F_i} is the Lebesgue–Stieltjes measure determined by F_i, $(i = 1, 2)$.

We shall examine signed measures more generally in the next section. For the present, we note the following:

Example 7.29

When considering the measure $P_X(E) = \int_E f_X \, dm$ induced on $\mathbb{R}$ by a density f_X we restrict attention to $f_X \geq 0$ to ensure that P_X is non-negative. But for a measurable function $f : \mathbb{R} \longrightarrow \mathbb{R}$ we set (Definition 4.16)

$$\int_E f \, dm = \int_E f^+ \, dm - \int_E f^- \, dm \text{ whenever } \int_E |f| \, dm < \infty.$$

The set function ν defined by $\nu(E) = \int_E f \, dm$ then splits naturally into the difference of two measures, i.e. $\nu = \nu^+ - \nu^-$, where $\nu^+(E) = \int_E f^+ \, dm$ and $\nu^-(E) = \int_E f^- \, dm$. Restricting to a function f supported on $[a, b]$ and setting $F(x) = \int_a^x f \, dm$, we obtain $m_F = \nu$, and if $F = F_1 - F_2$ as in the above definition, then $m_{F_1} = \nu^+$, $m_{F_2} = \nu^-$ by the minimality properties of the splitting of F.

7.3.4 Signed measures

The above example and the definition of Lebesgue–Stieltjes measures generated by a BV function motivate the following abstract definition and the subsequent search for a similar decomposition into the difference of two measures. We proceed to outline briefly the structure of signed measures in the abstract setting, which provides a general context for the above development of Stieltjes integrals and distribution functions. Our results will enable us to define integrals of functions relative to signed measures by reference to the decomposition of the signed measure into 'positive and negative parts', exactly as above. We also obtain a more general Lebesgue decomposition and Radon–Nikodym theorem, thus completing the description of the structure of a bounded signed measure relative to a given σ-finite measure. This leads to the general version of the fundamental theorem of the calculus signalled earlier.

Definition 7.30

A *signed measure* on a measurable space $(\Omega, \mathcal{F})$ is a set function $\nu : \mathcal{F} \longrightarrow (-\infty, +\infty]$ satisfying

(i) $\nu(\emptyset) = 0$

(ii) $\nu(\bigcup_{i=1}^\infty E_i) = \sum_{i=1}^\infty \nu(E_i)$ if $E_i \in \mathcal{F}$ and $E_i \cap E_j = \emptyset$ for $i \neq j$.

We need to avoid ambiguities like $\infty - \infty$ by demanding that ν should take at most one of the values $\pm\infty$; therefore we consistently demand that $\nu(E) > -\infty$ for all sets E in its domain. Note also that in (ii) either both sides are $+\infty$, or they are both finite, so that the series converges in $\mathbb{R}$. Since the left side is unaffected by any re-arrangement of the terms of the series, it follows that the series converges absolutely whenever it converges, i.e. $\sum_{i=1}^{\infty} |\nu(E_i)| < \infty$ if and only if $|\nu(\bigcup_{i=1}^{\infty} E_i)| < \infty$. The convergence is clear in the motivating example, since for any $E \subseteq \mathbb{R}$ we have

$$|\nu(E)| = |\int_E f\,dm| \le \int_E |f|\,dm < \infty \text{ when } f \in L^1(\mathbb{R}).$$

Note that ν is finitely additive (let $E_i = \emptyset$ for all $i > n$ in (ii), then (i) implies $\nu(\bigcup_{i=1}^{n} E_i) = \sum_{i=1}^{n} \nu(E_i)$ if $E_i \in \mathcal{F}$ and $E_i \cap E_j = \emptyset$ for $i \neq j$, $i,j \le n$). Hence if $F \subseteq E$, $F \in \mathcal{F}$, and $|\nu(E)| < \infty$, then $|\nu(F)| < \infty$, since both sides of $\nu(E) = \nu(F) + \nu(E \setminus F)$ are finite and $\nu(E \setminus F) > -\infty$ by hypothesis.

Signed measures do not inherit the properties of measures without change: as a negative result we have:

Proposition 7.31

A signed measure ν defined on a σ-field $\mathcal{F}$ is monotone increasing ($F \subset E$ implies $\nu(F) \le \nu(E)$) if and only ν is a measure on $\mathcal{F}$.

Hint $\emptyset$ is a subset of every $E \in \mathcal{F}$.

On the other hand, a signed measure attains its bounds at some sets in $\mathcal{F}$. More precisely: given a signed measure ν on $(\Omega, \mathcal{F})$ one can find sets A and B in $\mathcal{F}$ such that $\nu(A) = \inf\{\nu(F) : F \in \mathcal{F}\}$ and $\nu(B) = \sup\{\nu(F) : F \in \mathcal{F}\}$.

Rather than prove this result directly we shall deduce it from the Hahn–Jordan decomposition theorem. This basic result shows how the set A and its complement can be used to define two (positive) measures ν^+, ν^- such that $\nu = \nu^+ - \nu^-$, with $\nu^+(F) = \nu(F \cap A^c)$ and $\nu^-(F) = -\nu(F \cap A)$ for all $F \in \mathcal{F}$. The decomposition is minimal: if $\nu = \lambda_1 - \lambda_2$ where the λ_i are measures, then $\nu^+ \le \lambda_1$ and $\nu^- \le \lambda_2$.

Restricting attention to bounded signed measures (which suffices for applications to probability theory), we can derive this decomposition by applying the Radon–Nikodym theorem. (Our account is a special case of the treatment given in [11], Ch. 6, for complex-valued set functions.) First, given a bounded signed measure $\nu : \mathcal{F} \longrightarrow \mathbb{R}$, we seek the smallest (positive) measure μ that dominates ν, i.e. satisfies $\mu(E) \ge |\nu(E)|$ for all $E \in \mathcal{F}$. Defining

$$|\nu|(E) = \sup\{\sum_{i=1}^{\infty} |\nu(E_i)| : \{E_i\} \subset \mathcal{F}, \ E = \bigcup_{i \ge 1} E_i, \ E_i \cap E_j = \emptyset \text{ if } i \neq j\}$$

produces a set function which satisfies $|\nu|(E) \geq |\nu(E)|$ for every E. The requirement $\mu(E_i) \geq |\nu(E_i)|$ for all i then yields

$$\mu(E) = \sum_{i=1}^{\infty} \mu(E_i) \geq \sum_{i=1}^{\infty} |\nu(E_i)|$$

for any measure μ dominating ν. Hence to prove that $|\nu|$ has the desired properties we only need to show that it is countably additive. We call $|\nu|$ the *total variation* of ν. Note that we use countable partitions of Ω here, just as we used sequences of intervals when defining Lebesgue measure in $\mathbb{R}$.

Theorem 7.32

The total variation $|\nu|$ of a bounded signed measure is a (positive) measure on $\mathcal{F}$.

Proof

Partitioning $E \in \mathcal{F}$ into sets (E_i), choose (a_i) in $\mathbb{R}^+$ such that $a_i < |\nu|(E_i)$ for all i. Partition each E_i in turn into sets $\{A_{ij}\}_j$, and by definition of sup we can choose these to ensure that $a_i < \sum_j |\nu(A_{ij})|$ for every $i \geq 1$. But the (A_{ij}) also partition E, hence $\sum_i a_i < \sum_{i,j} |\nu(A_{i,j})| < |\nu|(E)$. Taking the supremum over all sequences (a_i) satisfying these requirements ensures that $\sum_i |\nu|(E_i) = \sup \sum_i a_i \leq |\nu|(E)$.

For the converse inequality consider any partition (B_k) of E and note that for fixed k, $(B_k \cap E_i)_{i \geq 1}$ partitions B_k, while for fixed i, $(B_k \cap E_i)_{k \geq 1}$ partitions E_i. This means that

$$\sum_{k \geq 1} |\nu(B_k)| = \sum_{k \geq 1} |\sum_{i \geq 1} \nu(B_k \cap E_i)| \leq \sum_{k \geq 1} \sum_{i \geq 1} |\nu(B_k \cap E_i)|.$$

Since the terms of the double series are all non-negative, we can exchange the order of summation, so that finally

$$\sum_{k \geq 1} |\nu(B_k)| \leq \sum_{i \geq 1} \sum_{k \geq 1} |\nu(B_k \cap E_i)| \leq \sum_{i \geq 1} |\nu|(E_i).$$

But the partition (B_k) of E was arbitrary, so the estimate on the right also dominates $|\nu|(E)$. This completes the proof that $|\nu|$ is a measure. $\square$

We now define the *positive* (resp. *negative*) *variation* of the signed measure ν by setting:

$$\nu^+ = \frac{1}{2}(|\nu| + \nu), \quad \nu^- = \frac{1}{2}(|\nu| - \nu).$$

Clearly both are positive measures on $\mathcal{F}$, and we have

$$\nu = \nu^+ - \nu^- \text{ and } |\nu| = \nu^+ + \nu^-.$$

With these definitions we can immediately extend the Radon–Nikodym and Lebesgue decomposition theorems to the case where ν is a bounded signed measure (we keep the notation used in Section 7.3.2, so here μ remains positive!):

Theorem 7.33

Let μ be σ-finite (positive) measure and suppose that ν is a bounded signed measure. Then there is unique decomposition $\nu = \nu_a + \nu_s$, into two signed measures, with $\nu_a \ll \mu$ and $\nu_s \perp \mu$. Moreover, there is a unique (up to sets of μ-measure 0) $h \in L^1(\mu)$ such that $\nu_a(F) = \int_F h \, d\mu$ for all $F \in \mathcal{F}$.

Proof

Given $\nu = \nu^+ - \nu^-$ we wish to apply the Lebesgue decomposition and Radon–Nikodym theorems to the pairs of finite measures (ν^+, μ) and (ν^-, μ). First we need to check that for a signed measure $\lambda \ll \mu$ we also have $|\lambda| \ll \mu$ (for then clearly both $\lambda^+ \ll \mu$ and $\lambda^- \ll \mu$). But if $\mu(E) = 0$ and (F_i) partitions E, then each $\mu(F_i) = 0$, hence $\lambda(F_i) = 0$, so that $\sum_{i \geq 1} |\lambda(F_i)| = 0$. As this holds for each partition, $|\lambda(E)| = 0$.

Similarly, if λ is concentrated on a set A, and $A \cap E = \emptyset$, then for any partition (F_i) of E we will have $\lambda(F_i) = 0$ for every $i \geq 1$. Thus $|\lambda|(E) = 0$, so $|\lambda|$ is also concentrated on A. Hence if two signed measures are mutually singular, so are their total variation measures, and thus also their positive and negative variations. Applying the Lebesgue decomposition and Radon–Nikodym theorems to the measures ν^+ and ν^- provides (positive) measures $(\nu^+)_a, (\nu^+)_s, (\nu^-)_a, (\nu^-)_s$ such that $\nu^+ = (\nu^+)_a + (\nu^+)_s$, and $(\nu^+)_a(F) = \int_F h' \, d\mu$, while $\nu^- = (\nu^-)_a + (\nu^-)_s$ and $(\nu^-)_a(F) = \int_F h'' \, d\mu$, for non-negative functions $h', h'' \in L^1(\mu)$, and with the measures $(\nu^+)_s, (\nu^-)_s$ each mutually singular with μ. Letting $\nu_a = (\nu^+)_a - (\nu^-)_a$ we obtain a signed measure $\nu_a \ll \mu$, and a function $h = h' - h'' \in \mathcal{L}^1(\mu)$ with $\nu_a(F) = \int_F h \, d\mu$ for all $F \in \mathcal{F}$. The signed measure $\nu_s = (\nu^+)_s - (\nu^-)_s$ is clearly singular to μ, and h is unique up to μ-null sets, since this holds for h', h'' and the decomposition $\nu = \nu^+ - \nu^-$ is minimal. $\qquad\square$

Example 7.34

If $g \in L^1(\mu)$ then $\nu(E) = \int_E g \, d\mu$ is a signed measure and $\nu \ll \mu$. The Radon–Nikodym theorem shows that (with our conventions) all signed measures $\nu \ll \mu$

have this form.

We are nearly ready for the general form of the fundamental theorem of calculus. First we confirm, as may be expected from the proof of the Radon–Nikodym theorem, the close relationship between the derivative of the bounded variation function F induced by a bounded signed (Borel) measure ν on $\mathbb{R}$ and the derivative $f = F'$:

Theorem 7.35

If ν is a bounded signed measure on $\mathbb{R}$ and $F(x) = \nu((-\infty, x])$ then for any $a \in \mathbb{R}$, the following are equivalent:

(i) F is differentiable at a, and $F'(a) = L$.

(ii) given $\varepsilon > 0$ there exists $\delta > 0$ such that $|\frac{\nu(J)}{m(J)} - L| < \varepsilon$ if the open interval J contains a and $l(J) < \varepsilon$.

Proof

We may assume that $L = 0$; otherwise consider $\rho = \nu - Lm$ instead, restricted to a bounded interval containing a. If (i) holds with $L = 0$ and $\varepsilon > 0$ is given, we can find $\delta > 0$ such that

$$|F(y) - F(x)| < \varepsilon|y - x| \text{ whenever } |y - x| < \delta.$$

Let $J = (x, y)$ be an open interval containing a with $(y - x) < \delta$. For sufficiently large N we can ensure that $a > x + \frac{1}{N} > x$ and so for $k \geq 1$, $y_k = x + \frac{1}{N+k}$ is bounded above by a and decreases to x as $k \to \infty$. Thus

$$|\nu(y_k, y])| = |F(y) - F(y_k)| \leq |F(y) - F(a)| + |F(a) - F(y_k)|$$
$$\leq \varepsilon\{(y - a) + (a - y_k)\} < \varepsilon m(J).$$

But since $y_k \to x$, $\nu(y_k, y] \to \nu(x, y]$ and we have shown that $|\frac{\nu(J)}{m(J)}| < \varepsilon$. Hence (ii) holds. For the converse, let ε, δ be as in (ii), so that with $x < a < y$ and $y - x < \delta$, (ii) implies $|\nu(x, y + \frac{1}{n})| < \varepsilon(y + \frac{1}{n} - x)$ for all large enough n. But as $(x, y] = \bigcap_n (x, y + \frac{1}{n})$, we also have

$$|\nu(x, y]| < |F(y) - F(x)| < \varepsilon(y - x). \tag{7.3}$$

Finally, since (ii) holds, $|\nu(\{a\})| \leq |\nu(I)| < \varepsilon l(I)$ for any small enough open interval I containing a. Thus $F(a) = F(a-)$ and so F is continuous at a. Since $x < a < y < x + \delta$, we conclude that (7.3) holds with a instead of x, which shows that the right-hand derivative of F at a is 0, and with a instead of y, which shows the same for the left-hand derivative. Thus $F'(a) = 0$, and so (i) holds.   □

Theorem 7.36 (fundamental theorem of calculus)

Let F be absolutely continuous on $[a, b]$. Then F is differentiable m-a.e. and its Lebesgue–Stieltjes signed measure m_F has Radon–Nikodym derivative $\frac{dm_F}{dm} = F'$ m-a.e. Moreover, for each $x \in [a, b]$,

$$F(x) - F(a) = m_F[a, x] = \int_a^x F'(t)\,dt.$$

You should compare the following theorem with with the elementary version given in Proposition 4.32.

Proof

The Radon–Nikodym theorem provides $\frac{dm_F}{dm} = h \in L^1(m)$ such that $m_F(E) = \int_E h\,dm$ for all $E \in \mathcal{B}$. Choosing the partitions

$$\mathcal{P}_n = \{(t_i, t_{i+1}] : t_i = a + \frac{i}{2^n}(b - a), i \leq 2^n\},$$

we obtain, successively, each $\mathcal{P}_n$ as the smallest common refinement of the partitions $\mathcal{P}_1, \mathcal{P}_2, \ldots, \mathcal{P}_{n-1}$. Thus, setting $h_n(a) = 0$ and

$$h_n(x) = \sum_{i=1}^{2^n} \frac{m_F(t_i, t_{i+1}]}{m(t_i, t_{i+1}]} \mathbf{1}_{(t_i, t_{i+1}]} = \sum_{i=1}^{2^n} \frac{F(t_{i+1}) - F(t_i)}{t_{i+1} - t_i} \mathbf{1}_{(t_i, t_{i+1}]} \text{ for } a < x \leq b,$$

we obtain a sequence (h_n) corresponding to the sequence $(h_{\mathcal{Q}_n})$ constructed in Step 2 of the proof of the Radon–Nikodym theorem. It follows that $h_n(x) \to h(x)$ m-a.e. But for any fixed $x \in (a, b)$, condition (ii) in Theorem 4.6 applied to the function F on each interval (t_i, t_{i+1}) with length less than δ, and with $L = h(x)$, shows that $h = F'$ m-a.e. The final claim is now obvious from the definitions.   □

The following result is therefore immediate and it justifies the terminology 'indefinite integral' in this general setting.

Corollary 7.37

If F is absolutely continuous on $[a, b]$ and $F' = 0$ m-a.e. then F is constant.

A final corollary now completes the circle of ideas for distribution functions and their densities:

Corollary 7.38

If $f \in L^1([a,b])$ and $F(x) = \int_a^x f \, dm$ for each $x \in [a,b]$ then F is differentiable m-a.e. and $F'(x) = f(x)$ for almost every $x \in [a,b]$.

7.3.5 Hahn–Jordan decomposition

As a further application of the Radon–Nikodym theorem we derive the Hahn–Jordan decomposition of ν which was outlined earlier. First we need the following

Theorem 7.39

Let ν be a bounded signed measure and let $|\nu|$ be its total variation. Then we can find a measurable function h such that $|h(\omega)| = 1$ for all $\omega \in \Omega$ and $\nu(E) = \int_E h \, d|\nu|$ for all $E \in \mathcal{F}$.

Proof

The Radon–Nikodym theorem provides a measurable function h with $\nu(E) = \int_E h \, d|\nu|$ for all $E \in \mathcal{F}$ since every $|\nu|$-null set is ν-null ($\{E, \varnothing, \varnothing, \ldots\}$ is a partition of E). Let $C_\alpha = \{\omega : |h(\omega)| < \alpha\}$ for $\alpha > 0$. Then, for any partition $\{E_i\}$ of C_α,

$$\sum_{i \geq 1} |\nu(E_i)| = \sum_{i \geq 1} \left| \int_{E_i} h \, d|\nu| \right| \leq \sum_{i \geq 1} \alpha |\nu|(E_i) = \alpha |\nu|(C_\alpha).$$

As this holds for any partition, it holds for the supremum, i.e. $|\nu|(C_\alpha) \leq \alpha |\nu|(C_\alpha)$. For $\alpha < 1$ we must conclude that C_α is $|\nu|$-null, and hence also ν-null. Therefore $|h| \geq 1$ ν-a.e.

To show that $|h| \leq 1$ ν-a.e. we note that if E has positive $|\nu|$-measure, then, by definition of h,

$$\frac{\left| \int_E h \, d|\nu| \right|}{|\nu|(E)} = \frac{|\nu(E)|}{|\nu|(E)} \leq 1.$$

That this implies $|h| \leq 1$ ν-a.e. follows from the proposition below, applied with $\rho = |\nu|$. Thus the set where $|h| \neq 1$ is $|\nu|$-null, hence also ν-null, and we can redefine h there, so that $|h(\omega)| = 1$ for all $\omega \in \Omega$. □

Proposition 7.40

Given a finite measure ρ and a function $f \in L^1(\rho)$, suppose that for every $E \in \mathcal{F}$ with $\rho(E) > 0$ we have $|\frac{1}{\rho(E)} \int_E f \, d\rho| \le 1$. Then $|f(\omega)| \le 1$, ρ-a.e.

Hint Let $E = \{f > 1\}$. If $\rho(E) > 0$ consider $\int_E \frac{f}{\rho(E)} \, d\rho$.

We are ready to derive the Hahn–Jordan decomposition very simply:

Proposition 7.41

Let ν be a bounded signed measure. There are disjoint measurable sets A, B such that $A \cup B = \Omega$ and $\nu^+(F) = \nu(B \cap F)$, $\nu^-(F) = \nu(A \cap F)$ for all $F \in \mathcal{F}$. Consequently, if $\nu = \lambda_1 - \lambda_2$ for measures λ_1, λ_2 then $\lambda_1 \ge \nu^+$ and $\lambda_2 \ge \nu^-$.

Hint Since $d\nu = h \, d|\nu|$ and $|h| = 1$ let $A = \{h = -1\}$, $B = \{h = 1\}$. Use the definition of ν^+ to show that $\nu^+(F) = \frac{1}{2} \int_F (1 + h) \, d|\nu| = \nu(F \cap B)$ for every F.

Exercise 7.14

Let ν be a bounded signed measure. Show that for all F, $\nu^+(F) = \sup_{G \subset F} \nu(G)$, $\nu^-(F) = -\inf_{G \subset F} \nu(G)$, all the sets concerned being members of $\mathcal{F}$.

Hint $\nu(G) \le \nu^+(G) \le \nu(B \cap G) + \nu((B \cap F) \setminus (B \cap G)) = \nu(B \cap F)$.

Exercise 7.15

Show that when $\nu(F) = \int_F f \, d\mu$ where $f \in L^1(\mu)$, where μ is a (positive) measure, the Hahn decomposition sets are $A = \{f < 0\}$ and $B = \{f \ge 0\}$, and $\nu^+(F) = \int_F f^+ \, d\nu$, while $\nu^-(F) = \int_F f^- \, d\nu$.

We finally arrive at a general definition of integrals relative to signed measures:

Definition 7.42

Let μ be signed measure and f a measurable function on $F \in \mathcal{F}$. Define the integral $\int_F f \, d\mu$ by

$$\int_F f \, d\mu = \int_F f \, d\mu^+ - \int_F f \, d\mu^-$$

whenever both terms on the right are finite or are not of the form $\pm(\infty - \infty)$.

The function is sometimes called *summable* if the integral so defined is finite. Note that the earlier definition of a Lebesgue–Stieltjes signed measure fits into this general framework. We normally restrict attention to the case when both terms are finite, which clearly holds when μ is bounded.

Exercise 7.16

Verify the following: Let μ be a finite measure and define the signed measure ν by $\nu(F) = \int_F g \, d\mu$. Prove that $f \in L^1(\nu)$ if and only if $fg \in L^1(\mu)$ and $\int_E f \, d\nu = \int_E fg \, d\mu$ for all μ-measurable sets E.

7.4 Probability

7.4.1 Conditional expectation relative to a σ-field

Suppose we are given a random variable $X \in L^1(P)$, where $(\Omega, \mathcal{F}, P)$ is a probability space. In Chapter 5 we defined the conditional expectation $\mathbb{E}(X|\mathcal{G})$ of $X \in L^2(P)$ relative to a sub-σ-field $\mathcal{G}$ of $\mathcal{F}$ as the a.s. unique random variable $Y \in L^2(\mathcal{G})$ (meaning $\mathcal{G}$-measurable) satisfying the condition

$$\int_G Y \, dP = \int_G X \, dP \text{ for all } G \in \mathcal{G}. \tag{7.4}$$

The construction was a consequence of orthogonal projections in the Hilbert space L^2 with the extension to all integrable random variables undertaken 'by hand', which required a little care. With the Radon–Nikodym theorem at our disposal we can verify the existence of conditional expectations for integrable random variables very simply:

The (possibly signed) bounded measure $\nu(F) = \int_F X \, dP$ is absolutely continuous with respect to P. Restricting both measures to $(\Omega, \mathcal{G})$ maintains this relationship, so that there is a $\mathcal{G}$-measurable, P-a.s. unique random variable Y such that $\nu(G) = \int_G Y \, dP$ for every $G \in \mathcal{G}$. But by definition $\nu(G) = \int_G X \, dP$, so the defining equation (7.4) of $Y = \mathbb{E}(X|\mathcal{G})$ has been verified.

Remark 7.43

In particular, this shows that for $X \in L^2(\mathcal{F})$ its orthogonal projection onto $L^2(\mathcal{G})$ is a version of the Radon–Nikodym derivative of the measure $\nu : F \to \int_F X \, dP$.

We shall write $\mathbb{E}(X|\mathcal{G})$ instead of Y from now on, always keeping in mind that we have freedom to choose a particular 'version', i.e. as long as the results we seek demand only that relations concerning $\mathbb{E}(X|\mathcal{G})$ hold P-a.s., we can alter this random variable on a null set without affecting the truth of the defining equation:

Definition 7.44

A random variable $\mathbb{E}(X|\mathcal{G})$ is called the *conditional expectation* of X relative to a σ-field $\mathcal{G}$ if

(i) $\mathbb{E}(X|\mathcal{G})$ is $\mathcal{G}$-measurable,

(ii) $\int_G \mathbb{E}(X|\mathcal{G})\,\mathrm{d}P = \int_G X\,\mathrm{d}P$ for all $G \in \mathcal{G}$.

We investigate the properties of the conditional expectation. To begin with, the simplest are left for the reader as a proposition. In this and the subsequent theorem we make the following assumptions:

1. All random variables concerned are defined on a probability space $(\Omega, \mathcal{F}, P)$;

2. X, Y and all (X_n) used below are assumed to be in $L^1(\Omega, \mathcal{F}, P)$;

3. $\mathcal{G}$ and $\mathcal{H}$ are sub-σ-fields of $\mathcal{F}$.

The properties listed in the next proposition are basic, and are used time and again. Where appropriate we give verbal description of its 'meaning' in terms of information about X.

Proposition 7.45

The conditional expectation $\mathbb{E}(X|\mathcal{G})$ has the following properties:

(i) $\mathbb{E}(\mathbb{E}(X|\mathcal{G})) = \mathbb{E}(X)$

   (more precisely: any version of the conditional expectation of X has the same expectation as X).

(ii) If X is $\mathcal{G}$-measurable, then $\mathbb{E}(X|\mathcal{G}) = X$

   (if, given $\mathcal{G}$, we already 'know' X, our 'best estimate' of it is perfect).

(iii) If X is independent of $\mathcal{G}$, then $\mathbb{E}(X|\mathcal{G}) = \mathbb{E}(X)$

   (if $\mathcal{G}$ 'tells us nothing' about X, our best guess of X is its average value).

(iv) (Linearity) $\mathbb{E}((aX + bY)|\mathcal{G}) = a\mathbb{E}(X|\mathcal{G}) + b\mathbb{E}(Y|\mathcal{G})$ for any real numbers a, b

(note again that this is really says that each linear combination of versions of the right-hand side is a version of the left-hand side).

Theorem 7.46

The following properties hold for $\mathbb{E}(X|\mathcal{G})$ as defined above:

(i) If $X \geq 0$ then $\mathbb{E}(X|\mathcal{G}) \geq 0$-a.s.

(positivity).

(ii) If $(X_n)_{n\geq 1}$ are non-negative and increase a.s. to X, then $(\mathbb{E}(X_n|\mathcal{G}))_{n\geq 1}$ increase a.s. to $\mathbb{E}(X|\mathcal{G})$

('monotone convergence' of conditional expectations).

(iii) If Y is $\mathcal{G}$-measurable and XY is integrable, then $\mathbb{E}(XY|\mathcal{G}) = Y\mathbb{E}(X|\mathcal{G})$

('taking out a known factor').

(iv) If $\mathcal{H} \subset \mathcal{G}$ then $\mathbb{E}([\mathbb{E}(X|\mathcal{G})]|\mathcal{H}) = \mathbb{E}(X|\mathcal{H})$

(the tower property).

(v) If $\varphi : \mathbb{R} \to \mathbb{R}$ is a convex function and $\varphi(X) \in L^1(P)$, then

$$\mathbb{E}(\varphi(X)|\mathcal{G}) \geq \varphi(\mathbb{E}(X|\mathcal{G})).$$

(This is known as the conditional Jensen inequality – a similar result holds for expectations. Recall that a real function φ is *convex* on (a, b) if for all $x, y \in (a, b)$, $\varphi(px + (1 - p)y) \leq p\varphi(x) + (1 - p)\varphi(y)$; the graph of φ stays on or below the straight line joining $(x, \varphi(x)), (y, \varphi(y))$.)

Proof

(i) For each $k \geq 1$ the set $E_k = \{\mathbb{E}(X|\mathcal{G}) < -\frac{1}{k}\} \in \mathcal{G}$, so that

$$\int_{E_k} X \, dP = \int_{E_k} \mathbb{E}(X|\mathcal{G}) \, dP.$$

As $X \geq 0$, the left-hand side is non-negative, while the right-hand side is bounded above by $-\frac{1}{k}P(E_k)$. This forces $P(E_k) = 0$ for each k, hence also $P(\mathbb{E}(X|\mathcal{G}) < 0) = P(\bigcup_k E_k) = 0$. Thus $\mathbb{E}(X|\mathcal{G}) \geq 0$-a.s.

(ii) For each n let Y_n be a version of $\mathbb{E}(X_n|\mathcal{G})$. By (i) and as in Section 5.4.3, the (Y_n) are non-negative and increase a.s. Letting $Y = \limsup_n Y_n$ provides a $\mathcal{G}$-measurable random variable such that the real sequence $(Y_n(\omega))_n$ converges to $Y(\omega)$ for almost all ω. Corollary 4.14 then shows that $(\int_G Y_n \, dP)_{n \geq 1}$ increases to $\int_G Y \, dP$. But we have $\int_G Y_n \, dP = \int_G X_n \, dP$ for each n, and (X_n) increases pointwise to X. By the monotone convergence theorem it follows that $(\int_G X_n \, dP)_{n \geq 1}$ increases to $\int_G X \, dP$, so that $\int_G X \, dP = \int_G Y \, dP$. This shows that Y is a version of $\mathbb{E}(X|\mathcal{G})$ and therefore proves our claim.

(iii) We can restrict attention to $X \geq 0$, since the general case follows from this by linearity. Now first consider the case of indicators: if $Y = \mathbf{1}_E$ for some $E \in \mathcal{G}$, we have, for all $G \in \mathcal{G}$,

$$\int_G \mathbf{1}_E \mathbb{E}(X|\mathcal{G}) \, dP = \int_{E \cap G} \mathbb{E}(X|\mathcal{G}) \, dP = \int_{E \cap G} X \, dP = \int_G \mathbf{1}_E X \, dP.$$

so that $\mathbf{1}_E \mathbb{E}(X|\mathcal{G})$ satisfies the defining equation and hence is a version of the conditional expectation of the product XY. So $\mathbb{E}(XY|\mathcal{G}) = Y\mathbb{E}(X|\mathcal{G})$ has been verified when $Y = \mathbf{1}_E$ and $E \in \mathcal{G}$. By the linearity property this extends to simple functions, and for arbitrary $Y \geq 0$ we now use (ii) and a sequence (Y_n) of simple functions increasing to Y to deduce that, for non-negative X, $\mathbb{E}(XY_n|\mathcal{G}) = Y_n\mathbb{E}(X|\mathcal{G})$ increases to $\mathbb{E}(XY|\mathcal{G})$ on the one hand and to $Y\mathbb{E}(X|\mathcal{G})$ on the other. Thus if X and Y are both non-negative we have verified (iii). Linearity allows us to extend this to general $Y = Y^+ - Y^-$.

(iv) We have $\int_G \mathbb{E}(X|\mathcal{G}) \, dP = \int_G X \, dP$ for $G \in \mathcal{G}$ and $\int_H \mathbb{E}(X|\mathcal{H}) \, dP = \int_H X \, dP$ for $H \in \mathcal{H} \subset \mathcal{G}$. Hence for $H \in \mathcal{H}$ we obtain $\int_H \mathbb{E}(X|\mathcal{G}) \, dP = \int_H \mathbb{E}(X|\mathcal{H}) \, dP$. Thus $\mathbb{E}(X|\mathcal{H})$ satisfies the condition defining the conditional expectation of $\mathbb{E}(X|\mathcal{G})$ with respect to $\mathcal{H}$, so that $\mathbb{E}([\mathbb{E}(X|\mathcal{G})]|\mathcal{H}) = \mathbb{E}(X|\mathcal{H})$.

(v) A convex function can be written as the supremum of a sequence of affine functions, i.e. there are sequences $(a_n), (b_n)$ of reals such that $\varphi(x) = \sup_n(a_n x + b_n)$ for every $x \in \mathbb{R}$. Fix n, then since $\varphi(X(\omega)) \geq a_n X(\omega) + b_n$ for all ω, the positivity and linearity properties ensure that

$$\mathbb{E}(\varphi(X)|\mathcal{G})(\omega) \geq \mathbb{E}([a_n X + b_n]|\mathcal{G})(\omega) = a_n \mathbb{E}(X|\mathcal{G})(\omega) + b_n$$

for all $\omega \in \Omega \setminus A_n$ where $P(A_n) = 0$. Since $A = \bigcup_n A_n$ is also null, it follows that for all $n \geq 1$, $\mathbb{E}(\varphi(X)|\mathcal{G})(\omega) \geq a_n \mathbb{E}(X|\mathcal{G})(\omega) + b_n$ a.s. Hence the inequality also holds when we take the supremum on the right, so that $(\mathbb{E}(\varphi(X)|\mathcal{G})(\omega) \geq \varphi[(\mathbb{E}(X|\mathcal{G})(\omega)]$ a.s. This proves (v). $\square$

An immediate consequence of (v) is that the L^P-norm of $\mathbb{E}(X|\mathcal{G})$ is bounded by that of X for $p \geq 1$, since the function $\varphi(x) = |x|^p$ is then convex: we obtain

$$|\mathbb{E}(X|\mathcal{G})|^P = \varphi(\mathbb{E}(X|\mathcal{G})) \leq \mathbb{E}(\varphi(X)|\mathcal{G}) = \mathbb{E}(|X|^P|\mathcal{G}) \text{ a.s.}$$

so that

$$|\mathbb{E}(X|\mathcal{G})|^p_p = \mathbb{E}(|\mathbb{E}(X|\mathcal{G})|^P) \leq \mathbb{E}(\mathbb{E}(|X|^P|\mathcal{G})) = \mathbb{E}(|X|^P) = |X|^p_p,$$

where the penultimate step applies (i) of Proposition 7.45 to $|X|^p$. Take pth roots to have $|\mathbb{E}(X|\mathcal{G})|_p \leq |X|_p$.

Exercise 7.17

Let $\Omega = [0, 1]$ with Lebesgue measure and let $X(\omega) = \omega$. Find $\mathbb{E}(X|\mathcal{G})$ if

(a) $\mathcal{G} = \{[0, \frac{1}{2}], (\frac{1}{2}, 1], [0, 1], \varnothing\}$,

(b) $\mathcal{G}$ is generated by the family of sets $\{B \subset [0, \frac{1}{2}], \text{Borel}\}$.

7.4.2 Martingales

Suppose we wish to model the behaviour of some physical phenomenon by a sequence (X_n) of random variables. The value $X_n(\omega)$ might be the outcome of the nth toss of a 'fair' coin which is tossed 1000 times, with 'heads' recorded as 1, 'tails' as 0. Then $Y(\omega) = \sum_{n=1}^{1000} X_n(\omega)$ would record the number of times that the coin had landed 'heads'. Typically, we would perform this random experiment a large number of times before venturing to make statements about the probability of 'heads' for this coin. We could average our results, i.e. seek to compute $\mathbb{E}(Y)$. But we might also be interested in guessing what the value of $X_n(\omega)$ might be after $k < n$ tosses have been performed, i.e. for a fixed $\omega \in \Omega$, does knowing the values of $(X_i(\omega))_{i \leq k}$ give us any help in predicting the value of $X_n(\omega)$ for $n > k$? In an 'idealised' coin-tossing experiment it is assumed that it does not, that is, the successive tosses are assumed to be independent – a fact which often perplexes the beginner in probability theory.
    There are many situations where the (X_n) would represent outcomes where the past behaviour of the process being modelled can reasonably be taken to influence its future behaviour, e.g. if X_n records whether it rains on day n. We seek a mathematical description of the way in which our knowledge of past behaviour of (X_n) can be codified. A natural idea is to use the σ-field $\mathcal{F}_k = \sigma\{X_i : 0 \leq i \leq k\}$ generated by the sequence $(X_n)_{n \geq 0}$ as representing the knowledge gained from knowing the first k outcomes of our experiment. We call $(X_n)_{n \geq 0}$ a (discrete) *stochastic process* to emphasise that our focus is now

on the 'dynamics' of the sequence of outcomes as it unfolds. We include a 0th stage for notational convenience, so that there is a 'starting point' before the experiment begins, and then $\mathcal{F}_0$ represents our knowledge before any outcome is observed.

So the information available to us by 'time' k (i.e. after k outcomes have been recorded) about the 'state of the world' ω is given by the values $(X_i(\omega))_{0 \leq i \leq k}$ and this is encapsulated in knowing which sets of $\mathcal{F}_k$ contain the point ω. But we can postulate a sequence of σ-fields $(\mathcal{F}_n)_{n \geq 0}$ quite generally, without reference to any sequence of random variables. Again, our knowledge of any particular ω is then represented at stage $k \geq 1$ by knowing which sets in $\mathcal{F}_k$ contain ω. A simple example is provided by the binomial stock price model of Section 2.6.3 (see Exercise 2.13). Guided by this example, we turn this into a general definition.

Definition 7.47

Given a probability space $(\Omega, \mathcal{F}, P)$, a (discrete) *filtration* is an increasing sequence of sub-σ-fields $(\mathcal{F}_n)_{n \geq 0}$ of $\mathcal{F}$; i.e.

$$\mathcal{F}_0 \subset \mathcal{F}_1 \subset \mathcal{F}_2 \subset \cdots \subset \mathcal{F}_n \subset \cdots \subset \mathcal{F}.$$

We write $\mathbb{F} = (\mathcal{F}_n)_{n \geq 0}$. We say that the sequence $(X_n)_{n \geq 0}$ of random variables is *adapted* to the filtration $\mathbb{F}$ if X_n is $\mathcal{F}_n$-measurable for every $n \geq 0$. The tuple $(\Omega, \mathcal{F}, (\mathcal{F}_n)_{n \geq 0}, P)$ is called a *filtered probability space*.

We shall normally assume in our applications that $\mathcal{F}_0 = \{\emptyset, \Omega\}$, so that we begin with 'no information', and very often we shall assume that the 'final' σ-field generated by the whole sequence, i.e. $\mathcal{F}_\infty = \sigma(\bigcup_{n \geq 0} \mathcal{F}_n)$, is all of $\mathcal{F}$ (so that, by the end of the experiment, 'we know all there is to know'). Clearly (X_n) is adapted to its *natural* filtration $(\mathcal{F}_n)$, where $\mathcal{F}_n = \sigma(X_i : 0 \leq i \leq n)$ for each n, and it is adapted to every filtration which contains this one. But equally, if $\mathcal{F}_n = \sigma(X_i : 0 \leq i \leq n)$ for some process (X_n), it may be that for some other process (Y_n), each Y_n is $\mathcal{F}_n$-measurable, i.e. (Y_n) is adapted to $(\mathcal{F}_n)$. Recall that by Proposition 3.24 this implies that for each $n \geq 1$ there is a Borel-measurable function $f_n : \mathbb{R}^{n+1} \to \mathbb{R}$ such that $Y_n = f(X_0, X_1, X_2, \ldots, X_n)$.

We come to the main concept introduced in this section:

Definition 7.48

Let $(\Omega, \mathcal{F}, (\mathcal{F}_n)_{n \geq 0}, P)$ be a filtered probability space. A sequence of random variables $(X_n)_{n \geq 0}$ on $(\Omega, \mathcal{F}, P)$ is a *martingale* relative to the filtration $\mathbb{F} = (\mathcal{F}_n)_{n \geq 0}$, provided:

(i) (X_n) is adapted to $\mathbb{F}$,

(ii) each X_n is in $L^1(P)$,

(iii) for each $n \geq 0$, $\mathbb{E}(X_{n+1}|\mathcal{F}_n) = X_n$.

We note two immediate consequences of this definition which are used over and over again:

1. If $m > n \geq 0$ then $\mathbb{E}(X_m|\mathcal{F}_n) = X_n$. This follows from the tower property of conditional expectations, since (a.s.)

$$\mathbb{E}(X_m|\mathcal{F}_n) = \mathbb{E}(\mathbb{E}(X_m|\mathcal{F}_{m-1})|\mathcal{F}_n) = \mathbb{E}(X_{m-1}|\mathcal{F}_n) = \cdots = \mathbb{E}(X_{n+1}|\mathcal{F}_n) = X_n.$$

2. Any martingale (X_n) has constant expectation:

$$\mathbb{E}(X_n) = \mathbb{E}(\mathbb{E}(X_n|\mathcal{F}_0)) = \mathbb{E}(X_0)$$

holds for every $n \geq 0$, by 1) and by (i) in Proposition 7.45.

A martingale represents a 'fair game' in gambling: betting, for example, on the outcome of the coin tosses, our winnings in 'game n' (the outcome of the nth toss) would be $\Delta X_n = X_n - X_{n-1}$, that is the difference between what we had before and after that game. (We assume that $X_0 = 0$.) If the games are fair we would predict at time $(n-1)$, before the nth outcome is known, that $\mathbb{E}(\Delta X_n|\mathcal{F}_{n-1}) = 0$, where $\mathcal{F}_k = \sigma(X_i : i \leq k)$ are the σ-fields of the natural filtration of the process (X_n). This follows because our knowledge at time $(n-1)$ is encapsulated in $\mathcal{F}_{n-1}$ and in a fair game we would expect our incremental winnings at any stage to be 0 on average. Hence in this situation the (X_n) form a martingale.

Similarly, in a game favourable to the gambler we should expect that $\mathbb{E}(\Delta X_n|\mathcal{F}_{n-1}) \geq 0$, i.e. $\mathbb{E}(X_n|\mathcal{F}_{n-1}) \geq X_{n-1}$ a.s. We call a sequence satisfying this inequality (and (i), (ii) of Definition 7.48) a *submartingale*, while a game unfavourable to the gambler (hence favourable to the casino!) is represented similarly by a *supermartingale*, which has $\mathbb{E}(X_n|\mathcal{F}_{n-1}) \leq X_{n-1}$ a.s. for every n. Note that for a submartingale the expectations of the (X_n) increase with n, while for a supermartingale they decrease. Finally, note that the properties of these processes do not change if we replace X_n by $X_n - X_0$ (as long as $X_0 \in L^1(\mathcal{F}_0)$, to retain integrability and adaptedness), so that we can work without loss of generality with processes that start with $X_0 = 0$.

Example 7.49

The most obvious, yet in some ways quite general, example of a martingale consists of a sequence of conditional expectations: given a random variable

$X \in L^1(\mathcal{F})$ and a filtration $(\mathcal{F}_n)_{n \geq 0}$ of sub-σ-fields of $\mathcal{F}$, let $X_n = \mathbb{E}(X|\mathcal{F}_n)$ for every n. Then $\mathbb{E}(X_{n+1}|\mathcal{F}_n) = \mathbb{E}(\mathbb{E}(X|\mathcal{F}_{n+1})|\mathcal{F}_n) = \mathbb{E}(X|\mathcal{F}_n) = X_n$, using the tower property again. We can interpret this by regarding each X_n as giving us the information available at time n, i.e. contained in the σ-field $\mathcal{F}_n$, about the random variable X. (Remember that the conditional expectation is the 'best guess' of X, with respect to mean-square errors, when we work in L^2.) For a finite filtration $\{\mathcal{F}_n : 0 \leq n \leq N\}$ with $\mathcal{F}_N = \mathcal{F}$ it is obvious that $\mathbb{E}(X|\mathcal{F}_N) = X$. For an infinite sequence we might hope similarly that 'in the limit' we will have 'full' information about X, which suggests that we should be able to retrieve X as the limit of the (X_n) in some sense. The conditions under which limits exist require careful study — see e.g. [13] and [5] for details.

A second standard example of a martingale is:

Example 7.50

Suppose $(Z_n)_{n \geq 1}$ is a sequence of independent random variables with zero mean. Let $X_0 = 0, \mathcal{F}_0 = \{\varnothing, \Omega\}$, set $X_n = \sum_{k=1}^n Z_k$ and define $\mathcal{F}_n = \sigma(Z_k : k \leq n)$ for each $n \geq 1$. Then $(X_n)_{n \geq 0}$ is a martingale relative to the filtration $(\mathcal{F}_n)$. To see this recall that for each n, Z_n is independent of $\mathcal{F}_{n-1}$, so that $\mathbb{E}(Z_n|\mathcal{F}_{n-1}) = \mathbb{E}(Z_n) = 0$. Hence $\mathbb{E}(X_n|\mathcal{F}_{n-1}) = \mathbb{E}(X_{n-1}|\mathcal{F}_{n-1}) + \mathbb{E}(Z_n) = X_{n-1}$, since X_{n-1} is $\mathcal{F}_{n-1}$-measurable. (You should check carefully which properties of the conditional expectation we used here!)

A 'multiplicative' version of this example is the following:

Exercise 7.18

Let $Z_n \geq 0$ be a sequence of independent random variables with $\mathbb{E}(Z_n) = \mu = 1$. Let $\mathcal{F}_n = \sigma\{Z_k : k \leq n\}$ and show that, $X_0 = 1$, $X_n = Z_1 Z_2 \ldots Z_n$ $(n \geq 1)$ defines a martingale for $(\mathcal{F}_n)$, provided all the products are integrable random variables, which holds, e.g., if all $Z_n \in L^\infty(\Omega, \mathcal{F}, P)$.

Exercise 7.19

Let $(Z_n)_{n \geq 1}$ be a sequence of independent random variables with mean $\mu = \mathbb{E}(Z_n) \neq 0$ for all n. Show that the sequence of their partial sums $X_n = Z_1 + Z_2 + \cdots + Z_n$ is not a martingale for the filtration $(\mathcal{F}_n)_n$, where $\mathcal{F}_n = \sigma\{Z_k : k \leq n\}$. How can we 'compensate' for this by altering X_n?

7.4.3 Doob decomposition

Let $X = (X_n)_{n \geq 0}$ be a martingale for the filtration $\mathbb{F} = (\mathcal{F}_n)_{n \geq 0}$ (with our above conventions); briefly we simply refer to the martingale $(X, \mathbb{F})$. The quadratic function is convex, hence by Jensen's inequality (Theorem 7.46) we have $\mathbb{E}(X_{n+1}^2|\mathcal{F}_n) \geq (\mathbb{E}(X_{n+1}|\mathcal{F}_n))^2 = X_n^2$, so X^2 is a submartingale. We investigate whether it is possible to 'compensate', as in Exercise 7.19, to make the resulting process again a martingale. Note that the expectations of the X_n^2 are increasing, so we will need to subtract an increasing process from X^2 to achieve this.

In fact, the construction of this 'compensator' process is quite general. Let $Y = (Y_n)$ be any adapted process with each $Y_n \in L^1$. For any process Z write its increments as $\Delta Z_n = Z_n - Z_{n-1}$ for all n. Recall that in this notation the martingale property can be expressed succinctly as $\mathbb{E}(\Delta Z_n|\mathcal{F}_{n-1}) = 0$ – we shall use this repeatedly in what follows.

We define two new processes $A = (A_n)$ and $M = (M_n)$ with $A_0 = 0, M_0 = 0$, via their successive increments,

$$\Delta A_n = \mathbb{E}(\Delta Y_n|\mathcal{F}_{n-1}) \text{ and } \Delta M_n = \Delta Y_n - \Delta A_n \text{ for } n \geq 1.$$

We obtain $\mathbb{E}(\Delta M_n|\mathcal{F}_{n-1}) = \mathbb{E}([\Delta Y_n - \mathbb{E}(\Delta Y_n|\mathcal{F}_{n-1})]|\mathcal{F}_{n-1}) = \mathbb{E}(\Delta Y_n|\mathcal{F}_{n-1}) - \mathbb{E}(\Delta Y_n|\mathcal{F}_{n-1}) = 0$, as $\mathbb{E}(\Delta Y_n|\mathcal{F}_{n-1})$ is $\mathcal{F}_{n-1}$-measurable. Hence M is a martingale. Moreover, the process A is increasing if and only if $0 \leq \Delta A_n = \mathbb{E}(\Delta Y_n|\mathcal{F}_{n-1}) = \mathbb{E}(Y_n|\mathcal{F}_{n-1}) - Y_{n-1}$, which holds if and only if Y is a submartingale. Note that $A_n = \sum_{k=1}^n \Delta A_k = \sum_{k=1}^n [\mathbb{E}(Y_k|\mathcal{F}_{k-1}) - Y_{k-1}]$ is $\mathcal{F}_{n-1}$-measurable. Thus the value of A_n is 'known' by time $n - 1$. A process with this property is called *predictable*, since we can 'predict' its future values one step ahead. It is a fundamental property of martingales that they are *not* predictable: in fact, if X is a predictable martingale, then we have

$$X_{n-1} = \mathbb{E}(X_n|\mathcal{F}_{n-1}) = X_n \text{ a.s. for every } n$$

where the first equality is the definition of martingale, while the second follows since X_n is $\mathcal{F}_{n-1}$-measurable. Hence a predictable martingale is a.s. constant, and if it starts at 0 it will stay there. This fact gives the decomposition of an adapted process Y into the sum of a martingale and a predictable process a useful uniqueness property: first, since $M_0 = 0 = A_0$, we have $Y_n = Y_0 + M_n + A_n$ for the processes M, A defined above. If also $Y_n = Y_0 + M_n' + A_n'$, where M_n' is a martingale, and A_n' is predictable, then

$$M_n - M_n' = A_n' - A_n \text{ a.s.}$$

is a predictable martingale, 0 at time 0. Hence both sides are 0 for every n and so the decomposition is a.s. unique.

We call this the *Doob decomposition* of an adapted process. It takes on special importance when applied to the submartingale $Y = X^2$ which arises from a martingale X. In that case, as we saw above, the predictable process A is increasing, so that $A_n \leq A_{n+1}$ a.s. for every n, and the Doob decomposition reads:

$$X^2 = X_0^2 + M + A.$$

In particular, if $X_0 = 0$ (as we can assume without loss of generality), we have written $X^2 = M + A$ as the sum of a martingale M and a predictable increasing process A. The significance of this is revealed in a very useful property of martingales, which was a key component of the proof of the Radon–Nikodym theorem (see Step 1 (iv) of Theorem 7.5, where the martingale connection is well hidden!): for any martingale X we can write, with $(\Delta X_n)^2 = (X_n - X_{n-1})^2$:

$$
\begin{aligned}
\mathbb{E}(\Delta X_n)^2|\mathcal{F}_{n-1}) &= \mathbb{E}([X_n^2 - 2X_nX_{n-1} + X_{n-1}^2]|\mathcal{F}_{n-1}) \\
&= \mathbb{E}(X_n^2|\mathcal{F}_{n-1}) - 2X_{n-1}\mathbb{E}(\Delta X_n|\mathcal{F}_{n-1}) + X_{n-1}^2 \\
&= \mathbb{E}([X_n^2 - X_{n-1}^2]|\mathcal{F}_{n-1}),
\end{aligned}
$$

where in the second step we took out 'what is known', then used the martingale property and cancelled the resulting terms. Hence given the martingale X with $X_0 = 0$, the decomposition $X^2 = M + A$ yields, since M is also a martingale:

$$
\begin{aligned}
0 = \mathbb{E}(\Delta M_n|\mathcal{F}_{n-1}) &= \mathbb{E}((\Delta X_n)^2 - \Delta A_n)|\mathcal{F}_{n-1}) \\
&= \mathbb{E}([X_n^2 - X_{n-1}^2]|\mathcal{F}_{n-1}) - \mathbb{E}(\Delta A_n|\mathcal{F}_{n-1}).
\end{aligned}
$$

In other words, since A is predictable,

$$\mathbb{E}((\Delta X_n)^2|\mathcal{F}_{n-1}) = \mathbb{E}(\Delta A_n|\mathcal{F}_{n-1}) = \Delta A_n, \qquad (7.5)$$

which exhibits the process A as a conditional 'quadratic variation' process of the original martingale X. Taking expectations: $\mathbb{E}((\Delta X_n)^2) = \mathbb{E}(\Delta A_n)$.

Example 7.51

Note also that $\mathbb{E}(X_n^2) = \mathbb{E}(M_n) + \mathbb{E}(A_n) = \mathbb{E}(A_n)$ (why?), so that both sides are bounded for all n if and only if the martingale X is bounded as a sequence in $L^2(\Omega, \mathcal{F}, P)$. Since (A_n) is increasing, the a.s. limit $A_\infty(\omega) = \lim_{n\to\infty} A_n(\omega)$ exists, and the boundedness of the integrals ensures in that case that $\mathbb{E}(A_\infty) < \infty$.

Exercise 7.20

Suppose $(Z_n)_{n\geq 1}$ is a sequence of Bernoulli random variables, with each Z_n taking the values 1 and -1, each with probability $\frac{1}{2}$. Let $X_0 = 0$,

$X_n = Z_1 + Z_2 + \cdots + Z_n$, and let $(\mathcal{F}_n)$ be the natural filtration generated by the (Z_n). Verify that (X_n^2) is a submartingale, and find the increasing process (A_n) in its Doob decomposition. What 'unexpected' property of (A_n) can you detect in this example?

In the discrete setting we now have the tools to construct 'stochastic integrals' and show that they preserve the martingale property. In fact, as we saw for Lebesgue–Stieltjes measures, for discrete distributions the 'integral' is simply an appropriate linear combination of increments of the distribution function. If we wish to use a martingale X as an integrator, we therefore need to deal with linear combinations of the increments $\Delta X_n = X_n - X_{n-1}$. Since we are now dealing with stochastic processes (that it, functions of both n and ω) rather than real functions, measurability conditions will help determine what constitutes an 'appropriate' linear combination. So, if for $\omega \in \Omega$ we set $I_0(\omega) = 0$ and form sums

$$I_n(\omega) = \sum_{k=1}^{n} c_k(\omega)(\Delta X_k)(\omega) = \sum_{k=1}^{n} c_k(\omega)(X_k(\omega) - X_{k-1}(\omega)) \text{ for } n \geq 1,$$

we look for measurability properties of the process (c_n) which ensure that the new process (I_n) has useful properties. We investigate this when (c_n) is a bounded predictable process and X is a martingale for a given filtration $(\mathcal{F}_n)$. Some texts call the process (I_n) a *martingale transform* — we prefer the term *discrete stochastic integral*. We calculate the conditional expectation of I_n:

$$\mathbb{E}(I_n|\mathcal{F}_{n-1}) = \mathbb{E}([I_{n-1} + c_n\Delta X_n]|\mathcal{F}_{n-1}) = I_{n-1} + c_n\mathbb{E}(\Delta X_n|\mathcal{F}_{n-1}) = I_{n-1},$$

since c_n is $\mathcal{F}_{n-1}$-measurable and $\mathbb{E}(\Delta X_n|\mathcal{F}_{n-1}) = \mathbb{E}(X_n|\mathcal{F}_{n-1}) - X_{n-1} = 0$. Therefore, when the process $c = (c_n)$ which is integrated against the martingale $X = (X_n)$, is predictable, the martingale property is preserved under the discrete stochastic integral: $I = (I_n)$ is also a martingale with respect to the filtration $(\mathcal{F}_n)$. We shall write this stochastic integral as $c \cdot X$, meaning that for all $n \geq 0$, $I_n = (c \cdot X)_n$. The result has sufficient importance for us to record it as a theorem:

Theorem 7.52

Let $(\Omega, \mathcal{F}, (\mathcal{F}_n)_{n \geq 0}, P)$ be a filtered probability space. If X is a martingale and c is a bounded predictable process, then the discrete stochastic integral $c \cdot X$ is again a martingale.

Note that we use the boundedness assumption in order to ensure that $c_k \Delta X_k$ is integrable, so that its conditional expectation makes sense. For L^2-martingales (which are what we obtain in most applications) we can relax this condition and demand merely that $c_n \in L^2(\mathcal{F}_{n-1})$ for each n.

While the preservation of the martingale property may please mathematicians, it is depressing news for gamblers! We can interpret the process c as representing the size of the stake the gambler ventures in every game, so that c_n is the amount (s)he bets in game n. Note that c_n could be 0, which mean that the gambler 'sits out' game n and places no bet. It also seems reasonable that the size of the stake depends on the outcomes of the previous games, hence c_n is $\mathcal{F}_{n-1}$-measurable, and thus c is predictable.

The conclusion that $c \cdot X$ is then a martingale means that 'clever' gambling strategies will be of no avail when the game is fair. It remains fair, whatever strategy the gambler employs! And, of course, if it starts out unfavourable to the gambler, so that X is a supermartingale ($X_{n-1} \geq \mathbb{E}(X_n|\mathcal{F}_{n-1})$), the above calculation shows that, as long as $c_n \geq 0$ for each n, then $\mathbb{E}(I_n|\mathcal{F}_{n-1}) \leq I_{n-1}$, so that the game remains unfavourable, whatever non-negative stakes the gambler places (and negative bets seem unlikely to be accepted, after all ...). You will verify immediately, of course, that a submartingale X produces a submartingale $c \cdot X$ when c is a non-negative process. Sadly, such favourable games are hard to find in practice.

Combining the definition of (I_n) with the Doob decomposition of the submartingale X^2, we obtain the identity which illustrates why martingales make useful 'integrators'. We calculate the expected value of the square of $(c \cdot X)_n$ when $c = (c_n)$ is predictable and $X = (X_n)$ is a martingale:

$$\mathbb{E}((c \cdot X)_n^2) = \mathbb{E}([\sum_{k=1}^n c_k \Delta X_k]^2) = \mathbb{E}(\sum_{j,k=1}^n c_j c_k \Delta X_j \Delta X_k).$$

Consider terms in the double sum separately: when $j < k$ we have

$$\mathbb{E}(c_j c_k \Delta X_j \Delta X_k) = \mathbb{E}(c_j c_k \Delta X_j \Delta X_k|\mathcal{F}_{k-1}) = \mathbb{E}(c_j c_k \Delta X_j \mathbb{E}(\Delta X_k|\mathcal{F}_{k-1})) = 0$$

since the first three factors are all $\mathcal{F}_{k-1}$-measurable, while $\mathbb{E}(\Delta X_k|\mathcal{F}_{k-1}) = 0$ since X is a martingale. With j, k interchanged this also shows that these terms are 0 when $k < j$.

The remaining terms have the form

$$\mathbb{E}(c_k^2(\Delta X_k)^2) = \mathbb{E}(c_k^2 \mathbb{E}((\Delta X_k)^2|\mathcal{F}_{k-1})) = \mathbb{E}(c_k^2 \Delta A_k).$$

By linearity, therefore, we have the fundamental identity for stochastic integrals relative to martingales (also called the *Itô isometry*):

$$\mathbb{E}([\sum_{k=1}^n c_k \Delta X_k]^2) = \mathbb{E}(\sum_{k=1}^n c_k^2 \Delta A_k).$$

Remark 7.53

The sum inside the expectation sign on the right is a 'Stieltjes sum' for the increasing process, so that it is now at least plausible that this identity allows us to define martingale integrals in the continuous-time setting, using approximation of processes by simple processes, much as was done throughout this book for real functions. The Itô isometry is of critical importance in the definition of stochastic integrals relative to processes such as Brownian motion: in defining Lebesgue–Stieltjes integrals our integrators were of bounded variation. Typically, the paths of Brownian motion (a process we shall not discuss in detail in this book — see (e.g.) [3] for its basic properties) are not of bounded variation, but the Itô isometry shows that their quadratic variation can be handled in the (much subtler) continuous-time version of the above framework, and this enables one to define integrals of a wide class of functions, using Brownian motion (and more general martingales) as the 'integrator'.

We turn finally to the idea of *stopping* a martingale at a random time.

Definition 7.54

A random variable $\tau : \Omega \rightarrow \{0, 1, 2, \ldots, n, \ldots\} \cup \{\infty\}$ is a *stopping time* relative to the filtration $(\mathcal{F}_n)$ if for every $n \geq 1$, the event $\{\tau = n\}$ belongs to $\mathcal{F}_n$.

Note that we include the value $\tau(\omega) = \infty$, so that we need $\{\tau = \infty\} \in \mathcal{F}_\infty = \sigma(\bigcup_{n \geq 1} \mathcal{F}_n)$, the 'limit σ-field'. Stopping times are also called *random times*, to emphasise that the 'time' τ is a random variable.

For a stopping time τ the event $\{\tau \leq n\} = \bigcup_{k=0}^{n} \{\tau = k\}$ is in $\mathcal{F}_n$ since for each $k \leq n$, $\{\tau = k\} \in \mathcal{F}_k$ and the σ-fields increase with n. On the other hand, given that for each n the event $\{\tau \leq n\} \in \mathcal{F}_n$, then

$$\{\tau = n\} = \{\tau \leq n\} \backslash \{\tau \leq n - 1\} \in \mathcal{F}_n.$$

Thus we could equally well have taken the condition $\{\tau \leq n\} \in \mathcal{F}_n$ for all n as the definition of stopping time.

Example 7.55

A gambler may decide to stop playing after a random number of games, depending on whether the winnings X have reached a predetermined level L (or his funds are exhausted!). The time $\tau = \min\{n : X_n \geq L\}$ is the first time at which the process X hits the interval $[L, \infty)$; more precisely, for $\omega \in \Omega$, $\tau(\omega) = n$ if $X_n(\omega) \geq L$ while $X_k(\omega) < L$ for all $k < n$. Since $\{\tau = n\}$ is thus determined by the values of X and those of the X_k for $k < n$ it is now clear that τ is a stopping time.

Example 7.56

Similarly, we may decide to sell our shares in a stock S if its value falls below 75% of its current (time 0) price. Thus we sell at the random time $\tau = \min\{n : S_n < \frac{3}{4}S_0\}$, which is again a stopping time. This is an example of a 'stop-loss strategy', and is much in evidence in a bear market.

Quite generally, the *first hitting time* τ_A of a Borel set $A \subset \mathbb{R}$ by an adapted process X is defined by setting $\tau_A = \min\{n \geq 0 : X_n \in A\}$. For any $n \geq 0$ we have $\{\tau_A \leq n\} = \bigcup_{k \leq n}\{X_k \in A\} \in \mathcal{F}_n$. To cater for the possibility that X never hits A we use the convention $\min \emptyset = \infty$, so that $\{\tau_A = \infty\} = \Omega \backslash (\bigcup_{n \geq 0}\{\tau_A \leq n\})$ represents this event. But its complement is in $\mathcal{F}_\infty = \sigma(\bigcup_{n \geq 0} \mathcal{F}_n)$, thus so is $\{\tau_A = \infty\}$. We have proved that τ_A is a stopping time.

Returning to the gambling theme, we see that stopping is simply a particular form of gambling strategy, and it should thus come as no surprise that the martingale property is preserved under stopping (with similar conclusions for super- and submartingales). For any adapted process X and stopping time τ, we define the *stopped process* X^τ by setting $X_n^\tau(\omega) = X_{n \wedge \tau(\omega)}(\omega)$ at each $\omega \in \Omega$. (Recall that for real x, y we write $x \wedge y = \min\{x, y\}$.)

The stopped process X^τ is again adapted to the filtration $(\mathcal{F}_n)$, since $\{X_{\tau \wedge n} \in A\}$ means that either $\tau > n$ and $X_n \in A$, or $\tau = k$ for some $k \leq n$ and $X_k \in A$. Now the event $\{X_n \in A\} \cap \{\tau > n\} \in \mathcal{F}_n$, while for each k the event $\{\tau = k\} \cap \{X_k \in A\} \in \mathcal{F}_k$. For all $k \leq n$ these events therefore all belong to $\mathcal{F}_n$. Hence so does $\{X_{\tau \wedge n} \in A\}$, which proves that X^τ is adapted.

Theorem 7.57

Let $(\Omega, \mathcal{F}, (\mathcal{F}_n), P)$ be a filtered probability space, and let X be a martingale with $X_0 = 0$. If τ is a stopping time, the stopped process X^τ is again a martingale.

Proof

We use the preservation of the martingale property under discrete stochastic integrals ('gambling strategies'). Let $c_n = \mathbf{1}_{\{\tau \geq n\}}$ for each $n \geq 1$. This defines a bounded predictable process $c = (c_n)$, since it takes only the values $0, 1$ and $\{c_n = 0\} = \{\tau \leq n - 1\} \in \mathcal{F}_{n-1}$, so that also $\{c_n = 1\} = \Omega \backslash \{c_n = 0\} \in \mathcal{F}_{n-1}$. Hence by Theorem 7.52 the process $c \cdot X$ is again a martingale. But by construction $(c \cdot X)_0 = 0 = X_0 = X_0^\tau$, while for any $n \geq 1$,

$$(c \cdot X)_n = c_1(X_1 - X_0) + c_2(X_2 - X_1) + \cdots + c_n(X_n - X_{n-1}) = X_{\tau \wedge n},$$

hence X^τ is a martingale.   □

Since $c_n \geq 0$ as defined in the proof, it follows that the supermartingale and submartingale properties are also preserved under stopping. For a martingale we have, in addition, that expectation is preserved, i.e. (in general) $\mathbb{E}(X_{\tau \wedge n}) = \mathbb{E}(X_0)$. Similarly, expectations increase for stopped submartingales, and decrease for stopped supermartingales.

None of this, however, guarantees that the random variable X_τ defined by $X_\tau(\omega) = X_{\tau(\omega)}(\omega)$ has finite expectation – to obtain a result which relates its expectation to that of X_0 we generally need to satisfy much more stringent conditions. For bounded stopping times (where there is a uniform upper bound N with $\tau(\omega) \leq N$ for all $\omega \in \Omega$), matters are simple: if X is a martingale, $X_{\tau \wedge n}$ is integrable for all n, and by the above theorem $\mathbb{E}(X_{\tau \wedge n}) = \mathbb{E}(X_0)$. Now apply this with $n = N$, so that $X_{\tau \wedge n} = X_{\tau \wedge N} = X_\tau$. Thus we have $\mathbb{E}(X_\tau) = \mathbb{E}(X_0)$ whenever τ is a bounded stopping time. We shall not delve any further, but refer the reader instead to texts devoted largely to martingale theory, e.g. [13], [8]. Bounded stopping times suffice for many practical applications, for example in the analysis of discrete American options in finance.

7.4.4 Applications to mathematical finance

A major task and challenge for the theory of finance is to price assets and securities by building models consistent with market practice. This consistency means that any deviation from the theoretical price should be penalized by the market. Specifically, if a market player quotes a price different from the price provided by the model, she should be bound to lose money.

The problem is the unknown future which has somehow to be reflected in the price since market participants express their beliefs about the future by agreeing on prices. Mathematically, an ideal situation is where the price process $X(t)$ is a martingale. Then we would have the obvious pricing formula $X(0) = \mathbb{E}(X(T))$ and in addition we would have the whole range of formulae for the intermediate prices by means of conditional expectation based on information gathered.

However, the situation where the prices follow a martingale is incompatible with the market fact that money can be invested risk-free, which creates a benchmark for expectations for investors investing in risky securities. So we modify our task by insisting that the discounted values $Y(t) = X(t) \exp\{-rt\}$

form a martingale. The modification is by means of a deterministic constant, so it does not create a problem for asset valuation.

A particular goal is pricing derivative securities where we are given the terminal value (a claim) of the form $f(S(T))$, where f is known and the probability distributions of the values of the underlying asset $S(t)$ are assumed to be known by taking some mathematical model. The above martingale idea would solve the pricing problem by constructing a process $X(t)$ in such a way that $X(T) = f(S(T))$ (We call X a *replication* of the claim.)

We can summarise the tasks: build a model of the prices of the underlying security $S(t)$ such that

1. there is a replication $X(t)$ such that $X(T) = f(S(T))$,

2. the process $Y(t) = X(t) \exp\{-rt\}$ is a martingale, so $Y(0)$ is the price of the security described by f,

3. any deviation from the resulting prices leads to a loss.

Steps 1 and 2 are mathematical in nature, but Step 3 is related to real market activities.

We perform the task for the single-step binomial model.

Step 1. Recall that the prices in this model are $S(0)$, $S(1) = S(0)\eta$ where $\eta = U$ or $\eta = D$ with some probabilities. Let $f(x) = (x - K)^+$. We can easily find $X(0)$, so that $X(1) = (S(1) - K)^+$ by building a portfolio of n shares S and m units of bank account (see Section 6.5.5) after solving the system

$$nS(0)U + mR = (S(0)U - K)^+,$$
$$nS(0)D + mR = (S(0)D - K)^+.$$

Note that $X(1)$ is obtained by means of the data at time $t = 0$ and the model parameters.

Step 2. Write $R = \exp\{r\}$. The martingale condition we need is trivial: $X(0)R = \mathbb{E}(X(1))$. The task here is to find the probability measure (X is defined, we have no influence on R, so this is the only parameter we can adjust). Recall that we assume $D < R < U$. We solve

$$X(0)R = pX(0)U + (1 - p)X(0)D$$

which gives

$$p = \frac{R - D}{U - D}$$

and we are done. Hence the theoretical price of a call is

$$C = R^{-1}p(S(0)U - K) = \exp\{-r\}\mathbb{E}(S(1) - K)^+.$$

Step 3. To see that within our model the price is right suppose that someone is willing to buy a call for $C' > C$. We sell it, immediately investing the proceeds in the portfolio from Step 1. At exercise our portfolio matches the call payoff and has earned the difference $C' - C$. So we keep buying the call until the seller realises the mistake and raises the price. Similarly if someone is selling a call at $C' < C$ we generate cash by forming the $(-n, -m)$ portfolio, buy a call (which, as a result of replication, settles our portfolio liability at maturity) and have profit until the call prices quoted hit the theoretical price C.

The above analysis summarises the key features of a general theory. A straightforward extension of this trivial model to n steps gives the so-called CRR (Cox–Ross–Rubinstein) model, which for large n is quite adequate for realistic pricing. We evaluated the expectation in the binomial model to establish the CRR price of a European call option in Proposition 4.56.

For continuous time the task becomes quite difficult. In the Black–Scholes model $S(t) = S(0) \exp\{(r - \frac{1}{2}\sigma^2)t + \sigma w(t)\}$, where $w(t)$ is a stochastic process (called the *Wiener process* or *Brownian motion*) with $w(0) = 0$ and independent increments such that $w(t) - w(s)$ is Gaussian with zero mean and variance $t - s$, $s < t$.

Exercise 7.21

Show that $\exp\{-rt\}S(t)$ is a martingale.

Existence of the replication process $X(t)$ is not easy but can be proved, as well as the fact that the process $Y(t)$ is a martingale. (Recall that we proved the martingale property of the discounted prices in the binomial model.) This results in the same general pricing formula: $\exp\{-rT\}\mathbb{E}(f(S(T)))$. Using the density of $w(T)$, this number can be written in an explicit form for particular derivative securities (see Section 4.7.5 where the formulae for the prices of call and put options were derived).

Remark 7.58

The reader familiar with finance theory will notice that we are focused on pricing derivative securities and this results in considering the model where the discounted prices form a martingale. This model is a mathematical creation not necessarily consistent with real life, which requires a different probability space and different parameters within the same framework. The link between the real world and the martingale one is provided by a special application of Radon–Nikodym theorem which links the probability measures, but this is a story we shall not pursue here; we refer the reader to numerous books devoted to the subject (for example [5]).

7.5 Proofs of propositions

Proof (of Proposition 7.2)

Suppose the (ε, δ)-condition fails. With A as in the hint, we have $\mu(A) \leq \mu(\bigcup_{i \geq n} F_i) \leq \sum_{i=n}^{\infty} \frac{1}{2^i} = \frac{1}{2^{n-1}}$ for every $n \geq 1$. Thus $\mu(A) = 0$. But $\nu(F_n) \geq \varepsilon$ for every n, hence $\nu(E_n) \geq \varepsilon$, where $E_n = \bigcup_{i \geq n} F_i$. The sequence (E_n) decreases, so that, as $\nu(F_1)$ is finite, Theorem 2.19 (i) gives: $\nu(A) = \nu(\lim E_n) = \lim \nu(E_n) \geq \varepsilon$. Thus ν is not absolutely continuous with respect to μ. Conversely, if the (ε, δ)-condition holds, and $\mu(F) = 0$, then $\mu(F) < \delta$ for any $\delta > 0$, and so, for every given $\varepsilon > 0$, $\nu(F) < \varepsilon$. Hence $\nu(F) = 0$. So $\nu \ll \mu$. □

Proof (of Proposition 7.6)

We proceed as in Remark 4.18. Begin with g as the indicator function $\mathbf{1}_G$ for $G \in \mathcal{G}$. Then we have: $\int_\Omega g \, d\mu = \mu(G) = \int_G h_\mu \, d\varphi$ by construction of h_μ. Next, let $g = \sum_{i=1}^n a_i \mathbf{1}_{G_i}$ for sets $G_1, G_2, \ldots, G_n$ in $\mathcal{G}$, and reals $a_1, a_2, \ldots, a_n$; then linearity of the integrals yields

$$\int_\Omega g \, d\mu = \sum_{i=1}^n a_i \mu(G_i) = \sum_{i=1}^n a_i \left(\int_{G_i} h_\mu \, d\varphi \right) = \sum_{i=1}^n a_i \left(\int_\Omega \mathbf{1}_{G_i} h_\mu \, d\varphi \right) = \int_\Omega g h_\mu \, d\varphi.$$

Finally, any $\mathcal{G}$-measurable non-negative function g is approximated from below by an increasing sequence of $\mathcal{G}$-simple functions $g_n = \sum_{i=1}^n a_i \mathbf{1}_{G_i}$; its integral $\int_\Omega g \, d\mu$ is the limit of the increasing sequence $(\int_\Omega g_n h_\mu \, d\varphi)_n$. But since $0 \leq h_\mu \leq 1$ by construction, $(g_n h_\mu)$ increases to $g h_\mu$ pointwise, hence the sequence $(\int_\Omega g_n h_\mu \, d\varphi)_n$ also increases to $\int_\Omega g h_\mu \, d\varphi$, so the limits are equal. For integrable $g = g^+ - g^-$ apply the above to each of g^+, g^- separately, and use linearity. □

Proof (of Proposition 7.8)

If $\bigcup_{n \geq 1} A_n = \Omega$ and the A_n are not disjoint, replace them by E_n, where $E_1 = A_1$, $E_n = A_n \setminus (\bigcup_{i=1}^{n-1} E_i)$, $n > 1$. The same can be done for the B_m and hence we can take both sequences as disjoint. Now $\Omega = \bigcup_{m,n \geq 1} (A_n \cap B_m)$ is also a disjoint union, and ν, μ are both finite on each $A_n \cap B_m$. Re-order these sets into a single sequence $(C_n)_{n \geq 1}$ and fix $n \geq 1$. Restricting both measures to the σ-field $\mathcal{F}_n = \{F \cap C_n : F \in \mathcal{F}\}$ yields them as finite measures on $(\Omega, \mathcal{F}_n)$, so that the Radon–Nikodym theorem applies, and provides a non-negative $\mathcal{F}_n$-measurable function h_n such that $\nu(E) = \int_E h \, d\mu$ for each $E \in \mathcal{F}_n$. But any set $F \in \mathcal{F}$ has the form $F = \bigcup_n F_n$ for $F_n \in \mathcal{F}_n$, so we can define h by setting $h = h_n$ for every $n \geq 1$. Now $\nu(F) = \sum_{n=1}^{\infty} \int_{F_n} h_n \, d\mu = \int_F h \, d\mu$. The uniqueness is clear: if g has the same properties as h, then $\int_F (h - g) \, d\mu = 0$ for each $F \in \mathcal{F}$, so $h - g = 0$ a.e. by Theorem 4.22. $\square$

Proof (of Proposition 7.9)

(i) This is trivial, since $\varphi = \lambda + \nu$ is σ-finite and absolutely continuous with respect to μ, and we have, for $F \in \mathcal{F}$:

$$\int_F \frac{d\phi}{d\mu} \, d\mu = \varphi(F) = (\lambda + \nu)(F) = \lambda(F) + \nu(F) = \int_F [\frac{d\lambda}{d\mu} + \frac{d\nu}{d\mu}] \, d\mu.$$

The integrands on the right and left extremes are thus a.s. (μ) equal, so the result follows.

(ii) Write $\frac{d\lambda}{d\nu} = g$ and $\frac{d\nu}{d\mu} = h$. These are non-negative measurable functions and we need to show that, for $F \in \mathcal{F}$

$$\lambda(F) = \int_F gh \, d\mu.$$

First consider this when g is replaced by a simple function of the form $\phi = \sum_{i=1}^{n} c_i \mathbf{1}_{E_i}$. Then we obtain:

$$\int_F \phi \, d\nu = \sum_{i=1}^{n} c_i \nu(F \cap E_i) = \sum_{i=1}^{n} c_i \int_{F \cap E_i} h \, d\mu = \int_F \phi h \, d\mu.$$

Now let (ϕ_n) be a sequence of simple functions increasing pointwise to g. Then by monotone convergence theorem:

$$\lambda(F) = \int_F g \, d\nu = \lim \int_F \phi_n \, d\nu = \lim \int_F \phi_n h \, d\mu = \int_F gh \, d\mu,$$

since $(\phi_n h)$ increases to gh. This proves our claim. $\square$

Proof (of Proposition 7.11)

Use the hint: $\lambda_1 + \lambda_2$ is concentrated on $A_1 \cup A_2$, while μ is concentrated on $B_1 \cap B_2$. But $A_1 \cup A_2$ is disjoint from $B_1 \cap B_2$, hence the measures $\lambda_1 + \lambda_2$ and μ are mutually singular. This proves (i). For (ii), choose a set E for which $\mu(E) = 0$ while λ_2 is concentrated on E. Let $F \subset E$, so that $\mu(F) = 0$ and hence $\lambda_1(F) = 0$ (since $\lambda_1 \ll \mu$). This shows that λ_1 is concentrated on E^c, hence λ_1 and λ_2 are mutually singular. Finally, (ii) applied with $\lambda_1 = \lambda_2 = \nu$ shows that $\nu \perp \nu$, which can only happen when $\nu = 0$.   □

Proof (of Proposition 7.15)

Fix $x \in \mathbb{R}$. The set $A = \{F(y) : y < x\}$ is bounded above by $F(x)$, while $B = \{F(y) : x < y\}$ is bounded below by $F(x)$. Hence $\sup A = K_1 \leq F(x)$, and $\inf B = K_2 \geq F(x)$ both exist in $\mathbb{R}$ and for any $\varepsilon > 0$, we can find $y_1 < x$ such that $K_1 - \varepsilon < F(y_1)$ and $y_2 > x$ such that $K_2 + \varepsilon > F(y_2)$. But since F is increasing this means that $K_1 - \varepsilon < F(y) < K_1$ throughout the interval (y_1, x) and $K_2 < F(y) < K_2 + \varepsilon$ throughout (y, y_2). Thus both one-sided limits $F(x-) = \lim_{y \uparrow x} F(y)$ and $F(x+) = \lim_{y \downarrow x} F(y)$ are well defined and by their definition $F(x-) \leq F(x) \leq F(x+)$.

Now let $C = \{x \in \mathbb{R} : F \text{ is discontinuous at } x\}$. For any $x \in C$ we have $F(x-) < F(x+)$. Hence we can find a rational $r = r(x)$ in the open interval $(F(x-), F(x+))$. No two distinct x can have the same $r(x)$, since if $x_1 < x_2$ we obtain $F(x_1+) \leq F(x_2-)$ from the definition of these limits. Thus the correspondence $x \leftrightarrow r(x)$ defines a one–one correspondence between C and a subset of $\mathbb{Q}$, so C is at most countable. At each discontinuity we have $F(x-) < F(x+)$, so all discontinuities result from jumps of F.   □

Proof (of Proposition 7.21)

Fix $\varepsilon > 0$, and let a finite set of disjoint intervals $J_k = (x_k, y_k)$ be given. Let $E = \bigcup_k J_k$. Then

$$\sum_{k=1}^{n} |F(y_k) - F(x_k)| = \sum_{k=1}^{n} |\int_{x_k}^{y_k} f \, dm| \leq \sum_{k=1}^{n} \int_{x_k}^{y_k} |f| \, dm = \int_E |f| \, dm.$$

But since $f \in L^1$, the measure $\mu(G) = \int_G |f| \, dm$ is absolutely continuous with respect to Lebesgue measure m and hence, by Proposition 7.2, there exists $\delta > 0$ such that $m(F) < \delta$ implies $\mu(F) < \varepsilon$. But if the total length of the intervals J_k is less than δ, then $m(F) < \delta$, hence $\mu(F) < \varepsilon$. This proves that the function F is absolutely continuous.   □

Proof (of Proposition 7.24)

Use the functions defined in the hint: for any partition $(x_k)_{k \leq n}$ of $[a, x]$ we have

$$F(x) - F(a) = \sum_{k=1}^{n}[F(x_k) - F(x_{k-1})]$$
$$= \sum_{k=1}^{n}[F(x_k) - F(x_{k-1})]^+ - \sum_{k=1}^{n}[F(x_k) - F(x_{k-1})]^-$$
$$= p(x) - n(x),$$

so that $p(x) = n(x) + [F(b) - F(a)] \leq N_F(x) + [F(b) - F(a)]$ by definition of sup. This holds for all partitions, hence $P_F(x) = \sup p(x) \leq N_F(x) + [F(b) - F(a)]$. On the other hand, writing $n(x) = p(x) + [F(a) - F(b)]$ yields $N_F(x) \leq P_F(x) + [F(a) - F(b)]$. Thus $P_F(x) - N_F(x) = F(b) - F(a)$. Now for any fixed partition we have

$$T_F(x) \geq \sum_{k=1}^{n}|F(x_k) - F(x_{k-1})| = p(x) + n(x) = p(x) + \{p(x) - [F(b) - F(a)]\}$$
$$= 2p(x) - [P_F(x) - N_F(x)] = 2p(x) + N_F(x) - P_F(x).$$

Take the supremum on the right: $T_F(x) \geq 2P_F(x) + N_F(x) - P_F(x) = P_F(x) + N_F(x)$. But we can also write $\sum_{k=1}^{n}|F(x_k) - F(x_{k-1})| = p(x) + n(x) \leq P_F(x) + N_F(x)$ for any partition, so taking the sup on the left provides $T_F(x) \leq P_F(x) + N_F(x)$. So the two sides are equal.   □

Proof (of Proposition 7.25)

It will suffice to prove this for T_F, as the other cases are similar. If the partition $\mathcal{P}$ of $[a, b]$ produces the sum $t(\mathcal{P})$ for the absolute differences, and if $\mathcal{P}' = \mathcal{P} \cup \{c\}$ for some $c \in (a, b)$, then $t(\mathcal{P})[a, b] \leq t(\mathcal{P}')[a, c] + t(\mathcal{P}')[c, b]$ and this is bounded above by $T_F[a, c] + T_F[c, b]$ for all partitions. Thus it bounds $T_F[a, b]$ also. On the other hand, any partitions of $[a, c]$ and $[c, b]$ together make up a partition of $[a, b]$, so that $T_F[a, b]$ bounds their joint sums. So the two sides must be equal. In particular, fixing a, $T_F[a, c] \leq T_F[a, b]$ when $c \leq b$, hence $T_F(x) = T_F[a, x]$ is increasing with x. The same holds for P_F and N_F. The final statement is obvious.   □

Proof (of Proposition 7.27)

(i) Given $\varepsilon > 0$, find $\delta > 0$ such that $\sum_{i=1}^{n}(y_i - x_i) < \delta$ implies $\sum_{i=1}^{n}|F(y_i) - F(x_i)| < \frac{\varepsilon\delta}{b-a}$. Given a partition $(t_i)_{i \leq K}$ of $[a, b]$, we add further partition points, uniformly spaced and at a distance $\frac{b-a}{M}$ from each other, to ensure that the combined partition $(z_i)_{i \leq N}$ has all its points less than δ apart. To do this we simply need to choose M as the integer part of $T = \frac{b-a}{\delta} + 1$. Since the (t_j) form a subset of the partition points $(z_i)_{i=0,1,\ldots,N}$ it follows that

$$\sum_{i=1}^{K} |F(t_i) - F(t_{i-1})| \leq \sum_{i=1}^{N} |F(z_i) - F(z_{i-1})|.$$

The latter sum can be re-ordered into M groups of terms where each group begins and ends with two consecutive new partition points: the kth group then contains (say) m_k points altogether, and by their construction, the sum of their consecutive distances (i.e. the distance between the two new endpoints!) is less than δ, so for each $k \leq M$, $\sum_{i=1}^{m_k} |F(w_{i,k}) - F(w_{i-1,k})| < \frac{\varepsilon\delta}{b-a}$, where the $(w_{i,k})$ are the re-ordered points (z_i). Thus the whole sum is bounded by $M(\frac{\varepsilon\delta}{b-a}) \leq T(\frac{\varepsilon\delta}{b-a}) < \varepsilon$. This shows that $F \in BV[a,b]$, since the bound is independent of the original partition $(t_i)_{i \leq K}$.

For (ii), note first that, by (i), the function F has bounded variation on $[a,b]$, so that over any subinterval $[x_i, y_i]$ the total variation function $T_F[x_i, y_i]$ is finite. Again take ε, δ as given in the definition of absolutely continuous functions. If $(x_i, y_i)_{i \leq n}$ are subintervals with $\sum_{i=1}^{n} |y_i - x_i| < \delta$ then $\sum_{i=1}^{n} |F(y_i) - F(x_i)| < \varepsilon$. As in the previous proposition this implies that $T_F[x_i, y_i] \leq \varepsilon$. Thus both $P_F[x_i, y_i]$ and $N_F[x_i, y_i]$ are less than ε, so that the functions F_1 and F_2 are absolutely continuous. □

Proof (of Proposition 7.31)

Obviously $\nu(\varnothing) = 0$ and $\varnothing \subset E$ for any E. So if ν is monotone increasing, $\nu(E) \geq \nu(\varnothing) \geq 0$. Hence ν is a measure. Conversely, if ν is a measure, $F \subset E$, $\nu(E) = \nu(F) + \nu(E \setminus F) \geq \nu(E)$. □

Proof (of Proposition 7.40)

If $E = \{f > 1\}$ has $\rho(E) > 0$, $\frac{f}{\rho(E)}$ is well defined and $\int_E \frac{f}{\rho(E)} \, d\rho = \frac{1}{\rho(E)} \int_E f \, d\rho > 1$. This contradicts the hypothesis, so $\rho(E) = 0$. Similarly, $F = \{f < -1\}$ has $\rho(F) = 0$. Hence $|f| \leq 1$ ρ-a.e. □

Proof (of Proposition 7.41)

Choose h, A, B as in the hint. Recall that $\nu^+ = \frac{1}{2}(|\nu| + \nu)$, and note that $\frac{1}{2}(1 + h) = h\mathbf{1}_B$, so that, for $F \in \mathcal{F}$,

$$\nu^+(F) = \frac{1}{2} \int_F (1 + h) \, d|\nu| = \int_{F \cap B} h \, d|\nu| = \nu(F \cap B).$$

But then, since $B = A^c$,

$$\nu^-(F) = -[\nu(F) - \nu^+(F)] = -[\nu(F) - \nu(F \cap B)] = -\nu(F \cap A).$$

Finally, if $\nu = \lambda_1 - \lambda_2$ where the λ_i are measures, then $\nu \leq \lambda_1$, so that $\nu^+(F) = \nu(F \cap B) \leq \lambda_1(F \cap B) \leq \lambda_1(F)$ by monotonicity. This proves the final statement of the proposition. $\qquad\Box$

Proof (of Proposition 7.45)

(i) Is immediate, as $\int_\Omega \mathbb{E}(X|\mathcal{G})\, \mathrm{d}P = \int_\Omega X\, \mathrm{d}P$ by definition.

(ii) If both integrands are $\mathcal{G}$-measurable and $\int_G \mathbb{E}(X|\mathcal{G})\, \mathrm{d}P = \int_G X\, \mathrm{d}P$ for all $G \in \mathcal{G}$, then the integrands are a.s. equal by Theorem 4.22, and thus X is a version of $\mathbb{E}(X|\mathcal{G})$.

(iii) For any $G \in \mathcal{G}$, $\mathbf{1}_G$ and X are independent random variables, so that

$$\int_G X\, \mathrm{d}P = \mathbb{E}(X\mathbf{1}_G) = \mathbb{E}(X)\mathbb{E}(\mathbf{1}_G) = \int_G \mathbb{E}(X)\, \mathrm{d}P$$

Hence by definition $\mathbb{E}(X)$ is a version of $\mathbb{E}(X|\mathcal{G})$. But $\mathbb{E}(X)$ is constant, so the identity holds everywhere.

(iv) Use the linearity of integrals:

$$\int_G (aX + bY) = a \int_G X\, \mathrm{d}P + b \int_G Y\, \mathrm{d}P = a \int_G \mathbb{E}(X|\mathcal{G})\, \mathrm{d}P + b \int_G \mathbb{E}(Y|\mathcal{G})\, \mathrm{d}P$$
$$= \int_G [a\mathbb{E}(X|\mathcal{G})\, \mathrm{d}P + b\mathbb{E}(Y|\mathcal{G})]\, \mathrm{d}P,$$

so the result follows. $\qquad\Box$

8
Limit theorems

In this chapter we present the classical limit theorems of probability theory. The reader may wish to omit the more technically demanding proofs at a first reading, in order to gain an overview of the principal limit theorems for sequences of random variables. We put the spotlight firmly on probability to derive substantive applications of the preceding theory.

First, however we discuss some basic modes of convergence of sequences of functions of real variable. Then we move to the probabilistic setting to which this chapter is largely devoted.

8.1 Modes of convergence

Let E be a Borel subset of $\mathbb{R}^n$. For a given sequence (f_n) in $L^p(E)$, $p \geq 1$, we can express the statement '$f_n \to f$ as $n \to \infty$' in a number of distinct ways:

Definition 8.1

(i) $f_n \to f$ *uniformly on* E: given $\varepsilon > 0$, there exists $N = N(\varepsilon)$ such that, for all $n \geq N$,
$$|f_n - f|_\infty = \sup_{x \in E}(|f_n(x) - f(x)|) < \varepsilon.$$

(Note that we need $f_n \in L^\infty(E)$ for the sup to be finite in general.)

(ii) $f_n \to f$ *pointwise on* E: for each $x \in E$, given $\varepsilon > 0$, there exists $N = N(\varepsilon, x)$ such that $|f_n(x) - f(x)| < \varepsilon$ for all $n \geq N$.

(iii) $f_n \to f$ *almost everywhere* (a.e.) on E: there is a null set $F \subset E$ such that $f_n \to f$ pointwise on $E \setminus F$.

(iv) $f_n \to f$ *in L^p-norm* (in the pth mean): $|f_n - f|_p \to 0$ as $n \to \infty$, i.e. for given $\varepsilon > 0$, $\exists N = N(\varepsilon)$ such that

$$|f_n - f|_p = \left(\int_E |f_n - f|^p \, dm \right)^{\frac{1}{p}} < \varepsilon$$

for all $n \geq N$.

Handling these different modes of convergence requires some care; at this point we have not even shown that they are all genuinely different. Clearly, pointwise (and a.e.) limits are often easier to 'guess', however, we cannot always be certain that the limit function, if it exists, is again a member of $L^p(E)$. For mean convergence, however, this is ensured by the completeness of $L^p(E)$, and similarly for uniform convergence, which is just convergence in the L^∞-norm. Note that the conclusions of the dominated and monotone convergence theorems yield the mean convergence of a.e. convergent sequences (f_n), but only by imposing additional conditions on the (f_n).

Theorem 8.2

With (f_n) as above, the only valid implications are the following: (i) $\Rightarrow$ (ii) $\Rightarrow$ (iii). For finite measures, (i) $\Rightarrow$ (iv).

Proof

The above implications are obvious. It is also obvious that (iii) $\not\Rightarrow$ (ii).

To see that (ii) $\not\Rightarrow$ (i) take $E = [0, 1]$, $f_n = \mathbf{1}_{(0, \frac{1}{n})}$ which converges to $f = 0$ at all points but $\sup f_n = 1$ for all n.

For (iii) $\not\Rightarrow$ (iv) take $f_n = n\mathbf{1}_{(0, \frac{1}{n}]}$; f_n converges to 0 pointwise, but $\int_0^1 f_n^p \, dm = n^p \frac{1}{n} = n^{p-1} \geq 1$.

To see that (iv) $\not\Rightarrow$ (iii), let $E = [0, 1]$ and put

$$g_1 = \mathbf{1}_{[0, \frac{1}{2}]} \quad g_2 = \mathbf{1}_{[\frac{1}{2}, 1]}$$

$$g_3 = \mathbf{1}_{[0, \frac{1}{4}]} \quad g_4 = \mathbf{1}_{[\frac{1}{4}, \frac{2}{4}]} \quad g_5 = \mathbf{1}_{[\frac{2}{4}, \frac{3}{4}]} \quad g_6 = \mathbf{1}_{[\frac{3}{4}, 1]}$$

$$\ldots$$

Then

$$\int_0^1 g_n^p \, \mathrm{d}m = \int_0^1 g_n \, \mathrm{d}m \to 0$$

but for each $x \in [0,1]$, $g_n(x) = 1$ for infinitely many n, so $g_n(x)$ does not converge at any x in E. $\square$

Example 8.3

We investigate

$$h_n(x) = x^n$$

for convergence in each of these modes on $E = [0,1]$.

The sequence converges everywhere to the function $h(x) = 0$ for $x \in [0,1)$, $h(1) = 1$, and so it also converges almost everywhere.

It does not converge uniformly since $\sup_{[0,1]} |h_n(x) - h(x)| = 1$ for all n.

It converges in L^p for $p > 0$:

$$\int_0^1 |h_n(x) - h(x)|^p \, \mathrm{d}x = \int_0^1 x^{pn} \, \mathrm{d}x = \frac{1}{pn+1} x^{pn+1} \Big|_0^1 \to 0.$$

Remark 8.4

There are still other modes of convergence which can be considered for sequences of measurable functions, and the relations between these and the above are quite complex in general. Here we will not pursue this theme in general, but specialise instead to probability spaces, where we derive additional relationships between the different limit processes.

Exercise 8.1

For each of the following decide whether $f_n \to 0$ in L^p, uniformly, pointwise, a.e.:

(a) $f_n = \mathbf{1}_{[n,n+\frac{1}{n}]}$,

(b) $f_n = n\mathbf{1}_{[0,\frac{1}{n}]} - n\mathbf{1}_{[-\frac{1}{n},0]}$.

8.2 Probability

The remainder of this chapter is devoted to a discussion of the basic limit theorems for random variables in probability theory. The very definition of 'probabilities' relies on a belief in such results, i.e. that we can ascribe a meaning to the 'limiting average' of successes in a sequence of independent identically distributed trials. Then the purpose of the 'endless repetition' of tossing a coin is to use the 'limiting frequency' of heads as the definition of the probability of heads.

Similarly, the pivotal role ascribed in statistics to the Gaussian density has its origin in the famous central limit theorem (of which there are actually many versions), which shows this density to describe the limit distribution of a sequence of distributions under appropriate conditions. Convergence of distributions therefore provides yet a further important limit concept for sequences of random variables.

In both cases the concept of independence plays a crucial role and first of all we need to extend this concept to infinite sequences of random variables. In what follows $(X_n)_{n\geq 1}$ will denote a sequence of random variables defined on a probability space.

Definition 8.5

We say that random variables $(X_n)_{n\geq 1}$ are independent if for any $k \in \mathbb{N}$ the variables $X_1, \ldots, X_k$ are independent (see Definition 3.18).

An alternative is to demand that any finite collection of X_n be independent. Of course this condition implies independence since finite collections cover the initial segments of k variables.

Conversely, take any finite collection of X_n and let k be the greatest index of this finite collection. Now $X_1, \ldots, X_k$ are independent and then for each subset the collection of its elements is independent; this includes, in particular, the chosen one.

We study the following sequence:

$$S_n = X_1 + \cdots + X_n.$$

If all X_n have the same distribution (we say that they are *identically distributed*), then $\frac{S_n}{n}$ is the average value of X_1 (or any X_k, it does not matter) after n repetitions of the same experiment.

We study the behaviour of S_n as n goes to infinity. The two main questions we address are:

1. When do the random variables $\frac{S_n}{n}$ converge to a certain number? Here there is an immediate question of the appropriate mode of convergence. Positive answers to such questions are known as laws of large numbers.

2. When do the distributions of the random variables $\frac{S_n}{n}$ converge to a measure? Under what conditions is this limit measure Gaussian? The results

we obtain in response are known as central limit theorems.

8.2.1 Convergence in probability

Our first additional mode of convergence, convergence in probability, is sometimes termed 'convergence in measure'.

Definition 8.6

A sequence (X_n) converges to X *in probability* if for each $\varepsilon > 0$

$$P(|X_n - X| > \varepsilon) \to 0$$

as $n \to \infty$.

Exercise 8.2

Go back to the proof of Theorem 8.2 (with $E = [0, 1]$) to see which of the sequences of random variables constructed there converge in probability.

Exercise 8.3

Find an example of a sequence of random variables on $[0, 1]$ that does not converge to 0 in probability.

We begin by showing that convergence almost surely (i.e. almost everywhere) is stronger than convergence in probability. But first we prove an auxiliary result.

Lemma 8.7

The following conditions are equivalent:

(i) $Y_n \to 0$ almost surely,

(ii) for each $\varepsilon > 0$,

$$\lim_{N \to \infty} P\left(\bigcup_{n=N}^{\infty} \{ \omega : |Y_n(\omega)| \geq \varepsilon \} \right) = 0.$$

Proof

Convergence almost surely, expressed succinctly, means that

$$P(\{\omega : \forall \varepsilon > 0 \; \exists N \in \mathbb{N} : \forall n \geq N, |Y_n(\omega)| < \varepsilon\}) = 1.$$

Writing this set of full measure another way we have

$$P(\bigcap_{\varepsilon > 0} \bigcup_{N \in \mathbb{N}} \bigcap_{n \geq N} \{\omega : |Y_n(\omega)| < \varepsilon\}) = 1.$$

The probability of the outer intersection (over all $\varepsilon > 0$) is less then the probability of any of its terms, but being already 1, it cannot increase, hence for all $\varepsilon > 0$

$$P(\bigcup_{N \in \mathbb{N}} \bigcap_{n \geq N} \{\omega : |Y_n(\omega)| < \varepsilon\}) = 1.$$

We have a union of increasing sets, so

$$\lim_{N \to \infty} P(\bigcap_{n \geq N} \{\omega : |Y_n(\omega)| < \varepsilon\}) = 1,$$

thus

$$\lim_{N \to \infty} (1 - P(\bigcap_{n \geq N} \{\omega : |Y_n(\omega)| < \varepsilon\})) = 0;$$

but we can write $1 = P(\Omega)$, so that

$$P(\Omega) - P(\bigcap_{n \geq N} \{\omega : |Y_n(\omega)| < \varepsilon\}) = P(\Omega \setminus \bigcap_{n \geq N} \{\omega : |Y_n(\omega)| < \varepsilon\})$$

$$= P(\bigcup_{n \geq N} \{\omega : |Y_n(\omega)| \geq \varepsilon\})$$

by de Morgan's law. Hence (i) implies (ii). Working backwards, these steps also prove the converse.   □

Theorem 8.8

If $X_n \to X$ almost surely then $X_n \to X$ in probability.

Proof

For simplicity of notation consider the difference $Y_n = X_n - X$ and the problem reduces to the discussion of convergence of Y_n to zero. We have

$$\lim_{n \to \infty} P(\bigcup_{k=n}^{\infty} \{\omega : |Y_n(\omega)| \geq \varepsilon\}) \geq \lim_{n \to \infty} P(\{\omega : |Y_k(\omega)| \geq \varepsilon\})$$

and by Lemma 8.7 the limit on the left is zero, hence so is that on the right.   □

Note that the two sides of the inequality neatly summarize the difference between convergence a.s. and in probability. For convergence in probability we consider the probabilities that individual Y_n are at least ε away from the limit, while for almost sure convergence we need to consider the whole tail sequence $(Y_n)_{n \geq k}$ simultaneously.

The following example shows that the implication in the above theorem cannot be reversed and also shows that the convergence in L^p does not imply almost sure convergence.

Example 8.9

Consider the following sequence of random variables defined on $\Omega = [0, 1]$ with Lebesgue measure: $X_1 = \mathbf{1}_{[0,1]}$, $X_2 = \mathbf{1}_{[0,1/2]}$, $X_3 = \mathbf{1}_{[1/2,1]}$, $X_4 = \mathbf{1}_{[0,1/4]}$, $X_5 = \mathbf{1}_{[1/4,1/2]}$ and so on (like in the proof of Theorem 8.2). The sequence clearly converges to zero in probability and in L^p but for each $\omega \in [0, 1]$, $X_n(\omega) = 1$ for infinitely many n, so it fails to converge pointwise.

Convergence in probability has an additional useful feature:

Proposition 8.10

The function defined by $d(X, Y) = \mathbb{E}(\frac{|X-Y|}{1+|X-Y|})$ is a metric and convergence in d is equivalent to convergence in probability.

Hint If $X_n \to X$ in probability then decompose the expectation into $\int_A + \int_{\Omega \setminus A}$ where $A = \{\omega : |X_n(\omega) - X(\omega)| < \varepsilon\}$.

We now give a basic estimate of the probability of a non-negative random variable, taking values in a given set by means of the moments of this random variable.

Theorem 8.11 (Chebyshev's inequality)

If Y is a non-negative random variable, $\varepsilon > 0$, $0 < p < \infty$, then

$$P(Y \geq \varepsilon) \leq \frac{\mathbb{E}(Y^p)}{\varepsilon^p}. \tag{8.1}$$

Proof

This is immediate from basic properties of integral: let $A = \{\omega : Y(\omega) \geq \varepsilon\}$

and then

$$\mathbb{E}(Y^p) \geq \int_A Y^p \, dP \quad \text{(integration over a smaller set)}$$
$$\geq P(A)\varepsilon^p,$$

since $Y^p(\omega) > \varepsilon^p$ on A, which gives the result after dividing by ε^p.   □

Chebyshev's inequality will be used mainly with small ε. But let us see what happens if ε is large.

Proposition 8.12

Assume that $\mathbb{E}(Y^p) < \infty$. Then

$$\varepsilon^p P(Y \geq \varepsilon) \to 0 \quad \text{as } \varepsilon \to \infty.$$

Hint Write

$$\mathbb{E}(Y^p) = \int_{\{\omega:Y(\omega)\geq\varepsilon\}} Y^p \, dP + \int_{\{\omega:Y(\omega)<\varepsilon\}} Y^p \, dP$$

and estimate the first term as in the proof of Chebyshev's inequality.

Corollary 8.13

Let X be a random variable with finite expectation $\mathbb{E}(X) = \mu$ and variance σ^2. Let $0 < a < \infty$; then

$$P(|X - \mu| \geq a\sigma) \leq \frac{1}{a^2}.$$

Proof

Using Chebyshev's inequality with $Y = |X - \mu|$ and $p = 2$, $\varepsilon = a\sigma$, we find that

$$P(|X - \mu| \geq a\sigma) \leq \frac{\mathbb{E}(|X - \mu|^2)}{a^2\sigma^2} = \frac{1}{a^2}$$

as required.   □

Remark 8.14

Chebyshev's inequality also shows that convergence in L^p implies convergence in probability. For, let $Y_n = |X_n - X|$ and assume that $\|X_n - X\|_p \to 0$. This implies that $P(Y_n \geq \varepsilon) \to 0$. The converse is false in general, as the next example shows.

Example 8.15

Let $\Omega = [0,1]$ with Lebesgue measure and let $X_n = n\mathbf{1}_{[0,\frac{1}{n}]}$. The sequence X_n converges to 0 pointwise, so we take $X = 0$ and we see that $|X_n - X|_p = \int_0^{\frac{1}{n}} n^p \, dm = n^{p-1}$. If $p \geq 1$, then $|X_n - X|_p \not\to 0$, however as we have already seen (Exercise 8.2),

$$P(|X_n - X| \geq \varepsilon) = P(X_n = n) = \frac{1}{n} \to 0,$$

showing that $X_n \to X$ in probability.

8.2.2 Weak law of large numbers

The simplest 'law of large numbers' provides an L^2-convergence result:

Theorem 8.16

If (X_n) are independent, $\mathbb{E}(X_n) = \mu$, $\mathrm{Var}(X_n) \leq K < \infty$, then $\frac{S_n}{n} \to \mu$ in L^2 and hence in probability.

Proof

First note that $\mathbb{E}(S_n) = \mathbb{E}(X_1) + \cdots + \mathbb{E}(X_n) = n\mu$ by the linearity of expectation. Hence $\mathbb{E}(\frac{S_n}{n}) = \mu$ and so $\mathbb{E}(\frac{S_n}{n} - \mu)^2 = \mathrm{Var}(\frac{S_n}{n})$. By the properties of the variance (Proposition 5.40),

$$\mathrm{Var}(\frac{S_n}{n}) = \frac{1}{n^2}\mathrm{Var}(S_n) = \frac{1}{n^2}(\mathrm{Var}(X_1) + \cdots + \mathrm{Var}(X_n)) \leq \frac{1}{n^2}nK = \frac{K}{n} \to 0$$

as $n \to \infty$. This in turn implies convergence in probability, as we saw in Remark 8.14. $\qquad\square$

Exercise 8.4

Using Chebyshev's inequality, find a lower bound for the probability that the average number of 'heads' in 100 tosses of a coin differs from $\frac{1}{2}$ by less than 0.1.

Exercise 8.5

Find a lower bound for the probability that the average number shown on a die in 1000 tosses differs from 3.5 by less than 0.1.

We give some classical applications of the weak law of large numbers. The Weierstrass theorem says that every continuous function can be uniformly approximated by polynomials. The Chebyshev inequality provides an easy proof.

Theorem 8.17 (Bernstein–Weierstrass approximation theorem)

If $f : [0,1] \to \mathbb{R}$ is continuous then the sequence of Bernstein polynomials

$$f_n(x) = \sum_{k=0}^{n} \binom{n}{k} x^k (1-x)^{n-k} f(\frac{k}{n})$$

converges to f uniformly.

Proof

The number $f_n(x)$ has a probabilistic meaning. Namely, $f_n(x) = \mathbb{E}(f(\frac{S_n}{n}))$ where $S_n = X_1 + \cdots + X_n$,

$$X_i = \begin{cases} 1 & \text{with probability } x \\ 0 & \text{with probability } 1-x. \end{cases}$$

Then writing $\mathbb{E}_x$ instead of $\mathbb{E}$ to emphasise the fact that the underlying probability depends on x, we have

$$\sup_{x \in [0,1]} |f_n(x) - f(x)| \leq \sup_{x \in [0,1]} |\mathbb{E}_x(f(\frac{S_n}{n})) - f(x)|$$

$$\leq \sup_{x \in [0,1]} \mathbb{E}_x |f(\frac{S_n}{n}) - f(x)|.$$

Take any $\varepsilon > 0$ and find $\delta > 0$ such that if $|x - y| < \delta$ then $|f(x) - f(y)| < \frac{\varepsilon}{2}$ (this is possible since f is uniformly continuous).

$$\mathbb{E}_x |f(\frac{S_n}{n}) - f(x)| = \int_{\{\omega: |\frac{S_n}{n} - x| < \delta\}} |f(\frac{S_n}{n}) - f(x)| \, dP$$

$$+ \int_{\{\omega: |\frac{S_n}{n} - x| \geq \delta\}} |f(\frac{S_n}{n}) - f(x)| \, dP$$

$$\leq \frac{\varepsilon}{2} + 2 \sup_{y \in [0,1]} |f(y)| \cdot P(|\frac{S_n}{n} - x| \geq \delta).$$

The last term converges to zero by the law of large numbers since $x = \mathbb{E}(\frac{S_n}{n})$, and $\mathrm{Var}(\frac{S_n}{n})$ is finite. This convergence is uniform in x:

$$P(|\frac{S_n}{n} - x| \geq \delta) \leq \frac{\mathrm{Var}(\frac{S_n}{n})}{\delta^2} = \frac{x(1-x)}{n\delta^2} \leq \frac{1}{4n\delta^2}$$

(due to $4x(1-x) \leq 1$ and $\text{Var}(S_n) = n\text{Var}(X_i) = nx(1-x)$). So, for sufficiently large n the right-hand side is less than ε, which completes the proof. $\square$

In many practical situations it is impossible to compute integrals (in particular areas or volumes) directly. The law of large numbers is the basis of the so-called Monte Carlo method, which gives an approximate solution by random selection of points. The next two examples illustrate this method.

Example 8.18

We restrict ourselves to $F \subset [0,1] \times [0,1]$ for simplicity. Assume that F is Lebesgue-measurable and our goal is to find its measure. Let X_n, Y_n be independent, uniformly distributed in $[0,1]$. Let $M_n = \frac{1}{n}\sum_{k=1}^{n} \mathbf{1}_F(X_k, Y_k)$. If we draw the pairs of numbers n times, then this sum gives the number of hits of the set F, and $M_n \to m(F)$ in probability. To see this, first observe that $\mathbb{E}(\mathbf{1}_F(X_k, Y_k)) = P((X_k, Y_k) \in F) = m(F)$ by the assumption on the distribution of X_k, Y_k: independence guarantees that the distribution of the pair is two-dimensional Lebesgue measure restricted to the square. Then

$$P(|M_n - m(F)| \geq \varepsilon) \leq \frac{m(F)}{n\varepsilon} \to 0.$$

A similar example illustrates the use of the Monte Carlo method for computing integrals.

Example 8.19

Let f be an integrable function defined on $[0,1]$. With X_n independent uniformly distributed on $[0,1]$ we take

$$I_n = \frac{1}{n}\sum_{k=1}^{n} f(X_k)$$

and we show that

$$I_n \to \int_0^1 f(x)\,dx$$

in probability. First note that the distribution of X_k is Lebesgue measure on $[0,1]$, hence

$$\mathbb{E}(f(X_k)) = \int f(x)\,dP_{X_k}(x) = \int_0^1 f(x)\,dx,$$

and so $\mathbb{E}(I_n) = \int_0^1 f(x)\,dx$. The weak law provides the desired convergence.

$$P(|I_n - \int_0^1 f(x)\,dx| \geq \varepsilon) \to 0.$$

We return to considering further weak laws of large numbers for sequences of random variables. The assumption that $\mathbb{E}(X_n)$ and $\text{Var}(X_n)$ are finite can be relaxed. There is, however, a price to pay by imposing additional conditions.

First we introduce some convenient notation. For a given sequence of random variables (X_n) introduce the truncated random variables

$$X_n(L) = X_n \mathbf{1}_{\{\omega : |X_n(\omega)| \leq L\}}$$

and set $\mu_L = \mathbb{E}(X_1(L))$. Note that $\mu_L = \mathbb{E}(X_n(L))$ for all $n \geq 1$ since the distributions of all X_n are the same. Also write

$$S_n(L) = X_1(L) + \cdots + X_n(L)$$

for all $n \geq 1$.

Theorem 8.20

If X_n are independent identically distributed random variables such that

$$aP(|X_1| > a) \to 0 \quad \text{as } a \to \infty, \tag{8.2}$$

then

$$\frac{S_n}{n} - \mu_n \to 0 \quad \text{in probability.} \tag{8.3}$$

We shall need the following lemma, which is of interest in itself and will be useful in what follows.

Lemma 8.21

If $Y \geq 0$, $Y \in L^p$, $0 < p < \infty$, then

$$\mathbb{E}(Y^p) = \int_0^\infty p y^{p-1} P(Y > y) \, dy.$$

In particular ($p = 1$),

$$\mathbb{E}(Y) = \int_0^\infty P(Y > y) \, dy.$$

Proof of the lemma

This is a simple application of Fubini's theorem:

$$\int_0^\infty py^{p-1} P(Y > y) \, dy$$

$$= \int_0^\infty \int_\Omega py^{p-1} \mathbf{1}_{\{\omega:Y(\omega)>y\}}(\omega) \, dP(\omega) \, dy$$

$$= \int_\Omega \int_0^\infty py^{p-1} \mathbf{1}_{\{\omega:Y(\omega)>y\}}(\omega) \, dy \, dP(\omega) \quad \text{(by Fubini)}$$

$$= \int_\Omega \int_0^{Y(\omega)} py^{p-1} \, dy \, dP(\omega)$$

$$= \int_\Omega Y^p(\omega) \, dP(\omega) \quad \text{(computing the inner integral, } \omega \text{ fixed)}$$

$$= \mathbb{E}(Y^p)$$

as required. $\qquad\qquad\qquad\qquad\qquad\qquad\qquad\qquad\qquad\qquad\qquad$ □

Proof of the theorem

Take $\varepsilon > 0$ and obviously

$$P(|\frac{S_n}{n} - \mu_n| \geq \varepsilon) \leq P(|\frac{S_n(n)}{n} - \mu_n| \geq \varepsilon) + P(S_n(n) \neq S_n).$$

We estimate the first term

$$P(|\frac{S_n(n)}{n} - \mu_n| \geq \varepsilon) \leq \frac{\mathbb{E}(|\frac{S_n(n)}{n} - \mu_n|^2)}{\varepsilon^2} \quad \text{(by Chebyshev)}$$

$$= \frac{\mathbb{E}(|\sum_{k=1}^n X_k(n) - n\mu_n|^2)}{n^2\varepsilon^2}$$

$$= \frac{\text{Var}(\sum_{k=1}^n X_k(n))}{n^2\varepsilon^2}.$$

Note that the truncated random variables are independent, being functions of the original ones, hence we may continue the estimation:

$$\leq \frac{\sum_{k=1}^n \text{Var}(X_k(n))}{n^2\varepsilon^2}$$

$$= \frac{\text{Var}(X_1(n))}{n\varepsilon^2} \quad \text{(as } \text{Var}(X_k(n)) \text{ are the same)}$$

$$\leq \frac{\mathbb{E}(X_1^2(n))}{n\varepsilon^2} \quad \text{(by } \text{Var}(Z) = \mathbb{E}(Z^2) - (\mathbb{E}Z)^2 \leq \mathbb{E}(Z^2))$$

$$= \frac{1}{n\varepsilon^2} \int_0^\infty 2y P(|X_1(n)| > y) \, dy \quad \text{(by the lemma for } p = 2)$$

$$= \frac{1}{n\varepsilon^2} \int_0^n 2y P(|X_1| > y) \, dy.$$

The function $y \mapsto 2yP(|X_1| > y)$ converges to 0 as $y \to \infty$ by hypothesis, hence for given $\delta > 0$ there is y_0 such that for $y \geq y_0$ this quantity is less than $\frac{1}{2}\delta\varepsilon^2$, and we have

$$
= \frac{1}{n\varepsilon^2} \int_0^{y_0} 2yP(|X_1| > y)\, \mathrm{d}y + \frac{1}{n\varepsilon^2} \int_{y_0}^n 2yP(|X_1| > y)\, \mathrm{d}y
$$

$$
\leq \frac{1}{n\varepsilon^2} y_0 \max_{y \in [0,\infty]} \{yP(|X_1| > y)\} + \frac{1}{n\varepsilon^2} n \frac{\delta\varepsilon^2}{2}
$$

$$
\leq \delta,
$$

provided n is sufficiently large. So the first term converges to 0. Now we tackle the second term:

$$
P(S_n(n) \neq S_n) \leq P(X_k(n) \neq X_k \text{ for some } k \leq n)
$$

$$
\leq \sum_{k=1}^n P(X_k(n) \neq X_k) \quad \text{(by subadditivity of } P\text{)}
$$

$$
= nP(X_1(n) \neq X_1) \quad \text{(the same distributions)}
$$

$$
= nP(|X_1| > n)
$$

$$
\to 0 \quad \text{by hypothesis.}
$$

This completes the proof.   □

Remark 8.22

Note that we cannot generalise the last theorem to the case of uncorrelated random variables since we made essential use of the independence. Although the identity

$$
\mathrm{Var}\left(\sum_{k=1}^n X_k(n) \right) = \sum_{k=1}^n \mathrm{Var}(X_k(n))
$$

holds for uncorrelated random variable, we needed the independence of the (X_k) – which implies that of the $(X_k(n))$– to conclude that the truncated random variables are uncorrelated.

Theorem 8.23

If X_n are independent and identically distributed, $\mathbb{E}(|X_1|) < \infty$, then (8.2) is satisfied, $\mu_n \to \mu = \mathbb{E}(X_1)$ and $\frac{S_n}{n} \to \mu$ in probability. (Note that we do not assume here that X_1 has finite variance.)

Proof

The finite expectation of $|X_1|$ gives condition (8.2):

$$aP(|X_1| > a) = a \int_\Omega \mathbf{1}_{\{\omega:|X_1(\omega)|>a\}} \, \mathrm{d}P$$

$$\leq \int_\Omega |X_1| \mathbf{1}_{\{\omega:|X_1(\omega)|>a\}} \, \mathrm{d}P$$

$$= \int_{\{\omega:|X_1(\omega)|>a\}} |X_1| \, \mathrm{d}P$$

$$\to 0$$

as $a \to \infty$ by the dominated convergence theorem. Hence $\frac{S_n}{n} - \mu_n \to 0$ but $\mu_n = \mathbb{E}(X_1 \mathbf{1}_{\{\omega:|X_1(\omega)|\leq n\}}) \to \mathbb{E}(X_1)$ as $n \to \infty$, so the result follows. $\qquad\square$

8.2.3 The Borel–Cantelli lemmas

The idea that a point $\omega \in \Omega$ belongs to 'infinitely many' events of a given sequence (A_n) of elements of $\mathcal{F}$ can easily be made precise: for every $n \geq 1$ we need to be able to find an $m_n \geq n$ such that $\omega \in A_{m_n}$. This identifies a subsequence (m_n) of indices such that for each $n \geq 1$, $\omega \in A_{m_n}$, i.e. $\omega \in \bigcup_{m \geq n} A_m$ for every $n \geq 1$. Thus we say that $\omega \in A_n$ *infinitely often*, and write $\omega \in A_n$ i.o., if $\omega \in \bigcap_{n=1}^\infty \bigcup_{m=n}^\infty A_m$. We call this set the upper limit of the sequence (A_n) and write it as

$$\limsup_{n\to\infty} A_n = \bigcap_{n=1}^\infty \bigcup_{m=n}^\infty A_m.$$

Exercise 8.6

Find $\limsup_{n\to\infty} A_n$ for a sequence $A_1 = [0,1]$, $A_2 = [0,\frac{1}{2}]$, $A_3 = [\frac{1}{2},1]$, $A_4 = [0,\frac{1}{4}]$, $A_5 = [\frac{1}{4},\frac{1}{2}]$ etc.

Given $\varepsilon > 0$, a sequence of random variables (X_n) and a random variable X, for each n set $A_n = \{|X_n - X| > \varepsilon\}$. Then $\omega \in A_n$ i.o. precisely when for every $\varepsilon > 0$, $|X_n(\omega) - X(\omega)| > \varepsilon$ occurs for all elements of an infinite subsequence (m_n) of indices, which means that (X_n) fails to converge to X. Hence it follows that

$$X_n \to X \text{ a.s. } (P) \iff \forall \varepsilon > 0 \; P(\limsup_{n\to\infty} A_n) = P(|X_n - X| > \varepsilon \text{ i.o.}) = 0.$$

Similarly, define

$$\lim_{n \to \infty} \inf A_n = \bigcup_{n=1}^{\infty} \bigcap_{m=n}^{\infty} A_n.$$

(We say that this set is the *lower limit* of the sequence (A_n).)

Proposition 8.24

(i) We have $\omega \in \liminf_{n \to \infty} A_n$ if and only if $\omega \in A_n$ except for finitely many n. (We say that $\omega \in A_n$ *eventually.*)

(ii) $P(X_n \longrightarrow X) = \lim_{\varepsilon \to 0} P(|X_n - X| < \varepsilon$ eventually).

(iii) If $A = \limsup_{n \to \infty} A_n$ then $A^c = \liminf_{n \to \infty} A_n^c$.

(iv) For any sequence (A_n) of events, $P(\{\omega \in A_n$ eventually$\}) \leq P(\{\omega \in A_n$ i.o.$\})$

Hint Use Fatou's lemma on the indicator functions of the sets in (iv).

The sets $\liminf_{n \to \infty} A_n$ and $\limsup_{n \to \infty} A_n$ are 'tail events' of the sequence (A_n): we can only determine whether a point belongs to them by knowing the whole sequence. It is frequently true that the probability of a tail event is either 0 or 1 – such results are known as 0 – 1 laws. The simplest of these is provided by combining the two Borel–Cantelli lemmas to which we now turn: together they show that for a sequence of independent events (A_n), $\limsup_{n \to \infty} A_n$ has either probability 0 or 1, depending on whether the series of their individual probabilities converges or diverges. In the first case, we do not even need independence, but can prove the result in general.

We have the following simple but fundamental fact.

Theorem 8.25 (Borel–Cantelli lemma)

If

$$\sum_{n=1}^{\infty} P(A_n) < \infty$$

then

$$P(\limsup_{n \to \infty} A_n) = 0,$$

i.e. '$\omega \in A_n$ for infinitely many n' occurs only with probability zero.

Proof

First note that $\limsup_{n\to\infty} A_n \subset \bigcup_{n=k}^{\infty} A_n$, hence

$$P(\limsup_{n\to\infty} A_n) \leq P(\bigcup_{n=k}^{\infty} A_n) \quad \text{(for all } k)$$

$$\leq \sum_{n=k}^{\infty} P(A_n) \quad \text{(by subadditivity)}$$

$$\to 0$$

since the tail of a convergent series converges to 0. $\qquad\qquad\qquad\square$

The basic application of the lemma provides a link between almost sure convergence and convergence in probability.

Theorem 8.26

If $X_n \to X$ in probability then there is a subsequence X_{k_n} converging to X almost surely.

Proof

We have to find a set of full measure on which a subsequence would converge. So the set on which the behaviour of the whole sequence is 'bad' should be of measure zero. For this we employ the Borel–Cantelli lemma whose conclusion is precisely that. So we introduce the sequence A_n encapsulating the 'bad' behaviour of X_n, which from the point of convergence is expressed by inequalities of the type $|X_n(\omega) - X(\omega)| > a$. Specifically, we take $a = 1$ and since $P(|X_n - X| > 1) \to 0$ we find k_1 such that

$$P(|X_n - X| > 1) < 1$$

for $n \geq k_1$. Next for $a = \frac{1}{2}$ we find $k_2 > k_1$ such that for all $n \geq k_2$,

$$P(|X_n - X| > \frac{1}{2}) < \frac{1}{4}.$$

We continue that process obtaining an increasing sequence of integers k_n with

$$P(|X_{k_n} - X| > \frac{1}{n}) < \frac{1}{n^2}.$$

The series $\sum_{n=1}^{\infty} P(A_n)$ converges, where $A_n = \{\omega : |X_{k_n}(\omega) - X(\omega)| > \frac{1}{n}\}$, hence $A = \limsup A_n$ has probability zero.

We observe that for $\omega \in \Omega \setminus A$, $\limsup X_{k_n}(\omega) = X(\omega)$. For, if $\omega \in \Omega \setminus A$, then for some k, $\omega \in \bigcap_{n=k}^{\infty}(\Omega \setminus A_n)$, so for all $n \geq k$, $|X_{k_n}(\omega) - X(\omega)| \leq \frac{1}{n}$, hence we have obtained the desired convergence. $\qquad \square$

The second Borel–Cantelli lemma partially completes the picture. Under the additional condition of independence it shows when the probability that infinitely many events occur is one.

Theorem 8.27

Suppose that the events A_n are independent. We have

$$\sum_{n=1}^{\infty} P(A_n) = \infty \quad \Rightarrow \quad P(\limsup_{n \to \infty} A_n) = 1.$$

Proof

It is sufficient to show that for all k

$$P(\bigcup_{n=k}^{\infty} A_n) = 1$$

since then the intersection over k will also have probability 1. Fix k and consider the partial union up to $m > k$. The complements of A_n are also independent, hence

$$P(\bigcap_{n=k}^{m} A_n^{\mathrm{c}}) = \prod_{n=k}^{m} P(A_n^{\mathrm{c}}) = \prod_{n=k}^{m}(1 - P(A_n)).$$

Since $1 - x \leq \mathrm{e}^{-x}$,

$$\prod_{n=k}^{m}(1 - P(A_n)) \leq \prod_{n=k}^{m} \mathrm{e}^{-P(A_n)} = \exp(-\sum_{n=k}^{m} P(A_n)).$$

The last expression converges to 0 as $m \to \infty$ by the hypothesis, hence

$$P(\bigcap_{n=k}^{m} A_n^{\mathrm{c}}) \to 0$$

but

$$P(\bigcap_{n=k}^{m} A_n^{\mathrm{c}}) = P(\Omega \setminus \bigcup_{n=k}^{m} A_n) = 1 - P(\bigcup_{n=k}^{m} A_n).$$

The sets $B_m = \bigcup_{n=k}^{m} A_n$ form an increasing chain with $\bigcup_{m=k}^{\infty} B_m = \bigcup_{n=k}^{\infty} A_n$ and so $P(B_m)$, which as we know converges to 1, converges to $P(\bigcup_{n=k}^{\infty} A_n)$. Thus this quantity is also equal to 1. $\qquad \square$

Exercise 8.7

Let $S_n = X_1 + X_2 + \cdots + X_n$ describe the position after n steps of a symmetric random walk on $\mathbb{Z}^d$. Using the asymptotic formula $n! \sim \left(\frac{n}{e}\right)^n \sqrt{2\pi n}$ and the Borel–Cantelli lemmas, show that the probability of $\{S_n = 0 \text{ i.o.}\}$ is 1 when $d = 1, 2$ and 0 for $d > 2$.

Below we discuss strong laws of large numbers, where convergence in probability is strengthened to almost sure convergence. But already we can observe some limitations of these improvements. Drawing on the second Borel–Cantelli lemma, we give a negative result.

Theorem 8.28

Suppose that $X_1, X_2, \ldots$ are independent identically distributed random variables and assume that $\mathbb{E}(|X_1|) = \infty$ (hence also $\mathbb{E}(|X_n|) = \infty$ for all n). Then

(i) $P(\{\omega : |X_n(\omega)| \geq n \text{ for infinitely many } n\}) = 1$,

(ii) $P(\lim_{n\to\infty} \frac{S_n}{n} \text{ exists and is finite}) = 0$.

Proof

(i) First,

$$\mathbb{E}(|X_1|) = \int_0^\infty P(|X_1| > x)\, dx \quad \text{(by Lemma 8.21)}$$

$$= \sum_{k=0}^\infty \int_k^{k+1} P(|X_1| > x)\, dx \quad \text{(countable additivity)}$$

$$\leq \sum_{k=0}^\infty P(|X_1| > k)$$

because the function $x \mapsto P(|X_1| > x)$ reaches its maximum on $[k, k+1]$ for $x = k$ since $\{\omega : |X_1(\omega)| > k\} \supset \{\omega : |X_1(\omega)| > x\}$ if $x \geq k$. By the hypothesis this series is divergent, but $P(|X_1| > k) = P(|X_k| > k)$ as the distributions are identical, so

$$\sum_{k=0}^\infty P(|X_k| > k) = \infty.$$

The second Borel–Cantelli lemma is applicable, yielding the claim.

(ii) Denote by A the set where the limit of $\frac{S_n}{n}$ exists (and is finite). Some elementary algebra of fractions gives

$$\frac{S_n}{n} - \frac{S_{n+1}}{n+1} = \frac{(n+1)S_n - nS_{n+1}}{n(n+1)} = \frac{S_n - nX_{n+1}}{n(n+1)} = \frac{S_n}{n(n+1)} - \frac{X_{n+1}}{n+1}.$$

For any $\omega_0 \in A$ the left-hand side converges to zero and also

$$\frac{S_n(\omega_0)}{n(n+1)} \to 0.$$

Hence also $\frac{X_{n+1}(\omega_0)}{n+1} \to 0$. This means that

$$\omega_0 \notin \{\omega : |X_k(\omega)| > k \text{ for infinitely many } k\} = B,$$

say, so $A \subset \Omega \setminus B$. But $P(B) = 1$ by (i), hence $P(A) = 0$. $\square$

8.2.4 Strong law of large numbers

We shall consider several versions of the strong law of large numbers, first by imposing additional conditions on the moments of the sequence (X_n), and then by gradually relaxing these until we arrive at Theorem 8.32, which provides the most general positive result.

The first result is due to von Neumann. Note that we do not impose the condition that the X_n have identical distributions. The price we pay is having to assume that higher-order moments are finite. However, for many familiar random variables, Gaussian for example, this is not a serious restriction.

Theorem 8.29

Suppose that the random variables X_n are independent, $\mathbb{E}(X_n) = \mu$, and $\mathbb{E}(X_n^4) \le K$. Then

$$\frac{S_n}{n} = \frac{1}{n}\sum_{k=1}^{n} X_k \to \mu \quad \text{a.s.}$$

Proof

By considering $X_n - \mu$ we may assume that $\mathbb{E}(X_n) = 0$ for all n. This simplifies the following computation:

$$\mathbb{E}(S_n^4) = \mathbb{E}\left(\sum_{k=1}^{n} X_k\right)^4$$

$$= \mathbb{E}\left(\sum_{k=1}^{n} X_k^4 + \sum_{i\neq j} X_i^2 X_j^2 + \sum_{i\neq j} X_i X_j X_k^2\right.$$

$$\left. + \sum_{i,j,k,l \text{ distinct}} X_i X_j X_k X_l\right).$$

The last two terms vanish by independence:

$$\mathbb{E}(\sum_{i\neq j} X_i X_j X_k^2) = \sum_{i\neq j} \mathbb{E}(X_i X_j X_k^2) = \sum_{i\neq j} \mathbb{E}(X_i)\mathbb{E}(X_j)\mathbb{E}(X_k^2) = 0$$

and similarly for the term with all indices distinct

$$\mathbb{E}(\sum X_i X_j X_k X_l) = \sum \mathbb{E}(X_i X_j X_k X_l)$$

$$= \sum \mathbb{E}(X_i)\mathbb{E}(X_j)\mathbb{E}(X_k)\mathbb{E}(X_l) = 0.$$

The first term is easily estimated by the hypothesis

$$\mathbb{E}(\sum X_k^4) = \sum \mathbb{E}(X_k^4) \leq nK.$$

To the remaining term we first apply the Schwarz inequality

$$\mathbb{E}(\sum_{i\neq j} X_i^2 X_j^2) = \sum_{i\neq j} \mathbb{E}(X_i^2 X_j^2) \leq \sum_{i\neq j} \sqrt{\mathbb{E}(X_i^4)}\sqrt{\mathbb{E}(X_j^4)} \leq NK,$$

where N is the number of components of this kind. (We could do better by employing independence, but then we would have to estimate the second moments by the fourth and it would boil down to the same.)

To find N first note that the pairs of two distinct indices can be chosen in $\binom{n}{2} = \frac{n(n-1)}{2}$ ways. Having fixed i, j, the term $X_i^2 X_j^2$ arises in 6 ways, corresponding to possible arrangements of 2 pairs of 2 indices: (i,i,j,j), (i,j,i,j), (i,j,j,i), (j,j,i,i), (j,i,j,i), (j,i,i,j). So $N = 3n(n-1)$ and we have

$$\mathbb{E}(S_n^4) \leq K(n + 3n(n-1)) = K(n + 3n^2 - 3n) \leq 3Kn^2.$$

By Chebyshev's inequality,

$$P(|\frac{S_n}{n}| \geq \varepsilon) = P(|S_n| \geq n\varepsilon) \leq \frac{\mathbb{E}(S_n^4)}{(n\varepsilon)^4} \leq \frac{3K}{\varepsilon^4} \cdot \frac{1}{n^2}.$$

The series $\sum \frac{1}{n^2}$ converges and by Borel–Cantelli the set $\limsup A_n$ with $A_n = \{\omega : |\frac{S_n}{n}| \geq \varepsilon\}$ has measure zero. Its complement is the set of full measure we need on which the sequence $\frac{S_n}{n}$ converges to 0. To see this let $\omega \notin \limsup A_n$, which means that ω is in finitely many A_n. So for a certain n_0, all $n \geq n_0$, $\omega \notin A_n$, i.e. $\frac{S_n}{n} < \varepsilon$ (as observed before), and this is precisely what was needed for the convergence in question.   □

The next law will only require finite moments of order 2, not necessarily uniformly bounded.

We precede it by an auxiliary but crucial inequality due to Kolmogorov. It gives a better estimate than does the Chebyshev inequality. The latter says that

$$P(|S_n| \geq \varepsilon) \leq \frac{\text{Var}(S_n)}{\varepsilon^2}.$$

In the theorem below the left-hand side is larger, hence the result is stronger.

Theorem 8.30

If $X_1, \ldots, X_n$ are independent with 0 expectation and finite variances, then for any $\varepsilon > 0$,

$$P(\max_{1 \leq k \leq n} |S_k| \geq \varepsilon) \leq \frac{\text{Var}(S_n)}{\varepsilon^2},$$

where $S_n = X_1 + \cdots + X_n$.

Proof

We fix an $\varepsilon > 0$ and describe the first instance that $|S_k|$ exceeds ε. Namely, we write

$$\varphi_k = \begin{cases} 1 & \text{if } |S_1| < \varepsilon, \ldots, |S_{k-1}| < \varepsilon, |S_k| \geq \varepsilon \\ 0 & \text{if all } |S_i| < \varepsilon. \end{cases}$$

For any ω at most one of the numbers $\varphi_k(\omega)$ may be 1, the remaining ones being 0, hence their sum is either 0 or 1. Clearly

$$\sum_{k=1}^{n} \varphi_k = 0 \quad \Leftrightarrow \quad \max_{1 \leq k \leq n} |S_k| < \varepsilon,$$

$$\sum_{k=1}^{n} \varphi_k = 1 \quad \Leftrightarrow \quad \max_{1 \leq k \leq n} |S_k| \geq \varepsilon.$$

Hence

$$P(\max_{1 \leq k \leq n} |S_k| \geq \varepsilon) = P(\sum_{k=1}^{n} \varphi_k = 1) = \mathbb{E}(\sum_{k=1}^{n} \varphi_k)$$

since the expectation is the integral of an indicator function:

$$\mathbb{E}(\sum_{k=1}^{n} \varphi_k) = \int_{\{\omega : \sum_{k=1}^{n} \varphi_k(\omega)=0\}} 0 \, dP + \int_{\{\omega : \sum_{k=1}^{n} \varphi_k(\omega)=1\}} 1 \, dP.$$

So it remains to show that

$$\mathbb{E}(\sum_{k=1}^{n} \varphi_k) \leq \frac{1}{\varepsilon^2} \mathrm{Var}(S_n) = \frac{1}{\varepsilon^2} \mathbb{E}(S_n^2),$$

the last equality because $\mathbb{E}(S_n) = 0$. We estimate $\mathbb{E}(S_n^2)$ from below:

$$\mathbb{E}(S_n^2) \geq \mathbb{E}(\sum_{k=1}^{n} \varphi_k \cdot S_n^2) \quad (\text{since } \sum_{k=1}^{n} \varphi_k \leq 1)$$

$$= \mathbb{E}(\sum_{k=1}^{n} [S_k^2 + 2S_k(S_n - S_k) + (S_n - S_k)^2]\varphi_k) \quad (\text{simple algebra})$$

$$\geq \mathbb{E}(\sum_{k=1}^{n} [S_k^2 + 2S_k(S_n - S_k)]\varphi_k) \quad (\text{non-negative term deleted})$$

$$= \mathbb{E}(\sum_{k=1}^{n} S_k^2 \varphi_k) + 2\mathbb{E}(\sum_{k=1}^{n} S_k(S_n - S_k)\varphi_k).$$

We show that the last expectation is equal to 0. Observe that φ_k is a function of random variables $X_1, \ldots, X_k$, so it is independent of $X_{k+1}, \ldots, X_n$ and also S_k is independent of $X_{k+1}, \ldots, X_n$ for the same reason. We compute one component of this last sum:

$$\mathbb{E}(S_k(S_n - S_k)\varphi_k) = \mathbb{E}(S_k \varphi_k (\sum_{i=k+1}^{n} X_i)) \quad (\text{by the definition of } S_n)$$

$$= \mathbb{E}(S_k \varphi_k) \mathbb{E}(\sum_{i=k+1}^{n} X_i) \quad (\text{by independence})$$

$$= 0 \quad (\text{since } \mathbb{E}(X_i) = 0 \text{ for all } i).$$

In the remaining sum note that for each $k \leq n$, $\varphi_k S_k^2 \geq \varphi_k \varepsilon^2$ (this is $0 \geq 0$ if $\varphi_k = 0$ and $S_k^k \geq \varepsilon^2$ if $\varphi_k = 1$, both true), hence

$$\mathbb{E}(\sum_{k=1}^{n} S_k^2 \varphi_k) \geq \mathbb{E}(\varepsilon^2 \sum_{k=1}^{n} \varphi_k) = \varepsilon^2 \mathbb{E}(\sum_{k=1}^{n} \varphi_k),$$

which gives the desired inequality. $\qquad\square$

Theorem 8.31

Suppose that $X_1, X_2, \ldots$ are independent with $\mathbb{E}(X_n) = 0$ and

$$\sum_{n=1}^{\infty} \frac{1}{n^2} \operatorname{Var}(X_n) < \infty.$$

Then

$$\frac{S_n}{n} \to 0 \quad \text{almost surely.}$$

Proof

We introduce auxiliary random variables

$$Y_m = \max_{k \le 2^m} |S_k|$$

and for $2^{m-1} \le n \le 2^m$

$$\left|\frac{S_n}{n}\right| \le \frac{1}{n} \max_{k \le 2^m} |S_k| \le \frac{1}{2^{m-1}} Y_m.$$

It is sufficient to show that $\frac{Y_m}{2^m} \to 0$ almost surely, and by Lemma 8.7 it is sufficient to show that for each $\varepsilon > 0$,

$$\sum_{m=1}^{\infty} P(|\frac{Y_m}{2^m}| \ge \varepsilon) < \infty.$$

First take a single term $P(|Y_m| \ge 2^m \varepsilon)$ and estimate it by Kolmogorov's inequality (Theorem 8.30):

$$P(|Y_m| \ge 2^m \varepsilon) \le \frac{\operatorname{Var}(S_{2^m})}{\varepsilon^2 2^{2m}}.$$

The problem reduces to showing that

$$\sum_{m=1}^{\infty} \operatorname{Var}(S_{2^m}) \frac{1}{4^m} < \infty.$$

We rearrange the components:

$$\sum_{m=1}^{\infty} \mathrm{Var}(S_{2^m}) \frac{1}{4^m} = \sum_{m=1}^{\infty} \frac{1}{4^m} \sum_{k=1}^{2^m} \mathrm{Var}(X_k)$$

$$= \mathrm{Var}(X_1) \sum_{m=1}^{\infty} \frac{1}{4^m} + \mathrm{Var}(X_2) \sum_{m=1}^{\infty} \frac{1}{4^m}$$

$$+ \mathrm{Var}(X_3) \sum_{m=2}^{\infty} \frac{1}{4^m} + \mathrm{Var}(X_4) \sum_{m=2}^{\infty} \frac{1}{4^m}$$

$$+ \mathrm{Var}(X_5) \sum_{m=3}^{\infty} \frac{1}{4^m} + \cdots + \mathrm{Var}(X_8) \sum_{m=3}^{\infty} \frac{1}{4^m}$$

$$+ \cdots$$

since $\mathrm{Var}(X_1), \mathrm{Var}(X_2)$ appear in all components of the series in m $(1, 2 \leq 2^1)$, $\mathrm{Var}(X_3), \mathrm{Var}(X_4)$ appear in all except the first one $(2^1 < 3, 4 \leq 2^2)$, $\mathrm{Var}(X_5), \ldots, \mathrm{Var}(X_8)$ appear in all except the first two $(2^2 < 5, 6, 7, 8 \leq 2^3)$, and so on. We arrive at the series

$$\sum_{k=1}^{\infty} \mathrm{Var}(X_k) a_k$$

where

$$a_k = \sum_{\{m : 2^m > k\}} \frac{1}{4^m}.$$

This is a geometric series with ratio $\frac{1}{4}$ and the first term $\frac{1}{4^j}$ where j is the least integer such that $2^j > k$. If we replace 2^j by k we increase the sum by adding more terms and the first term is then $\frac{1}{2^k}$:

$$a_k \leq \frac{\frac{1}{2^k}}{1 - \frac{1}{4}};$$

and by the hypothesis

$$\sum_{k=1}^{\infty} \mathrm{Var}(X_k) a_k \leq \frac{4}{3} \sum_{k=1}^{\infty} \mathrm{Var}(X_k) \frac{1}{2^k} < \infty,$$

which completes the proof. $\qquad\square$

Finally, we relax the conditions on moments even further; simultaneously we need to impose the assumption that the random variables are identically distributed.

Theorem 8.32

Suppose that $X_1, X_2, \ldots$ are independent, identically distributed, with $\mathbb{E}(X_1) = \mu < \infty$. Then

$$\frac{S_n}{n} \to \mu \quad \text{almost surely.}$$

Proof

The idea is to use the previous theorem where we needed finite variances. Since we do not have that here we truncate X_n:

$$Y_n = X_n(n) = X_n \mathbf{1}_{\{\omega : |X_n(\omega)| \le n\}}.$$

The truncated random variables have finite variances since each is bounded: $|Y_n| \le n$ (the right-hand side is forced to be zero if X_n dare upcross the level n). The new variables differ from the original ones if $|X_n| > n$. This, however, cannot happen too often, as the following argument shows. First,

$$\sum_{n=1}^{\infty} P(Y_n \ne X_n)$$

$$= \sum_{n=1}^{\infty} P(|X_n| > n)$$

$$= \sum_{n=1}^{\infty} P(|X_1| > n) \quad \text{(the distributions being the same)}$$

$$\le \sum_{n=1}^{\infty} \int_{n-1}^{n} P(|X_1| > x)\, \mathrm{d}x \quad \text{(as } P(|X_1| > x) \ge P(|X_1| > n))$$

$$\le \int_0^{\infty} P(|X_1| > x)\, \mathrm{d}x$$

$$= \mathbb{E}(|X_1|) \quad \text{(by Lemma 8.21)}$$

$$< \infty.$$

So by Borel–Cantelli, with probability one only finitely many events $X_n \ne Y_n$ happen; so, in other words, there is a set Ω' with $P(\Omega') = 1$ such that for $\omega \in \Omega'$, $X_n(\omega) = Y_n(\omega)$ for all except finitely many n. So, on Ω', if $\frac{Y_1 + \cdots + Y_n}{n}$ converges to some limit, the same holds for $\frac{S_n}{n}$.

To use the previous theorem we have to show the convergence of the series

$$\sum_{n=1}^{\infty} \frac{\mathrm{Var}(Y_n)}{n^2},$$

but since $\mathrm{Var}(Y_n) = \mathbb{E}(Y_n^2) - (\mathbb{E}Y_n)^2 \leq \mathbb{E}Y_n^2$ it is sufficient to show the convergence of

$$\sum_{n=1}^{\infty} \frac{\mathbb{E}(Y_n^2)}{n^2}.$$

To this end, note first:

$$\mathbb{E}(Y_n^2) = \int_0^{\infty} 2x P(|Y_n| > x)\,\mathrm{d}x \quad \text{(by Lemma 8.21)}$$

$$= \int_0^{n} 2x P(|Y_n| > x)\,\mathrm{d}x \quad \text{(since } P(|Y_n| > n) = 0\text{)}$$

$$= \int_0^{n} 2x P(|X_n| > x)\,\mathrm{d}x \quad \text{(if } |Y_n| \leq n \text{ then } Y_n = X_n\text{)}$$

$$= \int_0^{n} 2x P(|X_1| > x)\,\mathrm{d}x \quad \text{(identical distributions)}.$$

Next,

$$\sum_{n=1}^{\infty} \frac{\mathbb{E}(Y_n^2)}{n^2} = \sum_{n=1}^{\infty} \frac{1}{n^2} \int_0^{n} 2x P(|X_1| > x)\,\mathrm{d}x$$

$$= \sum_{n=1}^{\infty} \frac{1}{n^2} \int_0^{\infty} 2x \mathbf{1}_{[0,n)}(x) P(|X_1| > x)\,\mathrm{d}x$$

$$= 2 \int_0^{\infty} \sum_{n=1}^{\infty} \frac{1}{n^2} x \mathbf{1}_{[0,n)}(x) P(|X_1| > x)\,\mathrm{d}x.$$

We examine the function $x \mapsto \sum_{n=1}^{\infty} \frac{1}{n^2} x \mathbf{1}_{[0,n)}(x)$.

If $0 \leq x \leq 1$ then none of the terms in the series is killed and

$$\sum_{n=1}^{\infty} \frac{1}{n^2} x \mathbf{1}_{[0,n)}(x) \leq \sum_{n=1}^{\infty} \frac{1}{n^2} = \frac{\pi^2}{6} < 2,$$

as is well known.

If $x > 1$, then we have the sum only over $n \geq x$. Let $m = \mathrm{Int}(x)$ (the integer part of x). We have

$$\sum_{n=1}^{\infty} \frac{1}{n^2} x \mathbf{1}_{[0,n)}(x) = x \sum_{n=m+1}^{\infty} \frac{1}{n^2} \leq x \int_m^{\infty} \frac{1}{x^2}\,\mathrm{d}x = x \frac{1}{m} \leq 2.$$

In each case the function in question is bounded by 2, so

$$\sum_{n=1}^{\infty} \frac{\mathbb{E}(Y_n^2)}{n^2} \leq 4 \int_0^{\infty} P(|X_1| > x)\,\mathrm{d}x = 4\mathbb{E}(|X_1|) < \infty.$$

Consider $Y_n - \mathbb{E}(Y_n)$ and apply the previous theorem to get

$$\frac{1}{n}\sum_{k=1}^{n}(Y_k - \mathbb{E}(Y_k)) \to 0 \quad \text{almost surely.}$$

We have

$$\mathbb{E}(Y_k) = \mathbb{E}(X_k \mathbf{1}_{\{\omega:|X_k|\leq k\}}) = \mathbb{E}(X_1 \mathbf{1}_{\{\omega:|X_1(\omega)|\leq k\}}) \to \mu$$

since $X_1 \mathbf{1}_{\{\omega:|X_1(\omega)|\leq k\}} \to X_1$, the sequence being dominated by $|X_1|$ which is integrable. So we have by the triangle inequality

$$|\frac{1}{n}\sum_{k=1}^{n}Y_k - \mu| \leq |\frac{1}{n}\sum_{k=1}^{n}(Y_k - \mathbb{E}(Y_k))| + \frac{1}{n}\sum_{k=1}^{n}|\mathbb{E}(Y_k) - \mu| \to 0.$$

As observed earlier, this implies almost sure convergence of $\frac{S_n}{n}$. □

8.2.5 Weak convergence

In order to derive central limit theorems we first need to investigate the convergence of the distributions of the sequence (X_n) of random variables.

Definition 8.33

A sequence P_n of Borel probability measures on $\mathbb{R}^n$ converges *weakly* to P if and only their cumulative distribution functions F_n converge to the distribution function F of P at all points where F is continuous. If $P_n = P_{X_n}$, $P = P_X$ are the distributions of some random variables, then we say that the sequence (X_n) converges *weakly* to X.

The name 'weak' is justified, since this convergence is implied by the weakest we have come across so far, i.e. convergence in probability.

Theorem 8.34

If X_n converge in probability to X, then the distributions of X_n converge weakly.

Proof

Let $F(y) = P(X \leq y)$. Fix y, a continuity point of F, and $\varepsilon > 0$. The goal is to obtain

$$F(y) - \varepsilon < F_n(y) < F(y) + \varepsilon$$

for sufficiently large n. By continuity of F we can find $\delta > 0$ such that

$$P(X \leq y) - \frac{\varepsilon}{2} < P(X \leq y - \delta), \qquad P(X \leq y + \delta) < F(y) + \frac{\varepsilon}{2}.$$

By convergence in probability,

$$P(|X_n - X| > \delta) < \frac{\varepsilon}{2}.$$

Clearly, if $X_n \leq y$ and $|X_n - X| < \delta$, then $X < y + \delta$, so

$$P((X_n \leq y) \cap (|X_n - X| < \delta)) \leq P(X < y + \delta).$$

We can estimate the left-hand side from below:

$$P(X_n \leq y) - \frac{\varepsilon}{2} < P((X_n \leq y) \cap (|X_n - X| < \delta)).$$

Putting all these together we get

$$P(X_n \leq y) < P(X \leq y) + \varepsilon$$

and letting $\varepsilon \to 0$ we have achieved half of the goal. The other half is obtained similarly. $\qquad\qquad\square$

However, it turns out that weak convergence in a certain sense implies convergence almost surely. What we mean by 'in a certain sense' is explained in the next theorem: it also gives a central role to the Borel sets and Lebesgue measure in $[0, 1]$.

Theorem 8.35 (Skorokhod representation theorem)

If P_n converge weakly to P, then there exist X_n, X, random variables defined on the probability $([0, 1], \mathcal{B}, m_{[0,1]})$, such that $P_{X_n} = P_n$, $P_X = P$ and $X_n \to X$ a.s.

Proof

Take X_n^+, X_n^-, X^+, X^- corresponding to F_n, F, the distribution functions of P_n, P, as in Theorem 4.45. We have shown there that $F_{X^+} = F_{X^-} = F$, which implies $P(X^+ = X^-) = 1$. Fix an ω such that $X^+(\omega) = X^-(\omega)$. Let y be a continuity point of F such that $y > X^+(\omega)$. Then $F(y) > \omega$ and, by the weak convergence, for sufficiently large n we have $F_n(y) > \omega$. Then, by the construction, $X_n^+(\omega) \leq y$. This inequality holds for all except finitely many n, so it is preserved if we take the upper limit on the left:

$$\limsup X_n^+(\omega) \leq y.$$

Take a sequence y_k of continuity points of F converging to $X^+(\omega)$ from above (the set of discontinuity points of a monotone function is at most countable). For $y = y_k$ consider the above inequality and pass to the limit with k to get

$$\limsup X_n^+(\omega) \le X^+(\omega).$$

Similarly

$$\liminf X_n^-(\omega) \ge X^-(\omega),$$

so

$$X^-(\omega) \le \liminf X_n^-(\omega) \le \limsup X_n^+(\omega) \le X^+(\omega).$$

The extremes are equal a.s., so the convergence holds a.s.   □

The Skorokhod theorem is an important tool in probability. We will only need it for the following result, which links convergence of distributions to that of the associated characteristic functions.

Theorem 8.36

If P_{X_n} converge weakly to P_X then $\varphi_{X_n} \to \varphi_X$.

Proof

Take the Skorokhod representation Y_n, Y of the measures P_{X_n}, P_X. Almost sure convergence of Y_n to Y implies that $\mathbb{E}(e^{itY_n}) \to \mathbb{E}(e^{itY})$ by the dominated convergence theorem. But the distributions of X_n, X are the same as the distributions of Y_n, Y, so the characteristic functions are the same.   □

Theorem 8.37 (Helly's theorem)

Let F_n be a sequence of distribution functions of some probability measures. There exists F, the distribution function of a measure (not necessarily probability), and a sequence k_n such that $F_{k_n}(x) \to F(x)$ at the continuity points of F.

Proof

Arrange the rational numbers in a sequence: $\mathbb{Q} = \{q_1, q_2, \ldots\}$. The sequence $F_n(q_1)$ is bounded (the values of a distribution function lie in $[0, 1]$), hence it has a convergent subsequence,

$$F_{k_n^1}(q_1) \to y_1.$$

Next consider the sequence $F_{k_n^1}(q_2)$, which is again bounded, so for a subsequence k_n^2 of k_n^1 we have convergence

$$F_{k_n^2}(q_2) \to y_2.$$

Of course also

$$F_{k_n^2}(q_1) \to y_1.$$

Proceeding in this way we find k_n^3, k_n^4, ..., each term a subsequence of the previous one, with

$$F_{k_n^3}(q_m) \to y_m \quad \text{for } m \le 3,$$
$$F_{k_n^4}(q_m) \to y_m \quad \text{for } m \le 4,$$

and so on. The diagonal sequence $F_{k_n} = F_{k_n^n}$ converges at all rational points. We define $F_{\mathbb{Q}}$ on $\mathbb{Q}$ by

$$F_{\mathbb{Q}}(q) = \lim F_{k_n}(q)$$

and next we write

$$F(x) = \inf\{F_{\mathbb{Q}}(q) : q \in \mathbb{Q}, q > x\}.$$

We show that F is non-decreasing. Since F_n are non-decreasing, the same is true for $F_{\mathbb{Q}}$ ($q_1 < q_2$ implies $F_{k_n}(q_1) \le F_{k_n}(q_2)$, which remains true in the limit). Now let $x_1 < x_2$. $F(x_1) \le F_{\mathbb{Q}}(q)$ for all $q > x_1$, hence in particular for all $q > x_2$, so $F(x_1) \le \inf_{q > x_2} F_{\mathbb{Q}}(q) = F(x_2)$.

We show that F is right-continuous. Let $x_n \searrow x$. By the monotonicity of F, $F(x) \le F(x_n)$, hence $F(x) \le \lim F(x_n)$. Suppose that $F(x) < \lim F(x_n)$. By the definition of F there is $q \in \mathbb{Q}$, $x < q$, such that $F_{\mathbb{Q}}(q) < \lim F(x_n)$. For some n_0, $x \le x_{n_0} < q$, hence $F(x_{n_0}) \le F_{\mathbb{Q}}(q)$ again by the definition of F, thus $F(x_{n_0}) < \lim F(x_n)$, which is a contradiction.

Finally, we show that if F is continuous at x, then $F_{k_n}(x) \to F(x)$. Let $\varepsilon > 0$ be arbitrary and find rationals $q_1 < q_2 < x < q_3$ such that

$$F(x) - \varepsilon < F(q_1) \le F(x) \le F(q_3) < F(x) + \varepsilon.$$

Since $F_{k_n}(q_2) \to F_{\mathbb{Q}}(q_2) \ge F(q_1)$, then for sufficiently large n,

$$F(x) - \varepsilon < F_{k_n}(q_2).$$

But F_{k_n} is non-decreasing, so

$$F_{k_n}(q_2) \le F_{k_n}(x) \le F_{k_n}(q_3).$$

Finally, $F_{k_n}(q_3) \to F_{\mathbb{Q}}(q_3) \ge F(q_3)$, so for sufficiently large n,

$$F_{k_n}(q_3) < F(x) + \varepsilon.$$

Putting together the above three inequalities we get

$$F(x) - \varepsilon < F_{k_n}(x) < F(x) + \varepsilon,$$

which proves the convergence. $\square$

Remark 8.38

The limit distribution function need not correspond to a probability measure. Example: $F_n = \mathbf{1}_{[n,\infty)}$, $F_n \to 0$, so $F = 0$. This is a distribution function (non-decreasing, right-continuous) and the corresponding measure satisfies $P(A) = 0$ for all A. We then say informally that the mass escapes to infinity. The following concept is introduced to prevent this happening.

Definition 8.39

We say that a sequence of probabilities P_n on $\mathbb{R}^d$ is *tight* if for each $\varepsilon > 0$ there is M such that $P_n(\mathbb{R}^d \setminus [-M, M]) < \varepsilon$ for all n.

By an interval in $\mathbb{R}^n$ we understand the product of intervals: $[-M, M] = \{x = (x_1, \ldots, x_n) \in \mathbb{R}^n : x_i \in [-M, M] \text{ all } i\}$. It is important that the M chosen for ε is good for all n – the inequality is uniform. It is easy to find such an $M = M_n$ for each P_n separately. This follows from the fact that $P_n([-M, M]) \to 1$ as $M \to \infty$.

Theorem 8.40 (Prokhorov's theorem)

If a sequence P_n is tight, then it has a subsequence convergent weakly to some probability measure P.

Proof

By Helly's theorem a subsequence F_{k_n} converges to some distribution function F. All we have to do is to show that F corresponds to some probability measure P, which means we have to show that $F(\infty) = 1$ (i.e. $\lim_{y \to \infty} F(y) = 1$). Fix $\varepsilon > 0$ and find a continuity point such that $F_n(y) = P_n((-\infty, y]) > 1 - \varepsilon$ for all n (find M from the definition of tightness and take a continuity point of F which is larger than M). Hence $\lim_{n \to \infty} F_{k_n}(y) \geq 1 - \varepsilon$, but this limit is $F(y)$. This proves that $\lim_{y \to \infty} F(y) = 1$. $\square$

We need to extend the notion of the characteristic function.

Definition 8.41

We say that φ is the characteristic function of a Borel measure P on $\mathbb{R}$ if $\varphi(t) = \int e^{itx} \, dP(x)$.

In the case where $P = P_X$ we obviously have $\varphi_P = \varphi_X$, so the two definitions are consistent.

Theorem 8.42

Suppose (P_n) is tight and let P be the limit of a subsequence of (P_n) as provided by Prokhorov's theorem. If $\varphi_n(u) \to \varphi(u)$, where φ_n are the characteristic functions of P_n and φ is the characteristic function of P, then $P_n \to P$ weakly.

Proof

Fix a continuity point of F. For every subsequence F_{k_n} there is a subsequence (subsubsequence) l_{k_n}, l_n for brevity, such that F_{l_n} converge to some function H (Helly's theorem). Denote the corresponding measure by P'. Hence $\varphi_{l_n} \to \varphi_{P'}$, but on the other hand, $\varphi_{l_n} \to \varphi$. So $\varphi_{P'} = \varphi$ and consequently $P' = P$ (Corollary 6.25). The above is sufficient for the convergence of the sequence $F_n(y)$. $\square$

8.2.6 Central limit theorem

The following lemma will be useful in what follows. It shows how to estimate the 'weight' of the 'tails' of a probability measure, and will be a useful tool in proving tightness.

Lemma 8.43

If φ is the characteristic function of P, then

$$P(\mathbb{R} \setminus [-M, M]) \leq 7M \int_0^{\frac{1}{M}} [1 - \Re\varphi(u)]\, du.$$

Proof

$$M \int_0^{\frac{1}{M}} [1 - \Re\varphi(u)] \, du = M \int_0^{\frac{1}{M}} [1 - \Re \int_{\mathbb{R}} e^{ixu} \, dP(x)] \, du$$

$$= M \int_0^{\frac{1}{M}} [1 - \int_{\mathbb{R}} \cos(xu) \, dP(x)] \, du$$

$$= \int_{\mathbb{R}} M \int_0^{\frac{1}{M}} [1 - \cos(xu)] \, du \, dP(x) \quad \text{(by Fubini)}$$

$$= \int_{\mathbb{R}} (1 - \frac{\sin(\frac{x}{M})}{\frac{x}{M}}) \, dP(x)$$

$$\geq \int_{|t| \geq 1} (1 - \frac{\sin(\frac{x}{M})}{\frac{x}{M}}) \, dP(x)$$

$$\geq \inf_{|t| \geq 1} (1 - \frac{\sin t}{t}) \int_{|\frac{x}{M}| > 1} dP(x)$$

$$\geq \frac{1}{7} P(\mathbb{R} \setminus [-M, M])$$

since $1 - \frac{\sin t}{t} \geq 1 - \sin 1 \geq \frac{1}{7}$. $\qquad \square$

Theorem 8.44 (Levy's theorem)

Let φ_n be the characteristic functions of P_n. Suppose that $\varphi_n \to \varphi$ where φ is a function continuous at 0. Then φ is the characteristic function of a measure P and $P_n \to P$ weakly.

Proof

It is sufficient to show that P_n is tight (then $P_{k_n} \to P$ weakly for some P and k_n, $\varphi_{k_n} \to \varphi_P$, $\varphi = \varphi_P$, and by the previous theorem we are done).

Applying Lemma 8.43 we have

$$P_n(\mathbb{R} \setminus [-M, M]) \leq 7M \int_0^{\frac{1}{M}} [1 - \Re\varphi_n(u)] \, du.$$

Since $\varphi_n \to \varphi$, $|\varphi_n| \leq 1$, and for fixed M, this upper bound is integrable, so by the dominated convergence theorem

$$7M \int_0^{\frac{1}{M}} [1 - \Re\varphi_n(u)] \, du \to 7M \int_0^{\frac{1}{M}} [1 - \Re\varphi(u)] \, du.$$

If $M \to \infty$, then

$$7M \int_0^{\frac{1}{M}} [1 - \Re\varphi(u)] \, du \le 7M \frac{1}{M} \sup_{[0,\frac{1}{M}]} |1 - \Re\varphi(u)| \to 0$$

by continuity of φ at 0 (recall that $\varphi(0) = 1$). Now let $\varepsilon > 0$ and find M_0 such that $7 \sup_{[0,\frac{1}{M_0}]} |1 - \Re\varphi(u)| < \frac{\varepsilon}{2}$, and n_0 such that for $n \ge n_0$

$$\left| \int_0^{\frac{1}{M_0}} [1 - \Re\varphi_n(u)] \, du - \int_0^{\frac{1}{M_0}} [1 - \Re\varphi(u)] \, du \right| < \frac{\varepsilon}{2}.$$

Hence

$$P_n(\mathbb{R} \setminus [-M_0, M_0]) < \varepsilon$$

for $n \ge n_0$. Now for each $n = 1, 2, \ldots, n_0$ find M_n such that $P_n([-M_n, M_n]) > 1 - \varepsilon$ and let $M = \max\{M_0, M_1, \ldots, M_{n_0}\}$. Of course since $M \ge M_k$, $P_n([-M, M]) \ge P_n([-M_k, M_k]) > 1 - \varepsilon$ for each n, which proves the tightness of P_n. $\qquad\square$

For a sequence X_k with $\mu_k = \mathbb{E}(X_k)$, $\sigma_k^2 = \mathrm{Var}(X_k)$ finite, let $S_n = X_1 + \cdots + X_n$ as usual and consider the normalized random variables

$$T_n = \frac{S_n - \mathbb{E}(S_n)}{\sqrt{\mathrm{Var}(S_n)}}.$$

Clearly $\mathbb{E}(T_n) = 0$ and $\mathrm{Var}(T_n) = 1$ (by $\mathrm{Var}(aX) = a^2\mathrm{Var}(X)$). Write $c_n^2 = \mathrm{Var}(S_n)$ (if X_n are independent, then as we already know, $c_n^2 = \sum_{k=1}^n \sigma_k^2$). We state a condition under which the sequence of distributions of T_n converges to the standard Gaussian measure G (with the density $\frac{1}{\sqrt{2\pi}} e^{-\frac{1}{2}x^2}$):

$$\frac{1}{c_n^2} \sum_{k=1}^n \int_{\{x : |x - \mu_k| \ge \varepsilon c_n\}} (x - \mu_k)^2 \, dP_{X_k}(x) \to 0 \quad \text{as } n \to \infty. \tag{8.4}$$

In particular, if the distributions of X_n are the same, $\mu_k = \mu$, $\sigma_k = \sigma$, then this condition is satisfied. To see this, note that assuming independence we have $c_n^2 = n\sigma^2$ and then

$$\int_{\{x : |x - \mu| \ge \varepsilon\sigma\sqrt{n}\}} (x - \mu)^2 \, dP_{X_k}(x) = \int_{\{x : |x - \mu| \ge \varepsilon\sigma\sqrt{n}\}} (x - \mu)^2 \, dP_{X_1}(x),$$

hence

$$\frac{1}{n\sigma^2} \sum_{k=1}^n \int_{\{x : |x - \mu| \ge \varepsilon\sigma\sqrt{n}\}} (x - \mu)^2 \, dP_{X_k}(x)$$

$$= \frac{1}{\sigma^2} \int_{\{x : |x - \mu| \ge \varepsilon\sigma\sqrt{n}\}} (x - \mu)^2 \, dP_{X_1}(x) \to 0$$

as $n \to \infty$ since the set $\{x : |x - \mu| \geq \varepsilon\sigma\sqrt{n}\}$ decreases to $\emptyset$.

We are ready for the main theorem in probability. The proof is quite technical and advanced, and may be omitted at a first reading.

Theorem 8.45 (Lindeberg–Feller theorem)

Let X_n be independent with finite expectations and variances. If condition (8.4) holds, then $P_{T_n} \to G$ weakly.

Proof

Assume for simplicity that $\mu_n = 0$ (the general case follows by considering $X_n - \mu_n$). It is sufficient to show that the characteristic functions φ_{T_n} converge to the characteristic function of G, i.e. to show that

$$\varphi_{T_n}(u) \to e^{-\frac{1}{2}u^2}.$$

We compute

$$\begin{aligned}
\varphi_{T_n}(u) &= \mathbb{E}(e^{iuT_n}) \quad \text{(by the definition of } \varphi_{T_n}) \\
&= \mathbb{E}(e^{i\frac{u}{c_n}\sum_{k=1}^{n} X_k}) \\
&= \mathbb{E}(\prod_{k=1}^{n} e^{i\frac{u}{c_n}X_k}) \\
&= \prod_{k=1}^{n} \mathbb{E}(e^{i\frac{u}{c_n}X_k}) \quad \text{(by independence)} \\
&= \prod_{k=1}^{n} \varphi_{X_k}(\frac{u}{c_n}) \quad \text{(by the definition of } \varphi_{X_k}).
\end{aligned}$$

What we need to show is that

$$\log \prod_{k=1}^{n} \varphi_{X_k}(\frac{u}{c_n}) \to -\frac{1}{2}u^2.$$

We shall make use of the following formulae (particular cases of Taylor's formula for a complex variable):

$$\log(1 + z) = z + \theta_1 |z|^2 \quad \text{for some } \theta_1 \text{ with } |\theta_1| \leq 1,$$

$$e^{iy} = 1 + iy + \frac{1}{2}\theta_2 y^2 \quad \text{for some } \theta_2 \text{ with } |\theta_2| \leq 1,$$

$$e^{iy} = 1 + iy - \frac{1}{2}y^2 + \frac{1}{6}\theta_3 |y|^3 \quad \text{for some } \theta_3 \text{ with } |\theta_3| \leq 1.$$

So for fixed $\varepsilon > 0$,

$$
\begin{aligned}
\varphi_{X_k}(u) &= \int_{|x| \geq \varepsilon c_n} e^{iux}\, dP_{X_k}(x) + \int_{|x| < \varepsilon c_n} e^{iux}\, dP_{X_k}(x) \\
&= \int_{|x| \geq \varepsilon c_n} \left(1 + iux + \frac{1}{2}\theta_2 u^2 x^2\right) dP_{X_k}(x) \\
&\quad + \int_{|x| < \varepsilon c_n} \left(1 + iux - \frac{1}{2}u^2 x^2 + \frac{1}{6}\theta_3 |u|^3 |x|^3\right) dP_{X_k}(x) \\
&= 1 + \frac{1}{2}u^2 \int_{|x| \geq \varepsilon c_n} \theta_2 x^2\, dP_{X_k}(x) - \frac{1}{2}u^2 \int_{|x| < \varepsilon c_n} x^2\, dP_{X_k}(x) \\
&\quad + \frac{1}{6}|u|^3 \int_{|x| < \varepsilon c_n} \theta_3 |x|^3\, dP_{X_k}(x)
\end{aligned}
$$

since $\int x\, dP_{X_k}(x) = 0$ (this is $\mathbb{E}(X_k)$). For clarity we introduce the following notation:

$$
\alpha_{nk} = \int_{|x| \geq \varepsilon c_n} x^2\, dP_{X_k}(x),
$$

$$
\beta_{nk} = \int_{|x| < \varepsilon c_n} x^2\, dP_{X_k}(x).
$$

Observe that $\beta_{nk} \leq \varepsilon^2 c_n^2$ because on the set over which we take the integral we have $x^2 < \varepsilon^2 c_n^2$ and we integrate with respect to a probability measure. Condition (8.4) now takes the form

$$
\sum_{k=1}^{n} \frac{1}{c_n^2}\alpha_{nk} \to 0 \quad \text{as } n \to \infty. \tag{8.5}
$$

Since

$$
\sum_{k=1}^{n} (\alpha_{nk} + \beta_{nk}) = \sum_{k=1}^{n} \mathrm{Var}(X_k) = c_n^2
$$

we have

$$
\sum_{k=1}^{n} \frac{1}{c_n^2}\beta_{nk} \to 1 \quad \text{as } n \to \infty. \tag{8.6}
$$

The numbers α_{nk}, β_{nk} are positive, so the last convergence is monotone (the sequence is increasing). The above relations hold for each $\varepsilon > 0$.

We now analyse some terms in the expression for φ_{X_k}. Since

$$
\left| \int_{|x| \geq \varepsilon c_n} \theta_2 x^2\, dP_{X_k}(x) \right| \leq \int_{|x| \geq \varepsilon c_n} x^2\, dP_{X_k}(x),
$$

we have a θ_2' with $|\theta_2'| \leq 1$ such that

$$
\int_{|x| \geq \varepsilon c_n} \theta_2 x^2\, dP_{X_k}(x) = \theta_2' \int_{|x| \geq \varepsilon c_n} x^2\, dP_{X_k}(x) = \theta_2' \alpha_{nk}.
$$

Next

$$\left|\int_{|x|<\varepsilon c_n} \theta_3 |x|^3 \, dP_{X_k}(x)\right| \leq \int_{|x|<\varepsilon c_n} |x|^3 \, dP_{X_k}(x) \leq \int_{|x|<\varepsilon c_n} \varepsilon c_n x^2 \, dP_{X_k}(x)$$

(we replace one x by εc_n, leaving the remaining two), hence for some θ_3' with $|\theta_3'| \leq 1$,

$$\left|\int_{|x|<\varepsilon c_n} \theta_3 |x|^3 \, dP_{X_k}(x)\right| \leq \theta_3' \int_{|x|<\varepsilon c_n} \varepsilon c_n x^2 \, dP_{X_k}(x) = \theta_3' \varepsilon c_n \beta_{nk}.$$

We substitute this to the expression for φ_{X_k}, obtaining

$$\varphi_{X_k}(u) = 1 + \frac{1}{2}u^2 \theta_2' \alpha_{nk} - \frac{1}{2}u^2 \beta_{nk} + \frac{1}{6}|u|^3 \theta_3' \varepsilon c_n \beta_{nk}.$$

Replace u by $\frac{u}{c_n}$ to get

$$\varphi_{X_k}\left(\frac{u}{c_n}\right) = 1 + \frac{1}{2}u^2 \theta_2' \frac{1}{c_n^2} \alpha_{nk} - \frac{1}{2}u^2 \frac{1}{c_n^2} \beta_{nk} + \frac{1}{6}|u|^3 \theta_3' \varepsilon \frac{1}{c_n^2} \beta_{nk} = 1 + \gamma_{nk}$$

with

$$\gamma_{nk} = \frac{1}{2}u^2 \theta_2' \frac{1}{c_n^2} \alpha_{nk} - \frac{1}{2}u^2 \frac{1}{c_n^2} \beta_{nk} + \frac{1}{6}|u|^3 \theta_3' \varepsilon \frac{1}{c_n^2} \beta_{nk}.$$

The relations (8.5) and (8.6) give

$$\sum_{k=1}^{n} \gamma_{nk} \to -\frac{1}{2}u^2 + \frac{1}{6}|u|^3 \theta_3' \varepsilon. \tag{8.7}$$

Recall that what we are really after is

$$\log \prod_{k=1}^{n} \varphi_{X_k}\left(\frac{u}{c_n}\right) = \sum_{k=1}^{n} \log \varphi_{X_k}\left(\frac{u}{c_n}\right) = \sum_{k=1}^{n} (\gamma_{nk} + \theta_1 |\gamma_{nk}|^2),$$

where we introduced Taylor's formula for the logarithm, so we are not that far from the target. All we have to do is to show that

$$\left|\log \prod_{k=1}^{n} \varphi_{X_k}\left(\frac{u}{c_n}\right) + \frac{1}{2}u^2\right| \to 0$$

as $n \to \infty$. So let $\delta > 0$ be arbitrary and u fixed.

$$\left|\log \prod_{k=1}^{n} \varphi_{X_k}\left(\frac{u}{c_n}\right) + \frac{1}{2}u^2\right|$$

$$\leq \left|\sum_{k=1}^{n} \gamma_{nk} + \frac{1}{2}u^2\right| + |\theta_1| \sum_{k=1}^{n} |\gamma_{nk}|^2$$

$$\leq \left|\sum_{k=1}^{n} \gamma_{nk} + \frac{1}{2}u^2 - \frac{1}{6}|u|^3 \theta_3' \varepsilon\right| + \sum_{k=1}^{n} |\gamma_{nk}|^2 + |u|^3 \varepsilon |\theta_3'|.$$

The first term on the right converges to zero by (8.7), so it is less than $\frac{\delta}{2}$ for sufficiently large n. It remains to show that

$$\sum_{k=1}^{n} |\gamma_{nk}|^2 + |u|^3 \varepsilon < \frac{\delta}{2}$$

for large n. We choose ε so small that $|u|^3 \varepsilon < \frac{\delta}{4}$. It remains to show that

$$\sum_{k=1}^{n} |\gamma_{nk}|^2 < \frac{\delta}{4}$$

for large n. We have a formula for γ_{nk} and we know something about $\sum_{k=1}^{n} \gamma_{nk}$, hence we use the following trick:

$$\sum_{k=1}^{n} |\gamma_{nk}|^2 \leq \max_{k=1,\ldots,n} |\gamma_{nk}| \sum_{k=1}^{n} |\gamma_{nk}|.$$

The first factor has:

$$\max_{k=1,\ldots,n} |\gamma_{nk}| \leq \frac{1}{2} u^2 \max_{k=1,\ldots,n} |\frac{\alpha_{nk}}{c_n^2}| + \frac{1}{2} u^2 \varepsilon^2 + \frac{1}{6} |u|^3 \varepsilon^3$$

(using $\beta_{nk} \leq \varepsilon^2 c_n^2$) but

$$\max_{k=1,\ldots,n} |\frac{\alpha_{nk}}{c_n^2}| \leq \sum_{k=1}^{n} \frac{\alpha_{nk}}{c_n^2} \to 0.$$

The second factor satisfies

$$\sum_{k=1}^{n} |\gamma_{nk}| \leq \frac{1}{2} u^2 \sum_{k=1}^{n} \frac{1}{c_n^2} \alpha_{nk} + \frac{1}{2} u^2 + \frac{1}{6} |u|^3 \varepsilon,$$

where we used the fact $\sum \frac{\beta_{nk}}{c_n^2} \leq 1$. Writing $\sum_{k=1}^{n} \frac{1}{c_n^2} \alpha_{nk} = a_n$, $a_n \to 0$, for clarity we have, taking $\varepsilon \leq 1$,

$$\sum_{k=1}^{n} |\gamma_{nk}|^2 \leq (\frac{u^2}{2} a_n + \frac{u^2 \varepsilon^2}{2} + \frac{|u|^3 \varepsilon^3}{6})(\frac{u^2}{2} a_n + \frac{u^2}{2} + \frac{|u|^3 \varepsilon}{6}) \leq C a_n + D \varepsilon$$

for some numbers (depending only on u) C, D. So finally choose ε so that $D\varepsilon < \frac{\delta}{8}$ and then take n_0 so large that $C a_n < \frac{\delta}{8}$ for $n \geq n_0$. $\square$

As a special case we immediately deduce a famous result:

Corollary 8.46 (de Moivre–Laplace theorem)

Let X_n be identically distributed independent random variables with $P(X_n = 1) = P(X_n = -1) = \frac{1}{2}$. Then

$$P(a < \frac{S_n}{\sqrt{n}} < b) \to \frac{1}{\sqrt{2\pi}} \int_a^b e^{-\frac{1}{2}x^2} \, dx.$$

Exercise 8.8

Use the central limit theorem to estimate the probability that the number of 'heads' in 1000 independent tosses differs from 500 by less than 2%.

Exercise 8.9

How many tosses of a coin are required to have the probability at least 0.99 that the average number of 'heads' differs from 0.5 by less than 1%?

8.2.7 Applications to mathematical finance

We shall show that the Black–Scholes model is a limit of a sequence of suitably defined CRR models with option prices converging as well. For fixed T recall that in the Black–Scholes model the stock price is of the form

$$S(T) = S(0)\exp\{\xi(T)\},$$

where $\xi(T) = (r - \frac{\sigma^2}{2})T + \sigma w(T)$, r is the risk-free rate for continuous compounding. Randomness is only in $w(T)$, which is Gaussian with zero mean and variance T.

To construct the approximating sequence, fix n to decompose the time period into n steps of length $\frac{T}{n}$ and write

$$R_n = \exp\{r\frac{T}{n}\},$$

which is the risk-free growth factor for one step. We construct a sequence $\eta_n(i)$, $i = 1, \ldots, n$ of independent, identically distributed random variables, with binomial distribution, so that the price in the binomial model after n steps,

$$S_n(T) = S(0)\prod_{i=1}^{n}\eta_n(i),$$

converges to $S(T)$. Write

$$\eta_n(i) = \left\{ \begin{array}{l} U_n \\ D_n, \end{array} \right.$$

where each value is taken with probability $\frac{1}{2}$. Our task is to find U_n, D_n, assuming that $U_n > D_n$. The following condition

$$R_n = \frac{1}{2}(U_n + D_n) \qquad (8.8)$$

guarantees that $S_n(T)$ is a martingale (see Section 7.4.4). We look at the logarithmic returns:

$$\ln \frac{S_n(T)}{S(0)} = \ln(\prod_{i=1}^{n} \eta_n(i)) = \sum_{i=1}^{n} \ln \eta_n(i).$$

We wish to apply the Central Limit Theorem to the sequence $\ln \eta_n(i)$ so we adjust the variance. We want to have

$$\mathrm{Var}(\ln(\prod_{i=1}^{n} \eta_n(i))) = T\sigma^2.$$

On the left, using independence,

$$\mathrm{Var}(\sum_{i=1}^{n} \ln \eta_n(i)) = \sum_{i=1}^{n} \mathrm{Var}(\ln \eta_n(i)) = n\mathrm{Var}(\ln \eta_n(1)).$$

For the binomial distribution with $p = \frac{1}{2}$,

$$\mathrm{Var}(\ln \eta_n(1)) = \frac{1}{4}(\ln(U_n) - \ln(D_n))^2,$$

so the condition needed is

$$n\frac{1}{4}(\ln(U_n) - \ln(D_n))^2 = T\sigma^2.$$

Since $U_n > D_n$,

$$\ln(\frac{U_n}{D_n}) = 2\sigma\frac{\sqrt{T}}{\sqrt{n}},$$

so finally

$$U_n = D_n \exp\{2\sigma\sqrt{\frac{T}{n}}\}. \qquad (8.9)$$

We solve the system (8.8), (8.9) to get

$$D_n = e^{r\frac{T}{n}} \frac{2}{1 + e^{2\sigma\sqrt{T/n}}},$$

$$U_n = D_n e^{2\sigma\sqrt{T/n}}.$$

Consider the expected values

$$\mathbb{E}(\ln(\prod_{i=1}^{n} \eta_n(i))) = \mathbb{E}(\sum_{i=1}^{n} \ln \eta_n(i)) = n\mathbb{E}(\ln \eta_n(1))$$

$$= n\frac{1}{2}(\ln(U_n) + \ln(D_n)) = a_n,$$

say.

Exercise 8.10

Show that

$$a_n \to (r - \frac{1}{2}\sigma^2)T.$$

For each n we have a sequence of n independent identically distributed random variables $\xi_n(i) = \ln \eta_n(i)$ forming the so-called *triangular array*. We have the following version of central limit theorem. It can be proved in exactly the same way as Theorem 8.45 (see also [2]).

Theorem 8.47

If $\xi_n(i)$ is a triangular array, $\lambda_n = \sum \xi_n(i)$, $\mathbb{E}(\lambda_n) \to \mu = (r - \frac{1}{2}\sigma^2)T$, $\text{Var}(\lambda_n) \to \sigma^2 T$ then the sequence (λ_n) converges weakly to a Gaussian random variable with mean μ and variance $\sigma^2 T$.

The conditions stated in the theorem hold true with the values assigned above to U_n and D_n, as we have seen. As a result, $\ln S_n(T)$ converges to $\ln S(T)$ weakly. The value of put in the binomial model is given by

$$P_n(0) = \exp\{-rT\}\mathbb{E}(K - S_n(T))^+ = \exp\{-rT\}\mathbb{E}(g(\ln S_n(T)))$$

where $g(x) = (K - e^x)^+$ is a bounded, continuous function. Therefore

$$P_n(0) \to \exp\{-rT\}\mathbb{E}(g(\ln S(T))),$$

which, as we know (see Exercise 4.20), gives the Black–Scholes formula. The convergence of call prices follows immediately from the call–put parity, which holds universally, in each model.

8.3 Proofs of propositions

Proof (of Proposition 8.10)

Clearly $d(X, Y) = d(Y, X)$. If $d(X, Y) = 0$ then $\mathbb{E}(|X - Y|) = 0$, hence $X = Y$ almost everywhere so, $X = Y$ in L^1. The triangle inequality follows from the triangle inequality for the metric $\rho(x, y) = \frac{|x-y|}{1+|x-y|}$.

Next, assume that $X_n \to X$ in probability. Let $\varepsilon > 0$

$$d(X_n, X) = \int_{|X_n - X| < \frac{\varepsilon}{2}} \frac{|X_n - X|}{1 + |X_n - X|} \, dP$$
$$+ \int_{|X_n - X| \geq \frac{\varepsilon}{2}} \frac{|X_n - X|}{1 + |X_n - X|} \, dP$$
$$\leq \frac{\varepsilon}{2} + P(|X_n - X| \geq \frac{\varepsilon}{2})$$

since the integrand in the first term is less than ε (estimating the denominator by 1) and we estimate the integrand in the second term by 1. For n large enough the second term is less than $\frac{\varepsilon}{2}$.

Conversely, let $E_{\varepsilon,n} = \{\omega : |X_n(\omega) - X(\omega)| > \varepsilon\}$ and assume $0 < \varepsilon < 1$. Additionally, let $A_n = \{\omega : |X_n(\omega) - X(\omega)| < 1\}$ and write

$$d(X_n X) = \int_{A_n} \frac{|X_n - X|}{1 + |X_n - X|} \, dP + \int_{A_n^c} \frac{|X_n - X|}{1 + |X_n - X|} \, dP.$$

We estimate from below each of the two terms. First,

$$\int_{A_n} \frac{|X_n - X|}{1 + |X_n - X|} \, dP \geq \int_{A_n \cap E_{\varepsilon,n}} \frac{|X_n - X|}{1 + |X_n - X|} \, dP$$
$$\geq \frac{1}{2} \int_{A_n \cap E_\varepsilon} \varepsilon \, dP$$
$$= \frac{\varepsilon}{2} P(A_n \cap E_{\varepsilon,n})$$

since $\frac{a}{1+a} > \frac{a}{2}$ if $a < 1$. Second,

$$\int_{A_n^c} \frac{|X_n - X|}{1 + |X_n - X|} \, dP \geq \int_{A_n^c} \frac{1}{2} \, dP \geq \int_{A_n^c \cap E_{\varepsilon,n}} \frac{1}{2} \, dP \geq \frac{\varepsilon}{2} P(A_n \cap E_{\varepsilon,n})$$

since $\varepsilon < 1$. Hence, $d(X_n, X) \geq \frac{\varepsilon}{2} P(E_{\varepsilon,n}) \to 0$, so (X_n) converges to X in probability. $\qquad\square$

Proof (of Proposition 8.12)

First

$$\mathbb{E}(Y^p) = \int_{\{\omega : Y(\omega) \geq \varepsilon\}} Y^p \, dP + \int_{\{\omega : Y(\omega) < \varepsilon\}} Y^p \, dP$$

$$\geq \varepsilon^p P(Y \geq \varepsilon) + \int_{\{\omega : Y(\omega) < \varepsilon\}} Y^p \, dP.$$

Now if we let $\varepsilon \to \infty$, then the second term converges to $\int_\Omega Y^p \, dP = \mathbb{E}(Y^p)$ and the first term has no choice but to converge to 0.   □

Proof (of Proposition 8.24)

(i) Write $A = \liminf_{n \to \infty} A_n$. If $\omega \in A$ then there is an N such that $\omega \in A_n$ for all $n \geq N$, which implies $\omega \in A_n$ for all except finitely many n. Conversely, if $n \in A_n$ eventually, then there exists N such that $\omega \in A_n$ for all $n \geq N$ and $\omega \in A$.

(ii) Fix $\varepsilon > 0$. If $A_n^\varepsilon = \{|X_n - X| < \varepsilon\}$ then $\{|X_n - X| < \varepsilon \text{ e.v.}\} = \liminf_{n \to \infty} A_n^\varepsilon = A^\varepsilon$, say. But

$$\{X_n \to X\} = \bigcap_{\varepsilon > 0} \{|X_n - X| < \varepsilon \text{ ev.}\} = \bigcap_{\varepsilon > 0} A^\varepsilon$$

and the sets A^ε decrease as $\varepsilon \searrow 0$. Taking $\varepsilon = \frac{1}{n}$ successively shows that

$$P(X_n \to X) = P(\bigcap_{n=1}^{\infty} \liminf A_n^{1/n}) = \lim_{\varepsilon \to 0} A^\varepsilon = \lim_{\varepsilon \to 0} P(|X_n - X| < \varepsilon \text{ ev.}).$$

(iii) By de Morgan

$$(\bigcap_{n=1}^{\infty} \bigcup_{m=n}^{\infty} A_m)^c = \bigcup_{n=1}^{\infty} (\bigcup_{m=n}^{\infty} A_m)^c = \bigcup_{n=1}^{\infty} \bigcap_{m=n}^{\infty} A_m^c.$$

(iv) If $A = \liminf_{n \to \infty} A_n$ then $\mathbf{1}_A = \liminf_{n \to \infty} \mathbf{1}_{A_n}$ (since $\mathbf{1}_A(\omega) = 1$ if and only if $\mathbf{1}_{A_n} = 1$ eventually) and if $B = \limsup_{n \to \infty} A_n$, then $\mathbf{1}_B = \limsup_{n \to \infty} \mathbf{1}_{A_n}$ (since $\mathbf{1}_B(\omega) = 1$ if and only if $\mathbf{1}_{A_n} = 1$ i.o.). Fatou's lemma implies

$$P(A_n \text{ ev.}) = \int_\Omega \lim_{n \to \infty} \inf \mathbf{1}_{A_n} \, dP \leq \lim_{n \to \infty} \inf \int_\Omega \mathbf{1}_{A_n} \, dP,$$

that is

$$P(B) \leq \lim_{n \to \infty} \inf P(A_n) \leq \lim_{n \to \infty} \sup P(A_n).$$

But

$$\limsup_{n\to\infty} P(A_n) = \limsup_{n\to\infty} \int_\Omega \mathbf{1}_{A_n}\, dP \le \int_\Omega \limsup_{n\to\infty} \mathbf{1}_{A_n}\, dP$$

by Fatou in reverse (see the proof of Theorem 4.26), hence

$$\limsup_{n\to\infty} P(A_n) = \int_\Omega \mathbf{1}_A\, dP = P(A) = P(A_n \text{ i.o.}).$$

$\square$

Solutions

Chapter 2

2.1 If we can cover A by open intervals with prescribed total length, then we can also cover A by closed intervals with the same endpoints (closed intervals are bigger), and the total length is the same. The same is true for any other kind of intervals. For the converse, suppose that A is null in the sense of covering by closed intervals. Let $\varepsilon > 0$, take a cover $C_n = [a_n, b_n]$ with $\sum_n (b_n - a_n) < \frac{\varepsilon}{2}$, let $I_n = (a_n - \varepsilon \frac{1}{2^{n+2}}, b_n + \varepsilon \frac{1}{2^{n+2}})$. For each n, I_n contains C_n, so the I_n cover A; the total length is less than ε. In the same way we refine the cover by any other kind of intervals.

2.2 Write each element of C in ternary form. Suppose C is countable and so they can be arranged in a sequence. Define a number which is not in this sequence but has ternary expansion and so is in C, by exchanging 0 and 2 at the nth position.

2.3 For continuity at $x \in [0, 1]$ take any $\varepsilon > 0$ and find $F(x) - \varepsilon < a < F(x) < b < F(x) + \varepsilon$ of the form $a = \frac{k}{2^n}$, $b = \frac{m}{2^n}$. By the construction of F, these numbers are values of F taken on some intervals (a_1, a_2), (b_1, b_2), with ternary ends, and $a_1 < x < b_2$. Take a δ such that $a_1 < x - \delta < x + \delta < b_2$ to get the continuity condition.

2.4 $m^*(B) \le m^*(A \cup B) \le m^*(A) + m^*(B) = m^*(B)$ by monotonicity (Proposition 2.5) and subadditivity (Theorem 2.7). Thus $m^*(A \cup B)$ is squeezed between $m^*(B)$ and $m^*(B)$, so has little choice.

2.5 Since $A \subset B \cup (A \Delta B)$, $m^*(A) \le m^*(B) + m^*(A \Delta B) = m^*(B)$ (monotonicity and subadditivity). Reversing the roles of A and B gives the opposite inequality.

2.6 Since $A \cup B = A \cup (B \setminus A) = A \cup (B \setminus (A \cap B))$, using additivity and Proposition 2.15 we have $m(A \cup B) = m(A) + m(B) - m(A \cap B)$. Similarly $m(A \cup B \cup C) = m(A) + m(B) + m(C) - m(A \cap B) - m(A \cap C) - m(B \cap C) + m(A \cap B \cap C)$. Note, that the second part makes use of the first on the sets $A \cap C$ and $B \cap C$.

2.7 It is sufficient to note that

$$(a, b) = \bigcup_{n=1}^{\infty} (a, b - \frac{1}{n}], \quad (a, b) = \bigcup_{n=1}^{\infty} [a + \frac{1}{n}, b).$$

2.8 If E is Lebesgue-measurable, then existence of O and F is given by Theorems 2.17 and 2.29, respectively. Conversely, let $\varepsilon = \frac{1}{n}$, find O_n, F_n, and then $m^*(\bigcap O_n \setminus E) = 0$, $m^*(E \setminus \bigcup F_n) = 0$, so E is Borel up to null set. Hence E is in $\mathcal{M}$.

2.9 We can decompose A into $\bigcup_{i=1}^{\infty}(A \cap H_i)$; the components are pairwise disjoint, so

$$P(A) = \sum_{i=1}^{\infty} P(A \cap H_i) = \sum_{i=1}^{\infty} P(A|H_i) \cdot P(H_i)$$

using the definition of conditional probability.

2.10 Since $A^c \cap B = (\Omega \setminus A) \cap B = B \setminus (A \cap B)$, $P(A^c \cap B) = P(B) - P(A \cap B) = P(B) - P(A)P(B) = P(B)(1 - P(A)) = P(B)P(A^c)$.

2.11 There are 32 paths altogether. $S(5) = 524.88 = 500U^2D^3$, so there are $\binom{5}{2} = 10$ paths. $S(5) > 900$ in two cases: $S(5) = 1244.16 = 500U^5$ or $S(5) = 933.12 = 500U^4D$, so we have 6 paths with probability $0.5^5 = 0.03125$ each, so the probability in question is 0.1875.

2.12 There are 2^m paths of length m, $\mathcal{F}_m$ can be identified with the power set of the set of all such paths, thus it has 2^{2^m} elements.

2.13 $\mathcal{F}_m \subset \mathcal{F}_{m+1}$ since if the first $m+1$ elements of a sequence are identical, so are the first m elements.

2.14 It suffices to note that $P(A_m \cap A_k) = \frac{1}{4}$, which is the same as $P(A_m)P(A_k)$.

Chapter 3

3.1 If f is monotone (in the sense that $x_1 \leq x_2$ implies $f(x_1) \leq f(x_2)$), the inverse image of interval (a, ∞) is either $[b, \infty)$ or (b, ∞) with $b = \sup\{x : f(x) \leq a\}$, so it is obviously measurable.

3.2 The level set $\{x : f(x) = a\}$ is the intersection of $f^{-1}([a, +\infty))$ and $f^{-1}((-\infty, a])$, each measurable by Theorem 3.3.

3.3 If $b \geq a$, then $(f^a)^{-1}((-\infty, b]) = \mathbb{R}$, and if $b < a$, then $(f^a)^{-1}((-\infty, b]) = f^{-1}((-\infty, b])$; in each case a measurable set.

3.4 Let A be a non-measurable set and let $f(x) = 1$ if $x \in A$, $f(x) = -1$ otherwise. The set $f^{-1}([1, \infty))$ is non-measurable (it is A), so f is non-measurable, but $f^2 \equiv 1$ which is clearly measurable.

3.5 Let $g(x) = \limsup f_n(x)$, $h(x) = \liminf f_n(x)$; they are measurable by Theorem 3.9. Their difference is also measurable and the set where f_n converges is the level set of this difference: $\{x : f_n \text{ converges}\} = \{x : (h - g)(x) = 0\}$, hence is measurable.

3.6 If $\sup f = \infty$ then there is nothing to prove. If $\sup f = M$, then $f(x) \leq M$ for all x, so M is one of the z in the definition of ess sup (so the infimum of that set is at most M). Let f be continuous and suppose ess $\sup f < M$ finite. Then we take z between these numbers and by the definition of ess sup, $f(x) \leq z$ a.e. Hence $A = \{x : f(x) > z\}$ is null. However, by continuity A contains the set $f^{-1}((z, M))$ which is open – a contradiction. If $\sup f$ is infinite, then for each z the condition $f(x) \leq z$ a.e. does not hold. The infimum of the empty set is $+\infty$ so, we are done.

3.7 It is sufficient to notice that if $\mathcal{G}$ is any σ-field containing the inverse images of Borel sets, then $\mathcal{F}_X \subset \mathcal{G}$.

3.8 Let $f : \mathbb{R} \to \mathbb{R}$ be given by $f(x) = x^2$. The complement of the set $f(\mathbb{R})$ equal to $(-\infty, 0)$ cannot be of the form $f(A)$ since each of these is contained in $[0, \infty)$.

3.9 The payoff of a down-and-out European call is $f((S(0), S(1), \ldots, S(n)))$ with $f(x_0, x_1, \ldots, x_N) = (x_N - K)^+ \cdot \mathbf{1}_A$, where $A = \{(x_0, x_1, \ldots, x_N) : \min\{x_0, x_1, \ldots, x_N\} \geq L\}$.

Chapter 4

4.1 (a) $\int_0^{10} \text{Int}(x)\,dx = 0m([0, 1)) + 1m([1, 2)) + 2m([2, 3)) + \cdots + 9m([9, 10)) + 10m([10, 10]) = 45$.

(b) $\int_0^2 \text{Int}(x^2)\,dx = 0m([0, 1]) + 1m([1, \sqrt{2})) + 2m([\sqrt{2}, \sqrt{3})) + 3m([\sqrt{3}, 2)) = 5 - \sqrt{3} - \sqrt{2}$.

(c) $\int_0^{2\pi} \text{Int}(\sin x)\,dx = 0m([0, \frac{\pi}{2})) + 1m([\frac{\pi}{2}, \frac{\pi}{2}]) + 0m((\frac{\pi}{2}, \pi]) - 1m((\pi, 2\pi]) = -\pi$.

4.2 We have $f(x) = \sum_{k=1}^{\infty} k2^{k-1}\mathbf{1}_{A_k}(x)$, where A_k is the union of 2^{k-1} intervals of length $\frac{1}{3^k}$ each, that are removed from $[0,1]$ at the kth stage. The convergence is monotone, so

$$\int_{[0,1]} f \, dm = \lim_{n \to \infty} \sum_{k=1}^{n} k\frac{2^{k-1}}{3^k} = \frac{1}{3}\sum_{k=1}^{\infty} k\left(\frac{2}{3}\right)^{k-1}.$$

Since $\sum_{k=1}^{\infty} \alpha^k = \frac{1}{1-\alpha}$, differentiating term by term with respect to α we have $\sum_{k=1}^{\infty} k\alpha^{k-1} = \frac{1}{(1-\alpha)^2}$. With $\alpha = \frac{2}{3}$ we get $\int_{[0,1]} f \, dm = 3$.

4.3 The simple function $a\mathbf{1}_A$ is less than f, so its integral, equal to $am(A)$, is smaller than the integral of f. Next, $f \le b\mathbf{1}_A$, hence the second inequality.

4.4 Let $f_n = n\mathbf{1}_{(0,\frac{1}{n}]}$; $\lim f_n(x) = 0$ for all x but $\int f_n \, dm = 1$.

4.5 Let $\alpha \ne -1$. We have $\int x^\alpha \, dx = \frac{1}{\alpha+1}x^{\alpha+1}$ (indefinite integral). First consider $E = (0,1)$:

$$\int_0^1 x^\alpha \, dx = \frac{1}{\alpha+1} \left. x^{\alpha+1}\right|_0^1,$$

which is finite if $\alpha > -1$. Next $E = (1,\infty)$:

$$\int_1^\infty x^\alpha \, dx = \frac{1}{\alpha+1} \lim_{n \to \infty} \left. x^{\alpha+1}\right|_1^n$$

and for this to be finite we need $\alpha < -1$.

4.6 The sequence $f_n(x) = \frac{\sqrt{x}}{1+nx^3}$ converges to 0 pointwise, $\frac{\sqrt{x}}{1+nx^3} \le \frac{\sqrt{x}}{nx^3} \le \frac{1}{n}x^{-2.5} \le x^{-2.5}$ which is integrable on $[1,\infty)$, so the sequence of integrals converges to 0.

4.7 First $a = 0$. Substitute $u = nx$:

$$\int_0^\infty \frac{n^2 x e^{-n^2 x^2}}{1+x^2} \, dx = \int_0^\infty \frac{u e^{-u^2}}{1+(\frac{u}{n})^2} \, du.$$

The sequence of integrands converges to $u e^{-u^2}$ for all $u \ge 0$; it is dominated by $g(u) = u e^{-u^2}$, so

$$\lim \int_0^\infty f_n \, dm = \int_0^\infty \lim f_n \, dm = \int_0^\infty u e^{-u^2} \, du = \frac{1}{2}.$$

Now $a > 0$. After the same substitution we have

$$\int_a^\infty \frac{n^2 x e^{-n^2 x^2}}{1+x^2} \, dx = \int_{\mathbb{R}} \frac{u e^{-u^2}}{1+(\frac{u}{n})^2}\mathbf{1}_{[na,\infty)}(u) \, du = \int_{\mathbb{R}} f_n(u) \, du,$$

say, and $f_n \to 0$, $f_n(u) \le u e^{-u^2}$, so $\lim \int f_n \, dm = 0$.

4.8 The sequence $f_n(x)$ converges for $x \geq 0$ to e^{-x}. We find the dominating function. Let $n > 1$. For $x \in (0,1)$, $x^{\frac{1}{n}} \geq x^{\frac{1}{2}}$, $(1 + \frac{x}{n})^n \geq 1$, so $f_n(x) \leq \frac{1}{\sqrt{x}}$ which is integrable over $(0,1)$. For $x \in [1,\infty)$, $x^{-\frac{1}{n}} \leq 1$, so $f_n(x) \leq (1 + \frac{x}{n})^{-n}$. Next

$$\left(1 + \frac{x}{n}\right)^n = 1 + x + \frac{n(n-1)}{2!}\left(\frac{x}{n}\right)^2 + \cdots > x^2 \frac{n-1}{2n} \geq \frac{1}{4}x^2,$$

so $f_n(x) \leq \frac{4}{x^2}$ which is integrable over $[1,\infty)$.

Therefore, by the dominated convergence theorem,

$$\lim \int_0^\infty f_n \, dm = \int_0^\infty e^{-x} \, dx = 1.$$

4.9 (a) $\int_{-1}^1 |n^a x^n| \, dx = n^a \int_{-1}^1 |x|^n \, dx = 2n^a \int_0^1 x^n \, dx$ ($|x|^n$ is an even function) $= \frac{2n^a}{n+1}$. If $a < 0$, then the series $\sum_{n \geq 1} \frac{2n^a}{n+1}$ converges by comparison with $\frac{1}{n^{1-a}}$, we may apply the Beppo–Levi theorem and the power series in question defines an integrable function. If $a = 0$ the series is $\sum_{n \geq 1} x^n = \frac{x}{1-x}$, which is not integrable since $\int_{-1}^1 (\sum_{n \geq 1} x^n) \, dx = \sum_{n=1}^\infty \int_{-1}^1 x^n \, dx = \infty$. By comparison the series fails to give an integrable function if $a > 0$.

(b) Write $\frac{x}{e^x - 1} = x \frac{e^{-x}}{1 - e^{-x}} = \sum_{n \geq 1} x e^{-nx}$, $\int_0^\infty x e^{-nx} \, dx = x(-\frac{1}{n})e^{-nx}|_0^\infty - (-\frac{1}{n})\int_0^\infty e^{-nx} \, dx = \frac{1}{n^2}$ (integration by parts) and, as is well known, $\sum_{n=1}^\infty \frac{1}{n^2} = \frac{\pi^2}{6}$.

4.10 We extend f by putting $f(0) = 1$, so that f is continuous, hence Riemann-integrable on any finite interval. Let $a_n = \int_{n\pi}^{(n+1)\pi} f(x) \, dx$. Since f is even, $a_{-n} = a_n$ and hence $\int_{-\infty}^\infty f(x) \, dx = 2\sum_{n=0}^\infty a_n$. The series converges since $a_n = (-1)^n |a_n|$, $|a_n| \leq \frac{2}{n\pi}$ ($x \geq n\pi$, $|\int_{n\pi}^{(n+1)\pi} \sin x \, dx| = 2$). However for f to be in L^1 we would need $\int_{\mathbb{R}} |f| \, dm = 2\sum_{n=0}^\infty b_n$ finite, where $b_n = \int_{n\pi}^{(n+1)\pi} |f(x)| \, dx$. This is impossible due to $b_n \geq \frac{2}{(n+1)\pi}$.

4.11 Denote $\int_{-\infty}^\infty e^{-x^2} \, dx = I$; then

$$I^2 = \int\int_{\mathbb{R}^2} e^{-(x^2+y^2)} dx \, dy = \int_0^{2\pi} \int_0^\infty r e^{-r^2} dr \, d\alpha = \pi,$$

using polar coordinates and Fubini's theorem (Chapter 6).

Substitute $x = \frac{z-\mu}{\sqrt{2}\sigma}$ in I; $\sqrt{\pi} = \int_{-\infty}^\infty e^{-x^2} \, dx = \frac{1}{\sqrt{2}\sigma} \int_{-\infty}^\infty e^{-\frac{(z-\mu)^2}{2\sigma^2}} \, dz$, which completes the computation.

4.12 $\int_{\mathbb{R}} \frac{1}{1+x^2} \, dx = \arctan x|_{-\infty}^{+\infty} = \pi$, hence $\int_{-\infty}^\infty c(x) \, dx = 1$.

4.13 $\int_0^\infty e^{-\lambda x} \, dx = -\frac{1}{\lambda} e^{-\lambda x}|_0^\infty = \frac{1}{\lambda}$, hence $c = \lambda$.

4.14 Let $a_n \to 0$, $a_n \geq 0$. Then $P_X(\{y\}) = \lim_{n \to \infty} P_X((y - a_n, y]) = F_X(y) - \lim_{n \to \infty} F_X(y - a_n)$, which proves the required equivalence. (Recall that P_X is always right-continuous.)

4.15 (a) $F_X(y) = 1$ for $y \geq a$ and zero otherwise.

(b) $F_X(y) = 0$ for $y < 0$, $F_X(y) = 1$ for $y \geq \frac{1}{2}$, and $F_X(y) = 2y$ otherwise.

(c) $F_X(y) = 0$ for $y < 0$, $F_X(y) = 1$ for $y \geq \frac{1}{2}$, and $F_X(y) = 1 - (1 - 2y)^2$ otherwise.

4.16 In this case $g(x) = x^3$, $g^{-1}(y) = \sqrt[3]{y}$, $\frac{d}{dy} g^{-1}(y) = \frac{1}{3} y^{-\frac{2}{3}}$, hence $f_{X^3}(y) = \mathbf{1}_{[0,1]}(\sqrt[3]{y}) \frac{1}{3} y^{-\frac{2}{3}} = \frac{1}{3} y^{-\frac{2}{3}} \mathbf{1}_{[0,1]}(y)$.

4.17 (a) $\int_\Omega a \, dP = a P(\Omega) = a$ (constant function is a simple function).

(b) Using Exercise 4.15, $f_X(x) = 2 \times \mathbf{1}_{[0,\frac{1}{2}]}(x)$, so $\mathbb{E}(X) = \int_0^{\frac{1}{2}} 2x \, dx = \frac{1}{4}$.

(c) Again by Exercise 4.15, $f_X(x) = 4(1 - 2x) \times \mathbf{1}_{[0,\frac{1}{2}]}(x)$, $\mathbb{E}(X) = \int_0^{\frac{1}{2}} 4x(1 - 2x) \, dx = \frac{1}{6}$.

4.18 (a) With $f_X(x) = \frac{1}{b-a} \mathbf{1}_{[a,b]}$, $\mathbb{E}(X) = \frac{1}{b-a} \int_a^b x \, dx = \frac{1}{b-a} \frac{1}{2}(b^2 - a^2) = \frac{1}{2}(a + b)$.

(b) Consider the simple triangle distribution with $f_X(x) = x + 1$ for $x \in [-1, 0]$, $f_X(x) = -x + 1$ for $x \in (0, 1]$ and zero elsewhere. Then immediately $\int_{-1}^1 x f_X(x) \, dx = 0$. A similar computation for the density f_Y, whose triangle's base is $[a, b]$, gives $\mathbb{E}(X) = \frac{a+b}{2}$.

(c) $\lambda \int_0^\infty x e^{-\lambda x} \, dx = \frac{1}{\lambda}$ (integration by parts).

4.19 (a) $\varphi_X(t) = \frac{1}{(b-a)it}(e^{ibt} - e^{iat})$,

(b) $\varphi_X(t) = \lambda \int_0^\infty e^{(it - \lambda)x} \, dx = \frac{\lambda}{\lambda - it}$,

(c) $\varphi_X(t) = e^{i\mu t - \frac{1}{2}\sigma^2 t^2}$.

4.20 Using call–put parity, the formula for the call and symmetry of the Gaussian distribution we have $P = S(0)(N(d_1) - 1) - Ke^{-rT}(N(d_2) - 1) = -S(0)N(-d_1) + Ke^{-rT}N(-d_2)$

Chapter 5

5.1 First, $f \equiv f$ as $f = f$ everywhere. Second, if $f = g$ a.e., then of course $g = f$ a.e. Third, if $f = g$ on a full set $F_1 \subset E$ ($m(E \setminus F_1) = 0$) and $g = h$ on a full set $F_2 \subset E$, then $f = h$ on $F_1 \cap F_2$ which is full as well.

5.2 (a) $\|f_n - f_m\|_1 = 2$ if $m < n$, so the sequence is not Cauchy.

(b) $\|f_n - f_m\|_1 = \int_n^m \frac{1}{x}\,\mathrm{d}x = \log m - \log n$ (for simplicity assume that $n < m$), let $\varepsilon = 1$, take any N, let $n = N$, take m such that $\log m - \log N > 1$ ($\log m \to \infty$ as $m \to \infty$) – the sequence is not Cauchy.

(c) $\|f_n - f_m\|_1 = \int_n^m \frac{1}{x^2}\,\mathrm{d}x = -\frac{1}{x}\big|_n^m = \frac{1}{n} - \frac{1}{m}$ ($n < m$ as before), and for any $\varepsilon > 0$ take N such that $\frac{1}{N} < \frac{\varepsilon}{2}$ and for $n, m \geq N$, clearly $\frac{1}{n} - \frac{1}{m} < \varepsilon$ – the sequence is Cauchy.

5.3 $\|g_n - g_m\|_2^2 = \int_n^m \frac{1}{x^4}\,\mathrm{d}x = -\frac{1}{3x^3}\big|_n^m = \frac{1}{3}(\frac{1}{n^3} - \frac{1}{m^3})$ – the sequence is Cauchy.

5.4 (a) $\|f_n - f_m\|_2^2 = m - n$ if $n > m$, so the sequence is not Cauchy.

(b) $\|f_n - f_m\|_2^2 = \int_n^m \frac{1}{x^2}\,\mathrm{d}x = \frac{1}{n} - \frac{1}{m} \to 0$ – the sequence is Cauchy.

(c) $\|f_n - f_m\|_2^2 = \int_n^m \frac{1}{x^4}\,\mathrm{d}x = (\frac{1}{3n^3} - \frac{1}{3m^3})$ – the sequence is Cauchy.

5.5 $\|f + g\|^2 = 4$, $\|f - g\|^2 = 1$, $\|f\|^2 = 1$, $\|g\|^2 = 1$, and the parallelogram law is violated.

5.6 $\|f + g\|_1^2 = 0$, $\|f - g\|_1^2 = \frac{1}{4}$, $\|f\|_1^2 = \frac{1}{4}$, $\|g\|_1^2 = \frac{1}{4}$, which contradicts the parallelogram law.

5.7 Since $\sin nx \cos mx = \frac{1}{2}[\sin(n + m)x + \sin(n - m)x]$ and $\sin nx \sin mx = \frac{1}{2}[\cos(n - m)x + \cos(n + m)x]$, it is easy to compute the indefinite integrals. They are periodic functions, so the integrals over $[-\pi, \pi]$ are zero (for the latter we need $n \neq m$).

5.8 No: take any n, m (suppose $n < m$) and compute

$$\|g_n - g_m\|_4^4 = \int_{\frac{1}{m}}^{\frac{1}{n}} \left(\frac{1}{\sqrt{x}}\right)^4 \mathrm{d}x = -x^{-1}\Big|_{1/m}^{1/n} = (m - n) \geq 1,$$

so the sequence is not Cauchy.

5.9 Let $\Omega = [0, 1]$ with Lebesgue measure, $X(\omega) = \frac{1}{\sqrt{\omega}}$, $\mathbb{E}(X) = \int_0^1 X\,\mathrm{d}m = \int_0^1 \frac{1}{\sqrt{x}}\,\mathrm{d}x = 2$, $\mathbb{E}(X^2) = \int_0^1 \frac{1}{x}\,\mathrm{d}x = \infty$. If we take $X(\omega) = \frac{1}{\sqrt{\omega}} - 2$ then $\mathbb{E}(X) = 0$ and $\mathbb{E}(X^2) = \infty$.

5.10 $\mathrm{Var}(aX) = \mathbb{E}((aX)^2) - (\mathbb{E}(aX))^2 = a^2(\mathbb{E}(X^2) - (\mathbb{E}(X))^2) = a^2\mathrm{Var}(X)$.

5.11 Let $f_X(x) = \frac{1}{b-a}\mathbf{1}_{[a,b]}$, $\mathbb{E}(X) = \frac{a+b}{2}$, $\mathrm{Var}X = \mathbb{E}(X^2) - \frac{(a+b)^2}{4}$, $\mathbb{E}(X^2) = \frac{1}{b-a}\int_a^b x^2\,\mathrm{d}x = \frac{1}{b-a}\frac{1}{3}(b^3 - a^3)$ and simple algebra gives the result.

5.12 (a) $\mathbb{E}(X) = a$, $\mathbb{E}((X - a)^2) = 0$ since $X = a$ a.s.

(b) By Exercise 4.15, $f_X(x) = 2 \times \mathbf{1}_{[0,\frac{1}{2}]}(x)$ and by Exercise 4.17, $\mathbb{E}(X) = \frac{1}{4}$; so $\mathrm{Var}(X) = \int_0^{\frac{1}{2}} 2(x - \frac{1}{4})^2\,\mathrm{d}x = 2\int_{-\frac{1}{4}}^{\frac{1}{4}} x^2\,\mathrm{d}x = \frac{1}{48}$.

(c) By Exercise 4.15, $f_X(x) = (4 - 8x)\mathbf{1}_{[0,\frac{1}{2}]}(x)$, and by Exercise 4.17, $\mathbb{E}(X) = \frac{1}{6}$. So

$$\mathrm{Var}(X) = \int_0^{\frac{1}{2}} x^2(4 - 8x)\,\mathrm{d}x - (\frac{1}{6})^2 = \frac{1}{72}.$$

5.13 $\mathrm{Cov}(Y, 2Y+1) = \mathbb{E}((Y)(2Y+1)) - \mathbb{E}(Y)\mathbb{E}(2Y+1) = 2\mathbb{E}(Y^2) - 2(\mathbb{E}(Y))^2 =$
      $2\mathrm{Var}(Y)$, $\mathrm{Var}(2Y+1) = \mathrm{Var}(2Y) = 4\mathrm{Var}(Y)$, hence $\rho = 1$.

5.14 X, Y are uncorrelated by Exercise 5.7. Take $a > 0$ so small that the sets
      $A = \{\omega : \sin 2\pi\omega > 1 - a\}$, $B = \{\omega : \cos 2\pi\omega > 1 - a\}$ are disjoint. Then
      $P(A \cap B) = 0$ but $P(A)P(B) \neq 0$.

Chapter 6

6.1 The function
$$g(x,y) = \begin{cases} \frac{1}{x^2} & \text{if } 0 < y < x < 1 \\ -\frac{1}{y^2} & \text{if } 0 < x < y < 1 \\ 0 & \text{otherwise} \end{cases}$$

is not integrable since the integral of g^+ is infinite (the same is true for the
integral of g^-). However,

$$\int_0^1 \int_0^1 \left[\frac{1}{x^2}\mathbf{1}_{\{0<y<x<1\}}(x,y) - \frac{1}{y^2}\mathbf{1}_{\{0<x<y<1\}}(x,y) \right] \mathrm{d}x\,\mathrm{d}y$$

$$= \int_0^1 \left[\int_y^1 \frac{1}{x^2}\,\mathrm{d}x - \frac{1}{y^2}\int_0^y \mathrm{d}x \right] \mathrm{d}y = \int_0^1 \left[-1 + \frac{1}{y} - \frac{1}{y} \right] \mathrm{d}y = -1$$

while similarly

$$\int_0^1 \int_0^1 \left[\frac{1}{x^2}\mathbf{1}_{\{0<y<x<1\}}(x,y) - \frac{1}{y^2}\mathbf{1}_{\{0<x<y<1\}}(x,y) \right] \mathrm{d}y\,\mathrm{d}x$$

$$\int_0^1 \left[\frac{1}{x^2}\int_0^x \mathrm{d}y - \int_x^1 \frac{1}{y^2}\,\mathrm{d}y \right] \mathrm{d}x = 1,$$

which shows that the iterated integrals may not be equal if Fubini's theorem
condition is violated.

6.2 $\int \int_{[0,3]\times[-1,2]} x^2 y\,\mathrm{d}m_2 = \int_{-1}^2 \int_0^3 x^2 y\,\mathrm{d}x\,\mathrm{d}y = \int_{-1}^2 9y\,\mathrm{d}y = \frac{27}{2}$.

6.3 By symmetry it is sufficient to consider $x \geq 0$, $y \geq 0$, and hence the area
     is $4\frac{b}{a}\int_0^a \sqrt{a^2 - x^2}\,\mathrm{d}x = ab\pi$.

6.4 Fix $x \in [0,2]$, $\int_0^2 1_A(x,y)\,dy = m(A_x)$, hence $f_X(x) = x$ for $x \in [0,1]$, $f_X(x) = 2 - x$ for $x \in (1,2]$ and zero otherwise (triangle distribution). By symmetry, the same holds for f_Y.

6.5 $P(X + Y > 4) = P(Y > -X + 4) = \int \int_A f_{X,Y}(x,y)\,dx\,dy$ where $A = \{(x,y) : y > 4 - x\} \cap [0,2] \times [1,4]$, so $P(X + Y > 4) = \int_0^2 \int_{4-x}^4 \frac{1}{50}(x^2 + y^2)\,dy\,dx = \frac{1}{50}\int_0^2 (-4x^2 + \frac{4}{3}x^3 + 16x)\,dx = \frac{8}{15}$. Similarly $P(Y > X) = \int \int_A f_{X,Y}(x,y)\,dx\,dy$ where $A = \{(x,y) : y > x\} \cap [0,2] \times [1,4]$, so we get

$$P(Y > X) = \int_1^2 \int_0^y \frac{1}{50}(x^2 + y^2)\,dx\,dy + \int_2^4 \int_0^2 \frac{1}{50}(x^2 + y^2)\,dx\,dy$$
$$= \frac{1}{50}\int_1^2 \frac{4}{3}y^3\,dy + \frac{1}{50}\int_2^4 \left(\frac{8}{3} + 2y^2\right)\,dy = \frac{143}{150}$$

6.6

$$f_{X+Y}(z) = \int_R f_{X,Y}(x, z - x)\,dx = \begin{cases} 0 & z < 0 \\ z & 0 \le z \le 1 \\ 2 - z & 1 \le z \le 2 \\ 0 & 2 < z \end{cases}$$

6.7 By Exercise 6.4, $f_{X,Y}(x,y) \ne f_X(x)f_Y(y)$, so X, Y are not independent.

6.8 $f_{Y+(-X)}(z) = \int_{-\infty}^{+\infty} f_Y(y) f_{-X}(z - y)\,dy = \int_{-\infty}^{+\infty} \frac{1}{2} 1_{[0,2]}(y) 1_{[-1,0]}(z - y)\,dy$, so

$$f_{Y-X}(z) = \begin{cases} 0 & z < -1 \text{ or } 2 < z \\ \frac{1}{2}(z + 1) & -1 \le z \le 0 \\ \frac{1}{2} & 0 \le z \le 1 \\ \frac{1}{2}(2 - z) & 1 \le z \le 2 \end{cases}$$

$f_{X+Y}(z) = \int_{-\infty}^{+\infty} f_X(x) f_Y(z - x)\,dx = \int_{-\infty}^{+\infty} 1_{[0,1]}(x) \frac{1}{2} 1_{[0,2]}(z - x)\,dx$, hence

$$f_{X+Y}(z) = \begin{cases} 0 & z < 0 \text{ or } 3 < z \\ \frac{1}{2}z & 0 \le z \le 1 \\ \frac{1}{2} & 1 \le z \le 2 \\ \frac{1}{2}(3 - z) & 2 \le z \le 3 \end{cases}$$

$P(Y > X) = P(Y - X > 0) = \int_0^\infty f_{Y-X}(z)\,dz = \frac{1}{2} + \int_1^2 \frac{1}{2}(2 - z)\,dz = \frac{1}{2} + \frac{1}{4} = \frac{3}{4}$;

$P(X + Y > 1) = \int_1^\infty f_{X+Y}(z)\,dz = \frac{1}{2} + \int_2^3 \frac{1}{2}(3 - z)\,dz = \frac{1}{2} + \frac{1}{3} = \frac{3}{4}$.

6.9 $f_X(x) = \int_0^{-\frac{1}{2}x+1} 1_A\,dy = 1 - \frac{1}{2}x$, $h(y,x) = \frac{1_A(x,y)}{1 - \frac{1}{2}x}$ and $\mathbb{E}(Y|X = 1) = 2\int_0^{\frac{1}{2}} x\,dx = \frac{1}{4}$.

6.10 $f_Y(y) = \int_0^1 (x+y)\, dx = \frac{1}{2} + y$, $h(x|y) = \frac{x+y}{\frac{1}{2}+y}\mathbf{1}_A(x,y)$, $\mathbb{E}(X|Y = y) =$ $\int_0^1 x\frac{x+y}{\frac{1}{2}+y}\, dx = \frac{\frac{1}{3}+\frac{1}{2}y}{\frac{1}{2}+y}$.

Chapter 7

7.1 If $\mu(A) = 0$ then $\lambda_1(A) = 0$ and $\lambda_2(A) = 0$, hence $(\lambda_1 + \lambda_2)(A) = 0$.

7.2 Let $\mathcal{Q}$ be a finite partition of Ω which refines both $\mathcal{P}_1$ and $\mathcal{P}_2$. Each $A \in \mathcal{P}_1$ and each $B \in \mathcal{P}_2$ can be written as a disjoint union of in the form $A = \bigcup_{i \leq m} E_i$, $B = \bigcup_{j \leq n} F_j$, where the E_i and F_j are sets in $\mathcal{Q}$. Thus $A \cap B = \bigcup_{i,j}(E_i \cap F_j)$ is a finite disjoint union of sets in $\mathcal{Q}$, since in each case, $E_i \cap F_j$ equals either E_i or F_j (sets in $\mathcal{Q}$ are disjoint). Thus $\mathcal{Q}$ refines the partition $\mathcal{R} = \{A \cap B : A \in \mathcal{P}_1, B \in \mathcal{P}_2\}$. On the other hand, any set in $\mathcal{P}_1$ or $\mathcal{P}_2$ is obviously a finite union of sets in $\mathcal{R}$ (e.g. $A = A \cap \Omega = A \cap (\bigcup_{B \in \mathcal{P}_2} B)$ for $A \in \mathcal{P}_1$) as the partitions are finite.

7.3 We have to assume first that $m(A) \neq 0$. Then $B \subset A$ clearly implies that μ dominates ν. (In fact $m(B \setminus A) = 0$ is slightly more general.) Then consider the partition $\{B, A \setminus B, \Omega \setminus A\}$ to see that $h = \mathbf{1}_B$. To double check, $\nu(F) = m(F \cap B) = \int_{F \cap B} \mathbf{1}_B\, dm = \int_{F \cap B} \mathbf{1}_B\, d\mu$.

7.4 Clearly $\mu(\{\omega\}) \geq \nu(\{\omega\})$ is equivalent to μ dominating ν. For each ω we have $\frac{d\nu}{d\mu}(\omega) = \frac{\nu(\{\omega\})}{\mu(\{\omega\})}$.

7.5 Since $\nu(E) = \int_E g\, dm$ and we wish to have $\nu(E) = \int_E \frac{d\nu}{d\mu}\, d\mu = \int_E \frac{d\nu}{d\mu} f\, dm$ it is natural to aim at taking $\frac{d\nu}{d\mu}(x) = \frac{g(x)}{f(x)}$. Then a sufficient condition for this to work is that if $A = \{x : f(x) = 0\}$ then $\nu(A) = 0$, i.e. $g(x) = 0$ a.e. on A. Then we put $\frac{d\nu}{d\mu}(x) = \frac{g(x)}{f(x)}$ on A^c and 0 on A and we have $\nu(E) = \int_{E \cap A^c} g\, dm = \int_{E \cap A^c} \frac{d\nu}{d\mu} f\, dm = \int_E \frac{d\nu}{d\mu}\, d\mu$, as required.

7.6 Clearly $\nu \ll \mu$ is equivalent to $A = \{\omega : \mu(\{\omega\}) = 0\} \subset \{\omega : \nu(\{\omega\}) = 0\}$ and then $\frac{d\nu}{d\mu}(\omega) = \frac{\nu(\{\omega\})}{\mu(\{\omega\})}$ on A^c and zero on A.

7.7 Since $\nu \ll \mu$ we may write $h = \frac{d\nu}{d\mu}$, so that $\nu(F) = \int_F h\, d\mu$. As $\mu(F) = 0$ if and only if $\nu(F) = 0$, the set $\{h = 0\}$ is both μ-null and ν-null. Thus $h^{-1} = (\frac{d\nu}{d\mu})^{-1}$ is well-defined a.s., and we can use (ii) in Proposition 7.9 with $\lambda = \mu$ to conclude that $1 = h^{-1}h$ implies $\frac{d\mu}{d\nu} = h^{-1}$, as required.

7.8 $\delta_0((0, 25]) = 0$, but $\frac{1}{25}m|_{[0,25]}((0, 25]) = 1$; $\frac{1}{25}m|_{[0,25]}(\{0\}) = 0$ but $\delta_0(\{0\}) = 1$, so neither $P_1 \ll P_2$ nor $P_2 \ll P_1$. Clearly $P_1 \ll P_3$ with $\frac{dP_1}{dP_3}(x) = 2 \times \mathbf{1}_{\{0\}}(x)$ and $P_2 \ll P_3$ with $\frac{dP_2}{dP_3}(x) = 2 \times \mathbf{1}_{(0,25]}(x)$.

7.9 $\lambda_a = m|_{[2,3]}$, $\lambda_s = \delta_0 + m|_{(1,2)}$, and $h = \mathbf{1}_{[2,3]}$.

7.10 Suppose F is non-constant at a_i with positive jumps c_i, $i = 1, 2, \ldots$. Take $M \neq a_i$, with $-M \neq a_i$ and let $I = \{i : a_i \in [-M, M]\}$. Then

$$m_F([-M, M]) = F(M) - F(-M) = \sum_{i \in I} c_i = \sum_{i \in I} m_F(\{a_i\}),$$

which is finite since F is bounded on a bounded interval. So any $A \subset [-M, M] \setminus \bigcup_{i \in I}\{a_i\}$ is m_F-null, hence measurable. But $\{a_i\}$ are m_F-measurable, hence each subset of $[-M, M]$ is m_F-measurable. Finally, any subset E of $\mathbb{R}$ is a union of the sets of the form $E \cap [-M, M]$, so E is m_F-measurable as well.

7.11 m_F has density $f(x) = 2$ for $x \in [0, 1]$ and zero otherwise.

7.12 (a) $|x| = 1 + \int_{-1}^{x} f(y) \, dy$, where $f(y) = -1$ for $y \in [-1, 0]$, and $f(y) = 1$ for $y \in (0, 1]$.

(b) Let $1 > \varepsilon > 0$, take $\delta = \varepsilon^2$, $\sum_{k=1}^{n}(y_k - x_k) < \delta$, with $y_k \leq x_{k+1}$; then

$$\left(\sum_{k=1}^{n} |\sqrt{x_k} - \sqrt{y_k}|\right)^2 \leq (\sqrt{y_n} - \sqrt{x_1})^2 = y_n - 2\sqrt{y_n x_1} + x_1 < y_n - x_1$$

$$\leq \sum_{k=1}^{n}(y_k - x_k) < \varepsilon^2.$$

(c) The Lebesgue function f is a.e. differentiable with $f' = 0$ a.e. If it were absolutely continuous, it could be written as $f(x) = \int_0^x f'(y) \, dy = 0$, a contradiction.

7.13 (a) If F is monotone increasing on $[a, b]$, $\sum_{i=1}^{k} |F(x_i) - F(x_{i-1})| = F(b) - F(a)$ for any partition $a = x_0 < x_1 < \cdots < x_k = b$. Hence $T_F[a, b] = F(b) - F(a)$.

(b) If $F \in BV[a, b]$ we can write $F = F_1 - F_2$ where both F_1, F_2 are monotone increasing, hence have only countably many points of discontinuity. So F is continuous a.e. and thus Lebesgue-measurable.

(c) $f(x) = x^2 \cos \frac{\pi}{x^2}$ for $x \neq 0$ and $f(0) = 0$ is differentiable but does not belong to $BV[0, 1]$.

(d) If F is Lipschitz, $\sum_{i=1}^{k} |F(x_i) - F(x_{i-1})| \leq M \sum_{i=1}^{k} |x_i - x_{i-1}| = M(b - a)$ for any partition, so $T_F[a, b] \leq M(b - a)$ is finite.

7.14 Recall that $\nu^+(E) = \nu(E \cap B)$, where B is the positive set in the Hahn decomposition. As in the hint, if $G \subseteq F$, $\nu(G) \leq \nu^+(G \cap B) \leq \nu(G \cap B) + \nu((F \cap B) \setminus (G \cap B)) = \nu(F \cap B)$. Since the set $(F \cap B) \setminus (G \cap B) \subseteq B$, its ν-measure is non-negative. But $F \cap B \subseteq F$, so $\sup\{\nu(G) : G \subseteq F\}$ is attained and equals $\nu(F \cap B) = \nu^+(F)$. A similar argument shows $\nu^-(F) = \sup\{-\nu(G)\} = -\inf_{G \subset F}\{\nu(G)\}$.

7.15 For all $F \in \mathcal{F}$, $\nu^+(F) = \int_{B \cap F} f \, d\mu = \sup_{G \subset F} \int_G f \, d\mu$. If $f > 0$ on a set $C \subset A \cap F$ with $\mu(C) > 0$, then $\int_C f \, d\mu > 0$, so that $\int_C f \, d\mu + \int_{B \cap F} f \, d\mu > \sup_{G \subset F} \int_G f \, d\mu$. This is a contradiction since $C \cup (B \cap F) \subset F$. So $f \leq 0$-a.s. (μ) on $A \cap F$. We can take the set $\{f = 0\}$ into B, since it does not affect the integrals. Hence $\{f < 0\} \subset A$ and $\{f \geq 0\} \subset B$. But the two smaller sets partition Ω, so we have equality in both cases.

Hence $f^+ = f\mathbf{1}_B$ and $f^- = -f\mathbf{1}_A$, and for all $F \in \mathcal{F}$

$$\nu^+(F) = \nu(B \cap F) = \int_{B \cap F} f \, d\mu = \int_F f^+ \, d\mu,$$

$$\nu^-(F) = -\nu(A \cap F) = -\int_{A \cap F} f \, d\mu = \int_F f^- \, d\mu.$$

7.16 $f \in L^1(\nu)$ if and only if both $\int f^+ \, d\nu$ and $\int f^- \, d\nu$ are finite. Then $\int_E f^+ g \, d\mu$ and $\int_E f^- g \, d\mu$ are well defined and finite, and their difference is $\int_E fg \, d\mu$. So $fg \in L^1(\mu)$, as $\int_E (f^+ - f^-)|g| \, d\mu < \infty$. Conversely, if $fg \in L^1(\mu)$ then both $\int_E f^+|g| \, d\mu$ and $\int_E f^-|g| \, d\mu$ are finite, hence so is their difference $\int_E f \, d\nu$.

7.17 (a) $\mathbb{E}(X|\mathcal{G})(\omega) = \frac{1}{4}$ if $\omega \in [0, \frac{1}{2}]$, $\mathbb{E}(X|\mathcal{G})(\omega) = \frac{3}{4}$ otherwise.

  (b) $\mathbb{E}(X|\mathcal{G})(\omega) = \omega$ if $\omega \in [0, \frac{1}{2}]$, $\mathbb{E}(X|\mathcal{G})(\omega) = \frac{3}{4}$ otherwise.

7.18 $\mathbb{E}(X_n|\mathcal{F}_{n-1}) = \mathbb{E}(Z_1 Z_2 \ldots Z_n|\mathcal{F}_{n-1}) = Z_1 Z_2 \ldots Z_{n-1}\mathbb{E}(Z_n|\mathcal{F}_{n-1})$, and since Z_n is independent of $\mathcal{F}_{n-1}$, $\mathbb{E}(Z_n|\mathcal{F}_{n-1}) = \mathbb{E}(Z_n) = 1$, hence the result.

7.19 $\mathbb{E}(X_n) = n\mu \neq \mu = \mathbb{E}(X_1)$ so X_n is not a martingale. Clearly $Y_n = X_n - n\mu$ is a martingale.

7.20 Apply Jensen's inequality to $X_n = Z_1 + Z_2 + \cdots + Z_n$. We obtain

$$\mathbb{E}(X_n^2|\mathcal{F}_{n-1}) \leq [\mathbb{E}(X_n|\mathcal{F}_{n-1})]^2 = X_n^2,$$

so (X_n) is a submartingale. By (7.5) the compensator (A_n) satisfies $\Delta A_n = \mathbb{E}((\Delta X_n^2)|\mathcal{F}_{n-1})$ for each n. But here $\Delta X_n = Z_n$ and $Z_n^2 \equiv 1$. So $\Delta A_n = 1$ and hence $A_n = n$ is deterministic.

7.21 For $s < t$, since the increments are independent and $w(t) - w(s)$ has the same distribution as $w(t-s)$,

$$\mathbb{E}(\exp(-\sigma w(t) - \frac{1}{2}\alpha^2 t)) = e^{-\sigma w(s) - \frac{1}{2}\sigma^2 t}\mathbb{E}(\exp(-[\sigma(w(t) - w(s))])|\mathcal{F}_s)$$
$$= e^{-\sigma w(s) - \frac{1}{2}\sigma^2 t}\mathbb{E}(\exp(-[\sigma(w(t) - w(s))])$$
$$= e^{-\sigma w(s) - \frac{1}{2}\sigma^2 t}\mathbb{E}(\exp(-\sigma w(t-s))).$$

Now $\sigma w(t-s) \frown N(0, \sigma^2(t-s))$ so the expectation equals $\mathbb{E}(e^{-\sigma\sqrt{t-s}Z}) = e^{-\frac{1}{2}\sigma^2(t-s)}$ (where $Z \frown N(0,1)$) and so the result follows.

Chapter 8

8.1 (a) $f_n = \mathbf{1}_{[n,n+\frac{1}{n}]}$ converges to 0 in L^p, pointwise, a.e. but not uniformly.

(b) $f_n = n\mathbf{1}_{[0,\frac{1}{n}]} - n\mathbf{1}_{[-\frac{1}{n},0]}$ converges to 0 pointwise and a.e. It converges neither in L^p nor uniformly.

8.2 We have $\Omega = [0,1]$ with Lebesgue measure. The sequences $X_n = \mathbf{1}_{(0,\frac{1}{n})}$, $X_n = n\mathbf{1}_{(0,\frac{1}{n}]}$ converge to 0 in probability since $P(|X_n| > \varepsilon) \le \frac{1}{n}$ and the same holds for the sequence g_n.

8.3 There are endless possibilities, the simplest being $X_n(\omega) \equiv 1$ (but this sequence converges to 1) or, to make sure that it does not converge to anything, $X_n(\omega) \equiv n$.

8.4 Let $X_n = 1$ indicate the 'heads' and $X_n = 0$ the 'tails', then $\frac{S_{100}}{100}$ is the average number of 'heads' in 100 tosses. Clearly $\mathbb{E}(X_n) = \frac{1}{2}$, $\mathbb{E}(\frac{S_{100}}{100}) = \frac{1}{2}$, $\text{Var}(X_n) = \frac{1}{4}$, $\text{Var}(\frac{S_{100}}{100}) = \frac{1}{100^2}100 \cdot \frac{1}{4} = \frac{1}{400}$, so

$$P(|\frac{S_{100}}{100} - \frac{1}{2}| \ge 0.1) \le \frac{1}{0.1^2 400}$$

and

$$P(|\frac{S_{100}}{100} - \frac{1}{2}| < 0.1) \ge 1 - \frac{1}{0.1^2 400} = \frac{3}{4}.$$

8.5 Let X_n be the number shown on the die, $\mathbb{E}(X_n) = 3.5$, $\text{Var}(X_n) \approx 2.9167$.

$$P(|\frac{S_{1000}}{1000} - 3.5| < 0.1) \ge 0.2916.$$

8.6 The union $\bigcup_{m=n}^{\infty} A_m$ is equal to $[0,1]$ for all m and so is $\limsup_{n\to\infty} A_n$.

8.7 Let $d = 1$. There are $\binom{2n}{n}$ paths that return to 0, so $P(S_{2n} = 0) = \binom{2n}{n} \frac{1}{2^{2n}}$. Now

$$\frac{(2n)!}{(n!)^2} \sim \frac{(\frac{2n}{e})^{2n} \sqrt{2\pi 2n}}{(\frac{n}{e})^{2n} 2\pi n} = \frac{2^{2n}}{\sqrt{n\pi}},$$

so $P(S_{2n} = 0) \sim \frac{c}{\sqrt{n}}$ with $c = \sqrt{\frac{2}{\pi}}$. Hence $\sum_{n=1}^{\infty} P(A_n)$ diverges and Borel–Cantelli applies (as (A_n) are independent), so that $P(S_{2n} = 0 \text{ i.o.}) = 1$. Same for $d = 2$ since $P(A_n) \sim \frac{1}{n}$. But for $d > 2$, $P(A_n) \sim \frac{1}{n^{d/2}}$, the series converges and by the first Borel–Cantelli lemma $P(S_{2n} = 0 \text{ i.o.}) = 0$.

8.8 Write $S = S_{1000}$; $P(|S - 500| < 10) = P(\frac{|S-500|}{\sqrt{250}} < 0.63) = N(0.63) - N(-0.63) \approx 0.4729$, where N is the cumulative distribution function corresponding to the standard normal distribution.

8.9 The condition on n is $P(|\frac{S_n}{n} - 0.5| < 0.005) = P(\frac{|S_n - 0.5n|}{\sqrt{n/4}} < 0.01\sqrt{n}) \geq 0.99$, $N(0.01\sqrt{n}) \geq 0.995 = N(2.575835)$, hence $n \geq 66\,350$.

8.10 Write $x_n = e^{\sigma\sqrt{T/n}}$. Then

$$\frac{1}{2}(\ln U_n + \ln D_n) = \frac{1}{2}\ln(U_n D_n) = \frac{1}{2}\ln\frac{(2R_n x_n)^2}{(1+x_n^2)^2} = \ln e^{r\frac{T}{n}} - \ln(\frac{1+x_n^2}{2x_n}).$$

So it suffices to show that the last term on the right is $\frac{\sigma^2 T}{2n} + o(\frac{1}{n})$. But

$$\frac{1+x_n^2}{2x_n} = \frac{x_n^{-1} + x_n}{2} = \frac{e^{\sigma\sqrt{T/n}} + e^{\sigma\sqrt{T/n}}}{2} = \cosh(\sigma\sqrt{T/n})$$

$$= 1 + \frac{\sigma^2 T}{2n} + o(\frac{1}{n}),$$

so that

$$\ln(\frac{1+x_n^2}{2x_n}) = \ln(1 + \frac{\sigma^2 T}{2n} + o(\frac{1}{n})) = \frac{\sigma^2 T}{2n} + o(\frac{1}{n}).$$

Appendix

Existence of non-measurable and non-Borel sets

In Chapter 2 we defined the σ-field $\mathcal{B}$ of Borel sets and the larger σ-field $\mathcal{M}$ of Lebesgue-measurable sets, and all our subsequent analysis of the Lebesgue integral and its properties involved these two families of subsets of $\mathbb{R}$. The set inclusions

$$\mathcal{B} \subset \mathcal{M} \subset \mathcal{P}(\mathbb{R})$$

are trivial; however, it is not at all obvious at first sight that they are strict, i.e. that there are sets in $\mathbb{R}$ which are not Lebesgue-measurable, and as that there are Lebesgue-measurable sets which are not Borel sets. In this appendix we construct examples of such sets. Using the fact that $A \subset \mathbb{R}$ is measurable (resp. Borel-measurable) iff its indicator function $\mathbf{1}_A \in \mathcal{M}$ (resp. $\mathcal{B}$), it follows that we will automatically have examples of non-measurable (resp. measurable but not Borel) functions.

The construction of a non-measurable set requires some set-theoretic preparation. This takes the form of an axiom which, while not needed for the consistent development of set theory, nevertheless enriches that theory considerably. Its truth or falsehood cannot be proved from the standard axioms on which modern set theory is based, but we shall accept its validity as an axiom, without delving further into foundational matters.

The axiom of choice

Suppose that $\mathcal{A} = \{A_\alpha : \alpha \in \Gamma\}$ is a non-empty collection, indexed by some set Γ, of non-empty disjoint subsets of a fixed set Ω. There then exists a set $E \subset \Omega$ which contains precisely one element from each of the sets A_α, i.e. there is a *choice function* $f : \Gamma \to \mathcal{A}$.

Remark

The Axiom may seem innocuous enough, yet it can be shown to be independent of the (Zermelo–Fraenkel) axioms of sets theory. If the collection $\mathcal{A}$ has only finitely many members there is no problem in finding a choice function, of course.

To construct our example of a non-measurable set, first define the following equivalence relation on $[0,1]$: $x \sim y$ if $y - x$ is a rational number (which will be in $[-1,1]$). This relation is easily seen to be reflexive, symmetric and transitive. Hence it partitions $[0,1]$ into disjoint equivalence classes (A_α), where for each α, any two elements x, y of A_α differ by a rational, while elements of different classes will always differ by an irrational. Thus each A_α is countable, since $\mathbb{Q}$ is, but there are uncountably many different classes, as $[0,1]$ is uncountable.

Now use the axiom of choice to construct a new set $E \subset [0,1]$ which contains exactly one member a_α from each of the A_α. Now enumerate the rationals in $[-1,1]$: there are only countably many, so we can order them as a sequence (q_n). Define a sequence of translates of E by $E_n = E + q_n$. If E is Lebesgue-measurable, then so is each E_n and their measures are the same, by Proposition 2.15.

But the (E_n) are disjoint: to see this, suppose that $z \in E_m \cap E_n$ for some $m \neq m$. Then we can write $a_\alpha + q_m = z = a_\beta + q_n$ for some $a_\alpha, a_\beta \in E$, and their difference $a_\alpha - a_\beta = q_n - q_m$ is rational. Since E contains only one element from each class, $\alpha = \beta$ and therefore $m = n$. Thus $\bigcup_{n=1}^{\infty} E_n$ is a disjoint union containing $[0,1]$.

Thus we have $[0,1] \subset \bigcup_{n=1}^{\infty} E_n \subset [-1,2]$ and $m(E_n) = m(E)$ for all n. By countable additivity and monotonicity of m this implies:

$$1 = m([0,1]) \leq \sum_{n=1}^{\infty} m(E_n) = m(E) + m(E) + \cdots \leq 3.$$

This is clearly impossible, since the sum must be either 0 or ∞. Hence we must conclude that E is not measurable.

For an example of a measurable set that is not Borel, let C denote the Cantor set, and define the *Cantor function* $f : [0,1] \to C$ as follows: for $x \in [0,1]$ write $x = 0.a_1 a_2 \ldots$ in binary form, i.e. $x = \sum_{n=1}^{\infty} \frac{a_n}{2^n}$, where each $a_n = 0$ or 1 (taking non-terminating expansions where the choice exists). The function $x \mapsto a_n$ is determined by a system of finitely many binary intervals (i.e. the value of a_n is fixed by x satisfying finitely many linear inequalities) and so is measurable – hence so is the function f given by $f(x) = \sum_{n=1}^{\infty} \frac{2a_n}{3^n}$. Since all the terms of $y = \sum_{n=1}^{\infty} \frac{2a_n}{3^n}$ have numerators 0 or 2, it follows that the range R_f of f is a subset of C. Moreover, the value of y determines the sequence (a_n) and hence x, uniquely, so that f is invertible.

Now consider the image in C of the non-measurable set E constructed above, i.e. let $B = f(E)$. Then B is a subset of the null set C, hence by the completeness of m it is also measurable and null. On the other hand, $E = f^{-1}(B)$ is non-measurable. We show that this situation is incompatible with B being a Borel set.

Given a set $B \in \mathcal{B}$ and a measurable function g, then $g^{-1}(B)$ must be measurable. For, by definition of measurable functions, $g^{-1}(I)$ is measurable for every interval I, and we have

$$g^{-1}\left(\bigcup_{i=1}^{\infty} A_i\right) = \bigcup_{i=1}^{\infty} g^{-1}(A_i), \qquad g^{-1}(A^c) = (g^{-1}(A))^c$$

quite generally for any sets and functions. Hence the collection of sets whose inverse images under the measurable function g are again measurable forms a σ-field containing the intervals, and hence it also contains all Borel sets.

But we have found a measurable function f and a Lebesgue-measurable set B for which $f^{-1}(B) = E$ is *not* measurable. Therefore the measurable set B cannot be a Borel set, i.e. the inclusion $\mathcal{B} \subset \mathcal{M}$ is strict.

References

[1] T.M. Apostol, *Mathematical Analysis*, Addison-Wesley, Reading, MA, 1974.

[2] P. Billingsley, *Probability and Measure*, Wiley, New York, 1995.

[3] Z. Brzeźniak, T. Zastawniak, *Basic Stochastic Processes*, Springer, London, 1999.

[4] M. Capiński, T. Zastawniak, *Mathematics for Finance, An Introduction to Financial Engineering*, Springer, London, 2003.

[5] R.J. Elliott, P.E. Kopp, *Mathematics of Financial Markets*, Springer, New York, 1999.

[6] G.R. Grimmett, D.R. Stirzaker, *Probability and Random Processes*, Clarendon Press, Oxford, 1982.

[7] J. Hull, *Options, Futures, and Other Derivatives*, Prentice Hall, Upper Saddle River, NJ, 2000.

[8] P.E. Kopp, *Martingales and Stochastic Integrals*, Cambridge University Press. Cambridge, 1984.

[9] P.E. Kopp, *Analysis*, Edward Arnold, London, 1996.

[10] J. Pitman, *Probability*, Springer, New York, 1995.

[11] W. Rudin, *Real and Complex Analysis*, McGraw-Hill, New York, 1966.

[12] G. Smith, *Introductory Mathematics: Algebra and Analysis*, Springer, London, 1998.

[13] D. Williams, *Probability with Martingales*, Cambridge University Press, Cambridge, 1991.

Index